Global Occupational Safety and Health Management Handbook

Global Occupational Safety and Health Management Handbook

Edited by
Thomas P. Fuller

CRC Press
Taylor & Francis Group
Boca Raton London New York

CRC Press is an imprint of the
Taylor & Francis Group, an **informa** business

CRC Press
Taylor & Francis Group
6000 Broken Sound Parkway NW, Suite 300
Boca Raton, FL 33487-2742

Printed on acid-free paper

International Standard Book Number-13: 978-1-138-62672-0 (Hardback)

Library of Congress Cataloging-in-Publication Data

Names: Fuller, Thomas P., author.
Title: Global occupational safety and health management handbook / edited by Thomas P. Fuller.
Description: Boca Raton, FL : CRC Press/Taylor & Francis Group, 2019. |
Includes bibliographical references and index.
Identifiers: LCCN 2018049099 | ISBN 9781138626720 (hardback : acid-free paper) |
ISBN 9780429056475 (ebook)
Subjects: LCSH: Industrial safety—Handbooks, manuals, etc. | International business enterprises—Safety measures—Handbooks, manuals, etc.
Classification: LCC T55 .F853 2019 | DDC 658.3/82—dc23
LC record available at https://lccn.loc.gov/2018049099

Visit the Taylor & Francis Web site at
http://www.taylorandfrancis.com

and the CRC Press Web site at
http://www.crcpress.com

Contents

Preface

Occupational safety and health management is being increasingly practiced on a global scale. Therefore, it is important for professionals working in this field to be well informed about existing international programs, organizations, practices, and issues. This handbook is a starting point for someone interested in learning more about global health and safety topics. Occupational health and safety professionals with responsibilities for worker health and safety policy, planning, and operations internationally should be well versed in many of the subjects discussed in this handbook. Ready access to relevant guidelines and information on specific topics makes this handbook a valuable tool for the health and safety practitioner traveling abroad to work, or managing those that do. The aim of the handbook is to provide a reference and guide to those who, within organizations or governments, may hold key responsibilities for occupational safety and health of workers in foreign countries. Thus, its underlying purpose is to assist occupational hygienists and managers in attaining standards of professionalism that will enable them to deal with day-to-day health and safety management across international borders and in an expansive range of cultural, political, and regulatory situations.

The original focus of the handbook was primarily on existing international organizations and standards of practice. In identifying the fundamental international organizations, it directs the reader to the accepted international norms, agreements, and regulations. As the handbook was being written however, the subjects expanded to include diverse issues that are important to the discussion of what the term "global" occupational safety and health truly means. Certain chapters in the handbook expand the readers' understanding of what the concept of what "international" really means into areas that may not typically be considered the purview of worker health and safety. Some of these areas include such topics as child labor, informal workers, migrant workers, and modern slavery. Practicing occupational hygienists have a responsibility to understand the issues and concerns for these special types of workers for both ethical and practical reasons.

As an essential reference, the handbook is designed to cover the global occupational safety and health management field broadly and to offer guidelines on the major segments that constitute that field. The handbook does not attempt to deal with highly specialized aspects that are associated with detailed economic, technical, or regulatory issues. These are primarily the concern of existing specialists, who are normally available to provide relevant advice.

As the main author and editor of this book, I do not claim to have any specialized expertise in any one of the 18 topical chapters in this book. I have built upon my educational and work experiences of 38 years, working, studying, teaching, and serving in dozens of countries, to write about what seemed to me to be the most important issues and topics in each of the global subjects. My writing is built and based upon the work and research of many other dedicated professionals that have gone before me. I think that my work has been to clarify and consolidate available information on large and complex global health and safety topics.

Although I have been working on this book for over 4 years, and writing non-stop for two, as a first edition, I am aware of much additional information to be included in subsequent revisions. And there are easily several additional chapters to be included on global occupational health and safety topics in the next editions. As globalization continues, the need for greater understanding of global occupational safety and health issues and concerns will continue to grow. As countries continue to develop and grow, the need for occupational safety and health awareness will globally expand too. I hope this book is a step in the right direction, and towards a future where occupational safety and health is provided for all workers.

Acknowledgments

I thank Illinois State University President Larry Dietz, Provost Jan Murphy, and Health Sciences Department Chair Jeffery Clark for allowing me a one-semester sabbatical from my teaching duties to finish the 4-year process of writing this book. To those who do not understand or appreciate the value of having dedicated time for research and writing, I can attest that the benefits are broad and far-reaching.

I also thank my partner of the past 25 years, Dean Fournier, for the continuous support and encouragement, not only in this most recent project but also in all the others that came before, such as earning my doctorate, professional certifications, and even tenure at my university. I could not have achieved these goals without him.

I thank my mentors over the years: Evie Bain, Margaret Quinn, the now deceased Bill Charney. I thank my colleagues and cheering team at the American Industrial Hygiene Association, including Mary Ann Latko, Mellissa Cheszek, Larry Sloan, the Nonionizing Radiation Committee, and the International Affairs Committee (past and present members). I also thank all the new friends and colleagues I have made in the past few years in Workplace Health Without Borders, the Occupational Hygiene Training Association, and the International Occupational Hygiene Association. It is truly a privilege to work alongside and in collaboration with such an accomplished group of dedicated professionals.

I especially acknowledge the support in preparation of this book from a long list of expert reviewers of each of the chapters. Readers can trust that the book chapters were ultimately reviewed by some of the most respected and accomplished professionals in their fields of expertise on the given topics. I am profoundly thankful to the effort and time put in by all the reviewers, and their thoughtful comment and input to the content and direction of each section. This long list includes the following:

Nancy M. McClellan
Kul Garg
Richard Hirsh
Chris Laszcz-Davis
Deborah Nelson
Mary V. O'Reilly
Laurence Svirchev
Stephen Chiusano
David Zalk
Steven Verpaele
Christian Schumacher
Ulric Chung
Srinivas Durgam
Donna Heidel
Richard Olawoyin
Lorraine Brown
Susan Gunn, Garrett Brown, and Kevin Bulatek

Editor

Dr. Thomas P. Fuller has obtained a doctorate of science in industrial hygiene from the University of Massachusetts Lowell, Lowell, Massachusetts; a master of science in public health from the University of North Carolina, Chapel Hill, North Carolina; and a master's in business administration from Suffolk University, Boston, Massachusetts. He is currently an associate professor in the safety program at Illinois State University, Normal, Illinois. He is a certified industrial hygienist and certified safety professional with over 38 years of experience in occupational safety, radiation protection, emergency planning, industrial hygiene, infection control, environmental management, and public health. He has experience in health care, nuclear power plants, labor organizations, biopharmaceutical laboratories, manufacturing, and universities. Dr. Fuller is a member since 2002, and twice chair, of the American Industrial Hygiene Association (AIHA) Nonionizing Radiation Committee. He is also a member of the Health Care Working Group, Publications Committee, and a founding member of the AIHA Pandemic Planning Team. He is the past chair of the International Affairs Committee and the AIHA ambassador to France. He is also a member of Workplace Health Without Borders. Dr. Fuller is a contributing editor for the *American Journal of Nursing* and a member of the Editorial Advisory Board of the National Safety Council. He was recently admitted to the International Commission on Occupational Health and serves on the Industrial Hygiene Committee. Dr. Fuller is a past president of the Board of Directors of the *Journal of Occupational and Environmental Health*. He is the AIHA board representative of the International Occupational Hygiene Association and is on the board of the Occupational Hygiene Training Association.

Contributors

Marianne Levitsky is an industrial hygienist and was a founding president of the nonprofit Workplace Health Without Borders. She is a senior associate with Environmental Consulting Occupational Health (ECOH) in Mississauga, Canada. She was previously a director in the Ontario Workplace Safety and Insurance Board. Levitsky has been an occupational hygienist for the Ontario Ministry of Labour, a member of the Toronto Board of Health, and a founding board member of the Toronto Workers Health and Safety Legal Clinic in Toronto, Canada. She is an adjunct faculty of the University of Toronto, Toronto, Ontario, Canada; a recipient of the Hugh Nelson Award for Excellence in Occupational Hygiene; and an AIHA fellow.

Frederique Parrot is the global manager of occupational hygiene for Sanofi pharmaceutical company in Paris, France. She oversees occupational hygiene concerns at 120 production, research, and development sites around the world. She is also responsible for designing and developing standardized methods for evaluating and monitoring the workplaces in accordance with ISO 17025 standards for air, water, and soil.

Dr. Charles Redinger is the president/chief executive officer of the Institute for Advanced Risk Management in Harvard, Massachusetts. Since the 1980s, he has been at the forefront of developing innovative occupational health and safety (OHS) risk management solutions for public and private sector entities. His research and organizational engagements focus on unleashing organizational capacity that impacts global health and well-being. He has been a consultant to International Labor Organization and American National Standards Institute committees that developed global standards for risk management, and OHS management systems. He is a member of the National Institute of Occupational Safety and Health's Board of Scientific Counselors, and served on AIHA's board for 6 years. He is a past-chair of AIHA's Risk Assessment and Management Committees. Dr. Redinger received a PhD in industrial health from the University of Michigan, Ann Arbor, Michigan; an MPA in public policy from the University of Colorado, Denver, Colorado; and a BA in chemistry from the University of California, Oakland, California. He received a postdoctoral certificate in strategy and innovation from the MIT Sloan School of Management, Cambridge, Massachusetts.

James F. (Jim) Whiting is the principal risk engineer/consultant and MD of risk@workplaces Pty Ltd, Brisbane, Australia. A leading provider of expert advice and training in causal analysis and risk analysis systems. He has more than 40 years' experience in applications of engineering and science to risk management, health safety, and environment; incident/accident investigation; and root–cause analysis in energy, transport, petrochemical, mining, construction, health, and manufacturing industries. He has advised medium to large organizations and trained workers and managers in Australia and New Zealand, the United States, Canada, Singapore, China, Taiwan, Korea, Malaysia, and India. For 15 years, he was the national chief executive officer of the National Safety Council of Australia, Melbourne, Australia. Whiting is a past national chairman and honorary life member of the Risk Engineering Society of Australia, Barton ACT, Australia. He represented the National Safety Council and Engineers Australia on Standards Australia's various risk management sub-committees which developed AS/NZS 4801:2000 OHS Management Systems and the last edition of the Risk Management standard AS/NZS 4360: 2004 Risk Management, before it became the basis for the international version ISO 31000:2009, Risk management—Principles and guidelines.

1 Introduction to Global Occupational Safety and Health

Thomas P. Fuller
Illinois State University

CONTENTS

1.1 INTRODUCTION

Global Occupational Safety and Health (OSH) is the study of worldwide worker injury, illness, and fatality. It is the study of the factors that influence the well-being of workers internationally. It is also the study of the differences in occupational morbidity and mortality rates between countries, why those differences exist, and what can be done to improve working conditions in all nations and geographic regions. Global OSH is the analysis of complex intersections and interactions between economics, politics, culture, and science. It is more than the typical measurement and reporting of workplace exposures to risks, but an in-depth analysis of why and how the risks and hazards exist, and what social, political, economic, and cultural factors lead to those risks.

The term "globalization" represents the expanded breadth, intensity, and speed at which the world is connected. It has been represented by the following basic characteristics:

- Social, political, and economic activities commonly span across national, regional, and continental boundaries.
- The flow of trade, investment, wealth, people, and culture increases substantially.
- Global interactions of business, politics, and information are more rapid.
- Events that occur in one nation (e.g., policy changes, political instability, economic collapse, natural disasters) are felt more deeply and significantly in distant lands (Held, 1999).

In a globalized world, the actions and policies taken in one nation can have profound impacts on other nations. OSH impacts are not excluded from this premise. Globalization of business, politics, economics, and finance greatly influences OSH activities and approaches taken in individual countries around the world.

Globalization also includes the transfer of products, services, and technology. Each of these may come with their own inherent risks, including risks to workers using, manufacturing, handling, or disposing of the products. Globalization can mean the increased transfer of toxic chemicals, dangerous equipment and operations, and hazardous waste, each with their own significant and very real OSH concerns in the receiving countries. It can also bring the transfer of knowledge and education. This is particularly important for OSH, as it relates to the transfer of risks mentioned earlier.

A major catalyst for increased globalization occurs because of economic and financial benefits. And OSH has often been closely related to business and economics. Part of the original rationale for the creation of the U.S. Occupational Health and Safety Administration was to level the playing field between states that allowed different levels of safety within their operations that provided some with economic advantages. In international trade, the same incentives for businesses exist today globally, to find the country with the weakest environmental or safety regulations and set up operations there to take advantage of reduced operational or labor costs. Countries with governments that do not either understand or care about the risks to the environment or workers accept the hazardous activities, processes, and chemicals, for the purely economic benefit. Businesses then tend to seek those countries out to expand operations, and a viscous circle ensues. Countries compete for the business growth and reduce their national regulations further, which has resulted in a phenomenon coined "race to the bottom."

As economics play a special role in the national levels and norms of OSH, it is important to present some related concepts and terminology that will be used throughout this book. There is no clear international consensus on terminology to describe the economic or social status of nations, yet there is a need to have common reference words to describe conditions and make basic comparisons.

Some economists consider a developed economy to be one in which the average gross domestic product (GDP) per capita is US$12,000. Others may set this figure higher (Investopedia, 2018). However, many other factors can be considered in the definition including such parameters as access to safe drinking water, levels of sanitation and hygiene, food security, levels of air pollution, and climate vulnerability. The United Nations Human Development Index is a summary of social achievements of countries including such measures as economic growth, life expectancy, living standards, and expected years of schooling (UN, 2018). This index rates countries as very highly developed, highly developed, medium development, and low development (economically developing). Countries such as Norway, France, Spain, Japan, Canada, and Argentina are considered very highly developed, whereas countries such as Swaziland, Haiti, Yemen, and Afghanistan are considered in a state of low development.

The terminology is useful when discussing how social and economic activities impact OSH internationally. It is understandable that in the least developed nations, resources may be limited to fully implement desired OSH programs and protections. The ability to maintain adequate levels of OSH may also be related to the capacity to educate the workforce, or even supply adequate numbers of OSH professionals. Nutrition, sanitation, and availability of medical services, which are the important factors impacting the health and safety of the workforce, must all be considered when looking at OSH in a global context.

OSH and economics also overlap in terms of the discussion about supply and demand for products and services as they relate to the transfer of wealth. The costs and values of OSH are transferred and traded just as other services or "products." These are related to the status of nations or businesses. These factors also relate most basically to the differences that cultures or societies may place on the value of a life, value of longevity, or overall lifetime health. Just as there are economic disparities between nations, these closely relate through OSH to disparities in occupational injury, illness, and fatality rates, and ultimately average life span in a nation.

Many OSH professionals today feel an ethical responsibility to work to improve OSH globally, particularly in disadvantaged countries with little OSH capacity or infrastructure. Numerous volunteer opportunities exist to OSH professionals who want to share their expertise and services. There are opportunities to support OSH research to better understand the issues and controls needed to reduce worker risks globally. There are also opportunities to teach and work abroad in less developed countries. And there are roles for OSH professionals to play within their own organizations to improve the health and safety of workers in foreign countries, such as ensuring corporate supply chains follow international standards for worker safety, and the avoidance of slave and child labor in supply chains.

This edition of this book represents the first step in a broader discussion of the individual chapter topics in one resource. Many other important topics have not made it to this edition but can be included in the next revisions. For the most part, the issues and topics selected for this edition represent some of the more challenging and controversial subjects. They were selected partly as a means to begin to advance our conversations. Other chapters provide basic information on international agencies and organizations, in one location, that may be of use to OSH professionals working in international corporations or those that may be traveling abroad for work.

1.2 GROWTH AND GLOBALIZATION

Although international trade was known to occur between countries as early as 1900 BC, the need for locally grown food and limitations in transportation curtailed significant portions of societies' benefits from coming from distant lands. Mass migration occurred from time to time, but mostly by foot. Exotic materials such as spices or rare metals traveled long distances, but these represented limited parts of any given societies' trade (Stearns, 2001).

It was not until the latter half of the 19th century that industrialization and improvements in transportation brought large movements in people and products to more remote parts of the globe. Communications systems allowed for greater information exchange, societies became more urban rather than agrarian, and monetary systems based on the gold standard became more stable. By 1914, international trade had increased to 38% of the global GDP (Gun, 2017).

In the period after World War I, there was a sharp decline in trade as many countries instituted protectionist and isolationist strategies to protect their currencies and economies. The Great Depression brought further declines, and by the end of World War II, only 7% of the world's GDP was based on international trade.

During the massive reconstructions that took place after World War II, global trade grew rapidly. In 1944, the Bretton Woods accord led to numerous international agreements meant to stabilize and standardize international trade and the world economy. The International Monetary Fund was created that allowed member countries to regulate capital flows and make their currencies convertible for trade-related transactions. These monetary policies, the expansion of commercial aviation, and the development of container shipping over the next two decades led to the growth of global trade to 30% of the world GDP by 1980.

In 1995, the World Trade Organization (WTO) was created to provide a forum to negotiate international trade agreements on an agreed upon level playing field as a means to expand and promote global trade (WTO, 2017). With 164 member states, including China, the WTO has contributed to unprecedented international trade growth. Combined with the fall of the Soviet Union, government deregulation of banking systems, and significant advances in information technology, international trade has risen to 47% of global GDP (Gun, 2017).

As markets become more globally oriented, economically developing countries (EDCs) tend to increase their proportion of the world GDP, whereas those countries that are more developed lose percentage share of global GDP. Perhaps this explains the recent rise in protectionism in the G7 countries citing Brexit and the election of Donald Trump as indicators for popular nationalist and protectionist policies. And although discriminatory protectionist tariffs and barriers to free trade are on the rise and popular with some public and politicians, evidence indicates that overall, protectionist policies lead to lower total GDP growth for both developing and developed economies alike.

1.3 OSH AND GLOBALIZATION

With the expanded globalization of business and global value chains of products and workers, OSH activities must "Go Global" and reach across borders. Companies and other organizations with diverse interests and operations overseas need to expand OSH to provide adequate services to all facilities. Ex-patriot employees in foreign countries need safe working conditions, the same as in the home country. New employees hired in foreign countries have the right to safe and healthful working conditions. And various nations have differences in health and safety regulation development and enforcement. Upper managements of organizations need to consider OSH delivery and operations when expanding business abroad for mostly the same reasons that good OSH programs and implementation are required at home. In addition, organizations and countries can learn from their international counterparts or competitors about the latest trends in OSH and the values that they add to businesses and governments. That is to say, what is good for Denmark could very well also be good for Botswana with regard to worker safety. Countries can learn from one another. Companies in Botswana will need to have comparable OSH programs in place to compete in global markets in the long term.

Transfer of technology to developing countries due to the global expansion of trade has also led to transfer of hazards (ILO, 1988). Less developed countries are sometimes not in the position to handle additional hazards safely, due to the lack of adequate education, infrastructure, regulations, and availability of OSH professionals. There may be language barriers in the handling of complicated technical equipment for which operating manuals are only provided in English, or a few major languages. In Bolivia, for example, one study showed that farmers were overexposed to organophosphate pesticides because they were not aware of the hazards of exposure or benefit of protective measures (Jors, 2006).

In the absence of a concerted effort, technologies may be transferred without adequate consideration of such basic conditions as climate, sunlight, or power availability. Personal protective equipment (PPE) may not be available in adequate numbers, or workers may not know how to use it effectively. Machinery designed for healthy developed country workers may not be ergonomically correct for EDC country workers, or children and women. Nutritional deficiencies of developing country workers may exacerbate health effects from exposures to hazardous chemicals, biological, and physical agents, as well as extreme temperatures. In some developing nations, there is little awareness of the value of OSH, and therefore, little attention is paid to OSH training or research (Puplampu, 2012).

Enterprises transferring technology should consider the climate and location where the systems or equipment are to be used. Locations of plants in relation to geographic population centers should be considered. Other factors such as emergency response capabilities and environmental factors including rainfall or likelihood of flooding should also be considered. Potential effects of ongoing climate change should also be analyzed. Hazard analyses should be conducted to the same levels of awareness as would be done in developed countries, regardless of a lack of regulations or a legal impetus to do them. Hazardous chemical control, fire safety, and waste management all need to be considered prior to transferring technologies abroad. Companies and countries that are transferring technology need to maintain a constant level of adherence to international standards for OSH. The lack of regulations in a host country does not absolve organizations from a moral and ethical duty to provide the same level of care to the environment and workers as in a developed, home country.

The ethical standards for all companies and countries transferring technology to developing countries are all laid out in the International Labor Organization (ILO) 1988 document, "Safety, health and working conditions in the transfer of technology to developing countries—An ILO Code of Practice" (ILO, 1988). This document provides the criterion necessary to ensure operations within expected and accepted standards of practice. As large multinational companies continue to expand globally, they create larger numbers of global small to midsized enterprises (SMEs). These SMEs in EDCs are less likely to have the technical capacity and resources to fully address OSH issues (ILO, 2003).

1.4 NUMBERS OF INJURIES AND ILLNESSES

According to the ILO, 6,300 people die as the result of occupational accidents and illnesses each day (ILO, 2017), which supports a global total of 2.3 million occupational fatalities annually. But the numbers are not spread proportionately to populations; EDC injury and fatality rates exceed global averages and are more than double that of the most advanced nations. Occupational fatality rates for the African continent are 18.6 per 100,000 annually, compared with 6.9 for Europe and 13.7 for the United States (Hämäläinen, 2009). In addition, as a comparison, not only they are directly codependent, but the life expectancy rates for countries also vary significantly. The life expectancy is 65.7 in Botswana, 54.5 in Nigeria, and just under 50 in Swaziland. These are drastically lower than that in the United States (79.3) and Norway (81.8) (WHO, 2015).

Major causes of work-related deaths globally include cancer, communicable diseases, and circulatory diseases followed by accidents as shown in Table 1.1.

In the study by Nelson, 37% of occupational deaths were caused by chronic obstructive pulmonary disease (Nelson, 2005). Occupational diseases and injuries

TABLE 1.1
Major Causes of Death Worldwide

Major Causes of Occupational Fatality Globally (%)		
	Hämäläinen	**Takala**
Cancer	25	32
Circulatory diseases	21	23
Communicable diseases	28	17
Accidents	—	18

Source: Created from Takala (2014) and Hämäläinen (2009).

accounted for 24 million years of healthy life lost and 1.5% of all-cause mortality. The leading cause of years of healthy life lost was unintentional injuries (44%) followed by occupational hearing loss (18%). In developing countries, communicable diseases are a significant source of work-related disease, particularly in agriculture, food production, and health care (Hämäläinen, 2011). In more developed countries, such as the European Union (EU), the most common work-related injuries were musculoskeletal disorders at 60%, followed by depression, stress, and anxiety at 14% (EC Eurostat, 2009).

Although there are many good studies of injury and illness rates, there are still significant shortcomings in our true understanding of actual levels of OSH in many countries. Many countries do not require accident or injury reporting. In Pakistan, for example, which is the tenth most populated country in the world, there is no national system for recording occupational injuries or work-related casualties (Abbas, 2015). Those that do require record-keeping have large differences in systems and methods, which make it hard to compare results across borders (Takala, 2014). In the United States, for example, hundreds of thousands of government workers are not included in statistical databases and analyses.

Even when reporting structures are the same, there remain gross differences in various study assumptions and quantitative/qualitative measures of exposure. Synergistic health effects from exposures to multiple hazardous agents at the same time are also not accounted for in most reporting methods. In EDCs, employment in the informal sector reaches 70%, with the contribution to the GDP ranging from 10% to 60% (Rosenstock, 2006). Workers in the informal sector seldom have any means of recording and reporting injuries and illnesses. As a result of many of these shortcomings, the global burden of disease from occupational injury and exposure is greatly underestimated (Driscoll, 2005).

As EDCs become more industrialized, there is evidence that occupational exposures to toxic chemicals will increase, and with the increase, the risk of cancer and other illnesses (Purdue, 2015). Although these illnesses will increase, many of them will be unreported (Hämäläinen, 2009).

When developed countries move operations to less developed countries to take advantage of low labor rates and lax environmental/occupational safety laws, EDCs become more competitive and create the resulting "race to the bottom" where the least regulated country gets the larger share of business. In general, workers in the EDCs are expected to be less educated or experienced with toxic chemicals and hazardous operations, and have lower risk awareness and greater risk acceptance. Advanced countries sell/use obsolete equipment, processes, and chemicals to less developed nations (Stellman, 1998). These factors lead to potentially more unsafe and unhealthy working conditions and dangerous environmental consequences.

1.5 ECONOMIC AND SOCIAL COSTS

It is estimated that the annual economic burden of occupational injuries and disease is 4% of the global GDP (ILO, 2017). In the United States, the direct costs of disabling injuries to workers amounted to over US$53 billion This was 71% of the total compensation cost burden in 2008 (LMRI, 2010). In the United States, the direct and indirect costs of work-related injuries and illnesses are near US$250 billion (Leigh, 2011). Work-related accidents and injuries cost the EU 478 billion Euros each year. On average, worldwide work-related injuries and illnesses result in the loss of 2.9% of the world GDP. This amounts to a total global cost of 2,680 billion Euros (EU, 2017).

The social costs to families when an income earner is injured are not included in typical economic analyses. Individuals who are injured suffer significant financial consequences. The loss of income, in addition to medical expenses from the injuries, can lead to bankruptcy and forces many families into poverty as the result of a workplace injury or fatality. Many workers continue to work while injured, for both the income and fear of losing their job. This has a negative effect on the productivity of the enterprise (Boden, 2005).

1.6 LACK OF AWARENESS OF OSH PROGRAM BENEFITS

Many studies have shown a financial benefit from occupational health and safety programs, policies, and systems (Harms-Ringdahl, 1990; Lahiri, 2005; Amador-Rodenzo, 2005; Argiles-Bosch, 2014; Oxenburgh, 2005). When OSH professionals within organizations understand and can communicate the business value of OSH programs and systems, they contribute to better decision-making by management. Various methods for economic evaluation have been developed to assist managers and OSH professionals (Biddle, 2005; AIHA, 2011; Bergstrom, 2005; Tompa, 2010).

The business value of safety can be quantified in terms of return on investment (ROI) just as any other business endeavors, where the ROI is the reduction in financial losses due to injuries or illnesses, which is divided by the cost of the safety intervention or program. These direct costs typically include labor costs, turnover costs, productivity losses, costs of accident response, and higher insurance premiums (Lyon, 1997). There is often a payback period associated with the intervention that is compared to the initial financial outlays.

In addition to the direct financial benefits of OSH programs and control interventions, there are numerous indirect benefits to organizations that use OSH to reduce injury and illness rates. These include improvements to product quality, employee morale, reduced absenteeism, better corporate image, and fewer penalties for non-compliance with laws and regulations.

Studies have shown that top-level corporate financial decision-makers have favorable perceptions of OSH programs and interventions. The most beneficial safety modification as perceived by managers was worker OSH training (Huang, 2007, 2011). Yet many businesses, particularly small and midsized ones, have not fully recognized the business value of comprehensive OSH programs (DeArmond, 2010; Cagno, 2013).

What is good at the enterprise level has also been shown to be good at the national level. Good OSH programs and working conditions lead to improved productivity and greater competitiveness. In a study by the World Economic Forum and ILO, the relationships between economic competitiveness and the number of fatal accidents per 100,000 workers are shown clearly in Figure 1.1.

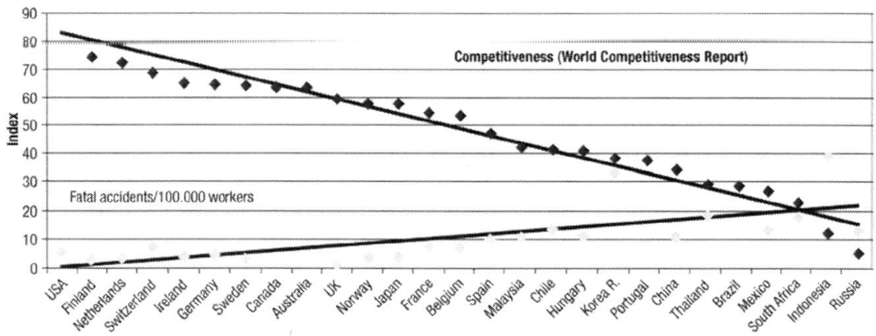

FIGURE 1.1 Competitiveness and safety. (*Source:* ILO, 2003.)

The most financially competitive and efficient business units tend to have the lowest fatality rates.

National governments are even less likely than businesses to thoroughly consider the financial benefits of OSH regulation. In addition, there has been a shortage of academic studies on the topic (Viscusi, 2006). Developing countries in the vacuum of clear data are reluctant to implement stringent OSH regulation (injury and illness reduction) for fear of stifling growth (Pouliakas, 2013). In one recent study in Ghana, it was shown that the implementation of a robust OSH management program could significantly improve the overall socioeconomic development of the country (Amponsah-Tawiah, 2013).

In addition to looking at OSH with an eye towards productivity and efficiency, several other international management principles are commonly used. The International Standards Organization (ISO) has published "ISO 45001 Occupational Health and Safety Management Systems," which provides the foundation for an OSH program, a system to monitor that all parts of the program are accomplished, and a format for continuous improvement. Other similar systems include the OHSAS 18001 Occupational Health and Safety Management Systems, ANSI/AIHA/ASSE Z10-2012 Occupational Health and Safety Management Systems, ILO-OSH Guidelines on OSH Management Systems, and the Australian/New Zealand document AS/NZS 4801:2001 Occupational Health and Safety Management Systems—specification with guidance for use.

1.7 MANAGEMENT AND BENCHMARKING

Despite the availability of numerous standardized and recognized OSH management systems, there remains a vacuum in the realm of OSH business strategy and research. Advances in corporate social responsibility reporting have begun to include OSH topics and performance indicators, but the practice is not pervasive. Advanced businesses and institutions may practice elaborate reporting and benchmarking between facilities, even those in foreign countries. But industry benchmarking in terms of OSH, particularly internationally, has several shortcomings in development, implementation, and reporting.

Reporting of injuries and illnesses, even if not required by the government, is the first step in understanding OSH issues and impacts in the workplace. Other "leading indicators" of performance are being developed and implemented to provide a better understanding of impacts and where program improvements would be most useful. Expanded benchmarking activities such as program reviews, audits, inspections, and the structured reporting of results would improve overall OSH. Benchmarks that allow comparisons between countries and continents would also be useful towards creating systems of continuous improvement.

1.8 CULTURAL, ECONOMIC, AND EDUCATIONAL DIFFERENCES

Global occupational hygiene cannot be fully understood or discussed without including the consideration of cultural differences between countries and peoples. Culture plays a large role in the judgment and acceptance of risk in general. Although there is a shortage of literature on the topic, it might be assumed that cultural awareness and responses to risk and hazards would flood over into the workplace.

In a country with a lower life expectancy, the people and workers themselves might be expected to be more complacent about workplace hazards and resulting injuries or fatalities. That is, if a long life is valued less in general, then it might be expected that there would be less concern over being killed at work. This is particularly true if the job is a necessity for survival. The economics of a person's condition can play a large role in his or her acceptance of risk in performing any hazardous job, in any culture.

In developing nations, women and children comprise a significantly higher percentage of the workforce. And women and children make up more of the workforce in informal economies, where workplace hazards are less well evaluated or controlled than in established economies. Children, in particular, are more susceptible to the harmful effects of hazardous workplaces. Physiological characteristics of children such as higher breathing and metabolic rates tend to increase their exposures and resulting doses to potentially toxic or carcinogenic chemicals. And their young age makes the exposure period or time for toxic actions to be well before the normal life expectancy. Children are typically less able to recognize, understand, or be aware of how or why to control occupational risks than adults from the same culture. Women are also less likely to be concerned about risks as they are often more dependent upon the job for survival of their families. And women have anatomic and physiological differences that may make them more susceptible than men to some hazards. For women, normal workplace protective controls, including PPE, may not be adequate since they are typically designed for men.

Workers in EDCs are more likely to be malnourished, dehydrated, and overworked, and are therefore inherently more "at risk" from exposure to hazardous agents or working conditions. Routine workplace protective controls and even PPE may not fit properly or be adequately protective. These workers may be more likely to be involved in an accident and may be more affected by working conditions such as intense thermal work environments.

1.9 EDUCATION, TRAINING, AND CREDENTIALING

1.9.1 Formal Education in OSH

Educational systems created to meet the demand for OSH professionals vary greatly between countries. Differences partly reflect the various modes of OSH practice in the country and also are a result of different levels of emphasis put on the value of OSH by governments and potential employers. In countries where the true preventive value of OSH practice has been realized, the educational system is typically more developed and streamlined to provide sufficient numbers of OSH graduates at the university, master's, and doctoral levels.

In a comprehensive study of OSH education in Canada, it was determined that the existing system of OSH education contained several gaps in program content and shortcomings in the ability to meet the need for OSH professionals at different levels, and that the OSH educational system needed significant development and enhancement to increase course offerings in certain sectors to ensure minimum levels of competencies on OSH issues (Curran, 2013). In the United States, the demand for qualified OSH professionals has historically outpaced the supply, and that continues to be the case today (Erickson, 1991). In Europe, OSH education systems vary greatly between countries, with most of the OSH courses offered within master's programs in engineering, applied sciences, or management (Arezes, 2012).

Globally, the demand for OSH professionals can also be expected to outpace the supply. As developing countries become more industrialized, the numbers and variety of occupational exposures to hazardous substances and working conditions lead to the need for more professionals capable to evaluate and control them. Policies and funding for OSH professional capacity building are needed in numerous countries to meet current and future demands (Harrison, 2016). Educational programs should, in addition, allow for education at the master's and doctoral levels to fully support the field of occupational hygiene and its development in the future (Vincent, 2005).

1.9.2 Global OSH Training

The shortage of formal educational development in OSH is mirrored by the need of OSH training by the workers. That is, the lack of competent qualified professionals to teach OSH courses in the workplace leaves workers in hazardous conditions where they are unaware of the risks or the appropriate controls needed to minimize the risks. OSH training is offered in many formats including online courses and seminars. Workers can obtain safety and health information for general awareness, or they can earn certificates in recognized accreditation programs. The Occupational Health Training Association (OHTA) is a nonprofit organization that has materials for several training courses available online for free. These course materials can be used by qualified OHTA trainers to allow students to sit for OHTA certificates in various OSH subject areas such as hazard control, ergonomics, and toxicology (OHTA, 2018).

Although Internet and web access training have the potential for reaching millions of workers, the true potential has yet to be achieved. The Multilingual Health and Safety Resource Guide to Worker Training Materials on the Web prepared by the State of California is an excellent source of contact information for OSH training (CA, 2014).

1.9.3 Mainstreaming OSH Education

A new approach for bringing OSH into the forefront of public awareness is the concept of mainstreaming. This is the introduction of basic safety concepts and principles into childhood and adolescent education programs with the belief that they will carry OSH concepts and principles into their working careers and daily lives as adults. It is a strategic community oriented approach being strongly promoted by the European Agency for Safety and Health at Work. Several EU countries have already developed elementary educational programs and training materials (EASHW, 2018).

The European Network Education and Training in Occupational Safety and Health (ENETOSH) is supported by the European Commission with the aim to maintain OSH education and training. The primary approach to training is through the mainstreaming of OSH into all facets and levels of education in society. ENETOSH has more than 80 partners from 33 countries that work together to collaborate on projects to achieve objectives (ENETOSH, 2018). The ENETOSH's belief is that the best way to ensure safe workplaces in the future for all workers is to ensure everyone is trained and receives an appropriate education of OSH topics throughout their education and ongoing careers.

1.9.4 Licensing and Credentialing

Government regulation for occupational licensure is required for many professions. Licenses may require a minimum number of hours of related training, age requirements, internships or apprenticeships, and/or passing an examination. Licensure has historically been associated with professions that impacted the public health in some way, such as nurses or physicians. In 2009, it was estimated that approximately 38% of workers required some sort of government certification or licensing (Kleiner, 2009). Licensure may also be associated with ensuring minimum levels of quality and protections for consumers, such as required training and licensing for hairdressers (Bryson, 2010). Unfortunately, in many other areas, professional licensure exists solely as a means of revenue for governments and/or a means to bar entry into a field, and raise wages to those already practicing. In one recent study, it was shown that the number of training hours required for emergency medical technicians (33 days) was significantly less than the number of training hours required for other professions much less related to public safety, such as cosmetologists (372 days) (Carpenter, 2015).

Although most would agree that occupational hygiene is closely related to public health and safety, few governments require licensure to practice as an occupational hygienist. When licensure is not required, the next best means available to document levels of expertise in a profession is through an organized and stepped system of credentialing. Different professional credentials or designations require various combinations of levels of education, experience, and examination (SIA, 2016). In lieu of licensure, certification can be a means for employers and the public to have a level of awareness about the stated capabilities of a person practicing occupational hygiene. Numerous OSH professional credentials have surfaced in several countries as a result of the need. In addition, international professional organizations such as the International Occupational Hygiene Association have begun to rate various national professional credentialing systems as a means to make comparisons between systems (IOHA, 2018).

In 2010, a new group of professional associations of safety professionals created the European Network of Safety and Health Professional Organizations. The goals of the group are to influence legislation, exchange information, and develop good standards of practice. In addition, they have created standardized certifications for two different levels of practice based on experience and education (ENSHPO, 2017).

1.10 INTERNATIONAL ORGANIZATIONS AND REGULATIONS

Several intergovernmental organizations (IGOs) exist to promote OSH in a variety of capacities. Some organizations such as the ILO are tripartite in nature, comprising representatives from government, workers, and employers. They create conventions and treaties that members can choose to ratify into national laws of standards if they so choose. Others have been created as subsets of larger IGOs, such as the World Health Organization. The basic information on these groups is provided solely as a simple introduction to their organizations and activities, and as a start point for more related information.

International nongovernmental organizations are presented, and their fundamental activities and objectives are provided. Many of these organizations have tens of member professional organizations from around the world. Individual national organizations have not been presented in this book, even though they may individually have large international memberships, in addition to international projects. For the most part, these organizations are represented by the international organizations to which they belong.

Chapter 11, on national profiles, represents only a sample of information on a small group of countries or regions. Future editions of the book will expand and balance the information provided on a larger number of countries and regions. The decision to include or exclude countries in this edition was based solely on the factors that impacted the ready availability of information on the given countries and regions.

1.11 SPECIAL TYPES OF WORKERS—CHILDREN AND INFORMAL WORKERS

1.11.1 Child Labor

The ILO estimates that globally, there are approximately 250 million children between the ages of 5 and 14 who work. And nearly half of them are working full-time (ILO, 2003). Many of these children work in hazardous or exploitive jobs, and despite being illegal in all countries, child labor continues and is supported financially by global supply chains. It is important for occupational hygienists to understand the issues surrounding the unethical and immoral use of the worst forms of child labor and to play a role in corporate social responsibility programs for informing organizations and governments.

1.11.2 Informal Workers

In many parts of the world such as Pakistan, Mali, Nepal, and Africa, more than 70% of the workforce does not work in formal workplace settings or industries (ILO, 2003). They do not have typical relationships with employers and are found in such jobs as waste collection, agriculture, fishing, transportation, and service industries. These workers often work from home and in cramped and unsafe workspaces. These workers lack legal protections from hazards, social services from governments, and medical support for injuries or illnesses incurred on the job. Informal workers face significant occupational risks and are often ill-prepared to protect themselves from them.

REFERENCES

Abbas, M., Trend of occupational injuries/diseases in Pakistan: Index value analysis of injured employed persons from 2001–02 to 2012–13, *Safety and Health at Work* (2015) Vol. 6, pp. 218–226.

AIHA (2011) *AIHA Business Value Strategy Manual*, the American Industrial Hygiene Association, Fairfax, VA ISBN: 978-1-935082-22-4, edited by Bernie Silverstein. American Industrial Hygiene Association.

Amador-Rodenzo, R., An overview to CERSSO's self evaluation of the cost-benefit on the investment in occupational safety and health in textile factories: A step by step methodology, *Journal of Safety Research* (2005) Vol. 36, pp. 215–29.

Amponsah-Tawiah, K., Occupational health and safety and sustainable development in Ghana, *International Journal of Business Administration* (2013) Vol. 4, No. 2, pp. 74–78.

Arezes, P., Swuste, P., Occupational health and safety post-graduation courses in Europe: A general overview, *Safety Science* (2012) Vol. 50, pp. 433–442.

Argiles-Bosch, J., Marti, J., Monllau, T., Garcia-Blandon, J., Urgell, T., Empirical analysis of the incidence of accidents in the workplace on firms' financial performance, *Safety Science* (2014) Vol. 70, pp. 123–132.

Bergstrom, M., The potential-method—An economic evaluation tool (2005), *ECON Proceedings* Vol. 36, pp. 237–240.

Biddle, E., Ray, T., Owusu-Edusei, K., Camm, T., Synthesis and recommendations of the economic evaluation of OHS interventions at the company level conference, *Journal of Safety Research* (2005) Vol. 36, pp. 261–267.

Boden, L., Running on empty: Families, time, and workplace injuries, *American Journal of Public Health* (2005) Vol. 95, No. 11, pp. 1894–1897.

Bryson, A., Kleiner, M., The regulation of occupations, *British Journal of Industrial Relations* (2010) Vol. 48, No. 4, pp. 670–675.

CA, California, *Multilingual Health & Safety Resources a Guide to Worker Training Materials on the Web* (2014) http://lohp.org/wp-content/uploads/2014/06/Multilingual-Guide-6th-edition-June-2014.pdf accessed July 29, 2018.

Cagno, E., Micheli, G., Masi, D., Jacinto, C., Economic evaluation of OSH and its way to SMEs: A constructive review, *Safety Science* (2013) Vol. 53, pp. 134–152.

Carpenter, D., Knepper, L., Erickson, A., Ross, J., Regulating work: Measuring the scope and burden of occupational licensure among low and moderate-income occupations in the United States, *Economic Affairs* (2015) Vol. 35, No. 1, pp. 3–20.

Curran, V., Hayward, M., Bornstein, S., Del Bianco, A., Demers, P., Bartlett, K., Davies, H., LeFort, S., MacKinnon, S., Miller, S. (2013). *Educational Offerings in Health and Safety in Canadian Post-Secondary Institutions: A Survey of Canadian Schools (RS2011-IG38).* Nova Scotia: WorkSafeBC, Workers' Compensation Board.

DeArmond, S., Huang, Y., Chen, P., Courtney, T., Corporate financial decision makers' perceptions of their company's safety performance, programs and personnel: Do company size and industry injury risk matter? *Work* (2010) Vol. 37, pp. 3–13.

Driscoll, T., Takala, J., Steenland, K., Corvalan, C., Fingerhut, M., Review of estimates of the global burden of injury and illness due to occupational exposures, *American Journal of Industrial Medicine* (2005) Vol. 48, pp. 491–502.

EASHW, European Agency for Safety and Health at Work, Homepage (2018) https://osha.europa.eu/en accessed July 29, 2018.

ENETOSH, European Network Education and Training in Occupational Safety and Health, Who We Are (2018) www.enetosh.net/webcom/show_article.php/_c-221/_nr-1/_lkm-175/i.html accessed July 29, 2018.

ENSHPO, European Network of Safety and Health Professional Organisations, Homepage (2017) http://enshpo.eu/home accessed July 29, 2018.

Erickson, J., Occupational safety and health professionals: A University of Southern California study, *Professional Safety* (1991) Vol. 36, No. 12, p. 33.

EU, European Union, Eurostat, (2009) *8.6% of Workers in the EU Experienced Work-Related Health Problems. Statistics in Focus 63/2009* (p. 12). European Communities http://ec.europa.eu/eurostat/documents/3433488/5283817/KS-SF-09-063-EN.PDF/10b62d3b-e4dd-403f-b337-af6ffd3de8de accessed July 1, 2016.

EU, EU-OSHA, Press Release-Work-Related Accidents and Injuries (2017) https://osha.europa.eu/en/about-eu-osha/press-room/eu-osha-presents-new-figures-costs-poor-workplace-safety-and-health-world accessed July 29, 2018.

Gun, D., Keller, C., Kochugovindan, S., Wieladek, T., The End of Globalisation As We Know It? (2017) www.investmentbank.barclays.com/our-insights/the-end-of-globalisation-as-we-know-it.html?cid=ppc_sc08e00v11m08WEpa00pv00&trid=43700022669873736 Barclays Bank accessed July 31, 2017.

Hämäläinen, P., Saarela, K. L., Takala, J., Global trend according to estimated number of occupational accidents and fatal work-related diseases at region and country level, *Journal of Safety Research* (2009) Vol. 40, pp. 125–139.

Hämäläinen, P., Saarela, K., Takala, J., Global estimates of fatal work-related diseases by region and disease group, *International Journal of Occupational and Environmental Health* (2011) Vol. 17, No. 1, pp. 49–56.

Harms-Ringdahl, L., On economic evaluation of systemic safety work at companies, *Journal of Occupational Accidents* (1990) Vol. 12, pp. 89–98.

Harrison, J., Dawson, L., Occupational health: Meeting the challenges of the next 20 years, *Safety and Health at Work* (2016) Vol. 7, pp. 143–149.

Held, D., McGrew, A., Goldblatt, D., Perraton, J. (1999). *Global Transformations Reader: Politics, Economics and Culture.* Cambridge: Polity Press.

Huang, Y., Leamon, T., Courtney, T., Chen, P., DeArmond, S., Corporate financial decision-makers' perceptions of workplace safety, *Accident Analysis and Prevention* (2007) Vol. 39, pp. 767–775.

Huang, Y., Leamon, T., Courtney, T., Chen, P., DeArmond, S., A comparison of workplace safety perceptions among financial decision-makers of medium- vs. large-size companies, *Accident Analysis and Prevention* (2011) Vol. 43, pp. 1–10.

ILO (1988). *Safety, Health and Working Conditions in the Transfer of Technology to Developing Countries—An ILO Code of Practice (1988).* Geneva: International Labor Organization. ISBN 92-2-106122-1.

ILO (2003). *Safety in Numbers: Pointers for a Global Safety Culture at Work.* Geneva: International Labor Organization. ISBN: 92-2-113741-4.

ILO, Safety and Health at Work (2017) www.ilo.org/global/topics/safety-and-health-at-work/lang--en/index.htm accessed August 8, 2017.

Investopedia, Top 25 Developed Countries (2018) www.investopedia.com/updates/top-developing-countries/ accessed July 28, 2018.

IOHA, International Occupational Hygiene Association, National Accreditation Recognition Committee (2018) https://ioha.net/national-accreditation-recognition-nar-committee/ accessed July 29, 2018.

Jors, E., Morant, R., Aguilar, G., Huici, O., Lander, F., Baelum, J., Konradsen, F., Occupational pesticide intoxications among farmers in Bolivia: A cross-sectional study, *Environmental Health* (2006) Vol. 5, p. 10.

Kleiner, M., Krueger, A. (2009), Analyzing the extent and influence of occupational licensing on the labor market, NBER Working Paper 14979, pp. 1–35. (From Bryson 2010).

Lahiri, S., Gold, J., Levenstein, C., Net-cost model for workplace interventions, *Journal of Safety Research* (2005) Vol. 36, pp. 241–255.

Leigh, J., Economic burden of occupational injury and illness in the United States, *Millbank Quarterly* (2011) Vol. 89, No. 4, pp. 728–772.

LMRI, Liberty Mutual Research Institute for Safety (2010). *Liberty Mutual Workplace Safety Index.* Hopkinton, MA: Liberty Mutual.

Lyon, B., Ergonomic benefit/cost analysis: Communicating the value of enhancements, *Professional Safety* (1997) Vol. 42, No. 3, p. 33.

Nelson, D., Concha-Barrientos, M., Driscoll, T., Steenland, K., Fingerhut, F., Punnett, L., Pruss-Ustun, A., Leigh, J., Corvalan, C., The global burden of selected occupational diseases and injury risks: Methodology and summary, *American Journal of Industrial Medicine* (2005) Vol. 48, pp. 400–418.

OHTA, The Occupational Hygiene Training Association homepage (2018) www.ohlearning.com/about-ohta/purpose-and-principles.aspx accessed June 24, 2018.

Oxenburgh, M., Marlow, P., The Productivity Assessment Tool: Computer-based cost benefit analysis model for the economic assessment of occupational health and safety interventions in the workplace, *Journal of Safety Research* (2005) Vol. 36, pp. 209–214.

Pouliakas, K., The economics of health and safety at work: An interdisciplinary review of the theory and policy, *Journal of Economic Surveys* (2013) Vol. 27, No. 1, pp. 167–208.

Puplampu, B., Quartey, S., Key issues on occupational health and safety practices in Ghana: A review, *International Journal of Business and Social Science* (2012) Vol. 3, No. 9, p. 2.

Purdue, M., Hutchings, S., Rushton, L., Silverman, D., The proportion of cancer attributable to occupational exposures, *Annals of Epidemiology* (2015) Vol. 25, pp. 188–192.

Rosenstock, L., Cullen, M., Fingerhut, M. (2006), Disease control priorities in developing countries. In D. T. Jamison, J. G. Breman, A. R. Measham, G. Alleyne, M. Claeson, D. B. Evans, P. Jha, A. Mills, P. Musgrove (Eds). *Occupational Health* (2nd ed., Chapter 60, p. 1127). New York: Oxford University Press www.ncbi.nlm.nih.gov/books/NBK11750/. ISBN-10: 0-8213-6179-1.

SIA, Safety Institute of Australia (SIA), Why is Credentialing Important? (2016) https://sia.org.au/certification/why-is-certification-important accessed August 12, 2017.

Stearns, P., Langer, W. (2001). *The Encyclopedia of World History: Ancient, Medieval, and Modern, Chronologically Arranged* (6th ed.). Boston: Houghton Mifflin Company. ISBN 0-395-65237-5.

Stellman, J., The ILO encyclopedia of occupational health and safety: A multidisciplinary challenge, *International Labour Review* (1998) Vol. 137, No. 3, pp. 410–418.

Takala, J., Hämäläinen, P., Saarela, K., Yun, L., Manickam, K., Jin, T., Heng, P., Tjong, C., Kheng, L., Lim, S., Lin, G., Global estimates of the burden of injury and illness at work in 2012, *Journal of Occupational and Environmental Hygiene* (2014) Vol. 11, No. 5, pp. 326–337.

Tompa, E., Verbeek, J., van Tulder, M., de Boer, A., Developing guidelines for good practice in the economic evaluation of occupational safety and health interventions, *Scandinavian Journal of Work, Environment & Health* (2010) Vol. 36, No. 4, pp. 313–318.

UN, Human Development Index (2018) http://hdr.undp.org/en/content/human-development-index-hdi accessed July 28, 2018.

Vincent, J., Graduate education in occupational hygiene: A rational framework, *Annals of Occupational Hygiene* (2005) Vol. 49, No. 8, pp. 649–659.

Viscusi (2006), Regulation of health, safety and environmental risk. Discussion paper 544, The Harvard John M. Olin Discussion Paper Series.

WHO, Life Expectancy at Birth Rates for 2015; Annex B Tables of Health Statistics by Country, Who Region and Globally (2015) www.who.int/gho/publications/world_health_statistics/2016/EN_WHS2016_AnnexB.pdf?ua=1 accessed August 4, 2017.

WTO, Overview (2017) www.wto.org/english/thewto_e/whatis_e/wto_dg_stat_e.htm accessed July 31, 2017.

2 Intergovernmental Occupational Safety and Health Organizations

Thomas P. Fuller
Illinois State University

CONTENTS

2.1 INTRODUCTION

Intergovernmental organizations (IGOs) are relationships and/or organizations created by treaties, charters, or other formal agreements between two or more nations to work towards a common goal or interest. Treaties that are formed by IGOs can be enforceable by international law. The main purpose of intergovernmental projects and agreements is typically to address a common set of economic or social problems. With the rise of increased globalization of business, communications, and science over the past few decades, the importance and roles of IGOs have grown significantly in terms of global governance and influence.

The standards and rules developed by IGOs can be useful and efficient means to identify and control risks. They are particularly of benefit for risks that may easily and potentially cross national or regional boundaries, such as those developed to address environmental hazards that may migrate via air or water to affect other countries downwind or downstream. IGO agreements may provide incentives to better control risks, communicate activities or releases, plan for emergencies, and develop safer or more sustainable systems and operations. Mutual assistance is more likely to flow from developed to less economically developed countries when there is an economic, financial, or political interest and benefit to the richer nations. In terms of environmental transboundary issues, businesses are well aware of the relationship between sound environmental judgment and financial and political stability (Kirchsteiger, 2005).

2.2 INTERNATIONAL LABOR ORGANIZATION

2.2.1 Background and History

The International Labor Organization (ILO) was founded in 1919, as part of a peace settlement in the Treaty of Versailles at the end of World War I. Portions of the treaty were founded on the notion that lasting peace could only be accomplished if society was fundamentally and universally just. And social justice included the concept of a minimum set of standards and accepted practice for humane work.

The original ILO constitution was created by a Labour Commission set up by the Versailles Peace Conference in 1919. This Labour Commission was a tripartite organization with representatives from government, labor, and employers. It had representatives from Belgium, Cuba, the former Czechoslovakia, France, Italy, Japan, Poland, the United Kingdom, and the United States (ILO, 2017a,b).

The original ILO constitution focused on the relationship between peace and harmony and the role that unfair and inhumane working conditions played in creating political dissent and unrest. The authors believed that nations with inadequate working conditions destabilized and placed in peril, other countries with better working conditions due to such activities as uncontrolled migration and open conflict.

Topics of concern in the preamble to the original constitution included reference to maximum working hours, provision of adequate and comparable wages for equal work, safe and healthy workplaces, special protections for women and children, provisions for old age and injured workers, basic freedoms for worker association, and the provision of training and worker development.

At the conclusion of World War II, the ILO constitution was revised to include the "Declaration of Philadelphia," which inserted a focus on universal social progress. It concluded that lasting peace cannot be attained locally, or globally, without universal social justice including economic security and equal opportunity. It held that poverty and severe inequality present a danger to prosperity and peace everywhere, and conscientious direct action must be taken to combat them internationally.

The ILO vision was that work should be a source of personal well-being and social integration. The Declaration of Philadelphia stated that "Labour is not a commodity." The organization is based on the principle that all people should have equal rights to beneficial work opportunities and conditions that prevent poverty and ensure security and longevity.

The ILO Constitution governance is based on a tripartite representation that allows for open discussion and democratic decision-making. Upon the creation of ILO Conventions and Recommendations regarding work, participating nation's governments adopt the policies for ratification or other action. A system of inspection is used to enforce commitments to conventions in national laws and regulations. The governance is, in addition, supported by a robust system of collaboration between international organizations to ensure that financial and economic programs continue to support the goals of sustained social progress and equality.

National commitments to ILO conventions and principles vary. And individual nation commitments may vary over time. The need for an international organization to protect workers and promote safe work conditions is based on the belief that not all individual countries can progress alone and that all nations are truly interdependent in their survival and success. Despite ebbs and flows of isolationism and nationalism, the world today truly functions as a global system and it is nearly unavoidable. The original ILO principles that require international collaboration and dialogue are perhaps more integral to how the world works today than at any time in the past.

2.2.2 Funding Sources

The ILO's biennial work program and budget funded by Member States is reviewed and approved every 2 years at the annual conference.

2.2.3 Regulatory Format and Legal Powers/Sources

The ILO Constitution establishes the purpose of the group and membership criteria, which includes charter members from 1945, Member States of the United Nations that chose to belong, and other nations that request membership and then are voted in by two-thirds delegates at the General Conference which meets at a minimum, once a year. The organization consists of a tripartite system that includes this General Conference and member representatives, a governing body composed of representatives from governments, employers, and employees, and an administrative ILO Office controlled by the governing body (ILO, 2017a,b).

Each group in the General Conference has voting powers in setting standards and policies. Within the General Conference, the government has two votes, and employers and employees each have one vote, on the adoption of ILO instruments and its agendas. The day-to-day governing body of the organization has 14 employer members, 14 employee members, and 28 members representing governmental bodies.

It is broadly believed that the tripartite system provides a means to present the interests of employers and workers in a more direct way than solely using government-appointed political representatives. In the tripartite system, enterprise viewpoints and worker rights have weight against various potentially contrary and conflicting governmental political interests. Employer and employee representatives have the additional power to transcend national boundaries and can create broad policies that umbrella over multiple governments since enterprise and worker concerns may be universal or multilateral in nature.

One shortfall of the tripartite system is that it still fails to represent a large segment of the working population throughout the world, specifically those that work in the informal economy. Some data show that up to 70% of the world's workforce work in neither the government nor formal businesses (ILO, 2003). In addition, labor organizations are losing power and numbers in many countries, so their representation is diminishing. There is now debate whether it might be appropriate for other nongovernmental organizations (NGOs) to be included in the governance of the ILO as a means to expand representation to a broader and underrepresented group.

2.2.4 Conventions and Recommendations Regarding Occupational Safety and Health

ILO standards are proposed and negotiated on a multinational tripartite forum by the International Labour Conference. Standards are set in either Conventions or Recommendations. Conventions that are adopted by the ILO are then ratified, or not, by member nations. The conventions are similar in nature to other international treaties, and member nations have the option not to ratify them in their national regulations or laws. Nations must consider the adoption of Conventions and Recommendations, and those that sign on to the Conventions must then promulgate and ratify laws in their own countries in order to meet their commitments to the ILO. ILO Recommendations are standards for good practice and provide guidance for work organization and conditions.

The International Labour Conference is held at least once a year in Geneva, Switzerland. The primary activities of each conference include creating and adopting new conventions or recommendations. In addition, the conference body reviews reports submitted by Member States regarding their status of compliance with their existing obligations.

Reports of the ILO Office on the topics of rights to collective bargaining and freedom of association, elimination of forced labor, abolition of child labor, and elimination of discrimination in the workplace are reviewed at each conference on an annually rotating basis. Other timely and important topics are discussed and debated at each conference, and plans for future directions and activities are determined.

The original 188 Recommendations and 188 Conventions adopted by the ILO in 1919 covered a broad range of labor law and the control of work. The governing body has identified eight fundamental conventions outlined in the ILO's Declaration on Fundamental Principles and Rights at Work (ILO, 1998). The eight fundamental rights are as follows:

1. Freedom of Association and Protection of the Right to Organise Convention, 1948 (No. 87)
2. Right to Organise and Collective Bargaining Convention, 1949 (No. 98)
3. Forced Labour Convention, 1930 (No. 29)
4. Abolition of Forced Labour Convention, 1957 (No. 105)
5. Minimum Age Convention, 1973 (No. 138)
6. Worst Forms of Child Labour Convention, 1999 (No. 182)
7. Equal Remuneration Convention, 1951 (No. 100)
8. Discrimination (Employment and Occupation) Convention, 1958 (No. 111)

The ILO has also designated four additional conventions as "governance conventions" because of their importance in the operation of the standards system. These conventions were listed in the Declaration on Social Justice for a Fair Globalization, and Member States are encouraged to follow them. The governance conventions are as follows (ILO, 2008):

1. Labour Inspection Convention, 1947 (No. 81)
2. Employment Policy Convention, 1964 (No. 122)
3. Labour Inspection (Agriculture) Convention, 1969 (No. 129)
4. Tripartite Consultation (International Labour Standards) Convention, 1976 (No. 144)

The Occupational Safety and Health (OSH) Convention C155 was adopted by the ILO in 1981 and put into force in 1983. It applies to all environments and workplaces of economic activity and public service and covers all employed persons. It requires Member States to design and implement policies, programs, and necessary laws and regulations to ensure safe working environments for employees free from hazardous working conditions, processes, or chemical, physical, and biological agents that pose an unacceptable risk. It requires the development of training and qualification for individuals responsible for the achievement of adequate levels of health and safety in the workplace.

The convention requires employers to determine the degree of hazard associated within various environments and processes, and implement effective controls to reduce them. Protocols must be developed for the response to and notification of accidents and any other occupationally related diseases or injuries to employees. The convention requires Member States to develop a system of enforcement of their regulations and penalties for employers who do not comply.

The Occupational Health Services Convention of 1985 establishes basic preventive functions for maintaining safe and healthy work environments. This convention requires Member States to design and implement programs that require employers to perform such activities as workplace risk assessments, surveillance of working conditions, active planning and program development to maintain worker health and safety on an ongoing basis, workplace safety training and communication, first aid and emergency treatment, and accident/illness investigation.

In 2009, the Promotional Framework for Occupational Safety and Health Convention number C187 went into force. This convention requires Member States to develop programs to ensure the continuous promotion of a national preventive OSH culture. It establishes requirements to progressively develop and periodically review programs to ensure continuous improvement. It also includes the use of research and statistics to review progress, collaboration with insurance and social security systems, and the expansion of training to include other aspects of capacity building.

Other miscellaneous OSH conventions and recommendations are provided in the following list:

P155—Protocol of 2002 to the Occupational Safety and Health Convention, 1981
R164—Occupational Safety and Health Recommendation, 1981 (No. 164)
R171—Occupational Health Services Recommendation, 1985 (No. 171)
R197—Promotional Framework for Occupational Safety and Health Recommendation, 2006 (No. 197)
R097—Protection of Workers' Health Recommendation, 1953 (No. 97)
R194—List of Occupational Diseases Recommendation, 2002 (No. 194)

Instrument with Interim Status
R031—Prevention of Industrial Accidents Recommendation, 1929 (No. 31)

Replaced Recommendations
R112—Occupational Health Services Recommendation, 1959 (No. 112)
C115—Radiation Protection Convention, 1960 (No. 115)
R114—Radiation Protection Recommendation, 1960 (No. 114)
C139—Occupational Cancer Convention, 1974 (No. 139)
R147—Occupational Cancer Recommendation, 1974 (No. 147)
C148—Working Environment (Air Pollution, Noise and Vibration) Convention, 1977 (No. 148)
R156—Working Environment (Air Pollution, Noise and Vibration) Recommendation, 1977 (No. 156)
C162—Asbestos Convention, 1986 (No. 162)
R172—Asbestos Recommendation, 1986 (No. 172)
C170—Chemicals Convention, 1990 (No. 170)
R177—Chemicals Recommendation, 1990 (No. 177)
C174—Prevention of Major Industrial Accidents Convention, 1993 (No. 174)
R181—Prevention of Major Industrial Accidents Recommendation, 1993 (No. 181)

2.2.5 Administrative Programs

One of the most important administrative activities of the ILO is the supervision of standards implementation. Each year participating nations must submit a report regarding ratified conventions and status of implementation to the ILO Committee of Experts. The governing body also has the right to require reports from Member States on their practices concerning unratified Conventions and Recommendations. This puts pressure on countries that have not ratified a certain convention to at least explain what they are doing to address the issue.

Complaints regarding a given country's adherence to its ILO commitments can be made by nation members, delegates to the International Labour Conference, and employer or worker representatives. Complaints are reviewed by the Governing Body Committee on Freedom of Association.

In some cases, the ILO creates other instruments to disseminate policies or programs. Declarations can be made to promote a certain orientation in programs. Another means is the creation of Codes of Practice, especially in the area of occupational health. Codes of Practice are not legally binding but provide guidance on complex technically or politically sensitive topics.

2.2.6 Completed Projects

In addition to the many conventions and recommendations for OSH, the ILO has completed numerous research projects and developed Codes of Practice to provide guidance on specific topics. The publications are typically available online free of charge. A partial list of topics is provided here:

- Guidelines on OSH management systems
- Safety and health in the use of machinery
- Safety and health in agriculture
- Safety and health in underground coal mines
- Safety and health in the iron and steel industry
- Safety and health in shipbreaking: Guidelines for Asian countries and Turkey
- Security in ports
- Safety and health in ports
- Workplace violence in services sectors and measures to combat this phenomenon
- Managing disability in the workplace
- Safety and health in the nonferrous metal industries
- Human immunodeficiency virus (HIV)/autoimmune deficiency syndrome (AIDS) and the world of work
- Safety in the use of synthetic vitreous fiber insulation wools
- Ambient factors in the workplace
- Safety and health in forestry work
- Recording and notification of occupational accidents and diseases
- Safety in the use of chemicals at work
- Safety and health in construction
- Prevention of major industrial accidents
- Safety, health, and working conditions in the transfer of technology to developing countries
- Radiation protection of workers
- Safety in the use of asbestos
- Occupational exposure to airborne substances harmful to health
- Safe design and use of chain saws

- Safe construction and operation of tractors
- Technical and ethical guidelines for worker's health surveillance

In general, these documents can provide a fundamental starting point for understanding hazardous conditions and how to remediate or control them. Unfortunately, some of the documents listed above are somewhat dated and in some cases may not have the latest relevant safety information. The guide on chain saws, for example, was written in 1978, and although it lists the use of leg protectors under safety equipment, it does not thoroughly describe the advances made in protective clothing to date. The guide for airborne exposure to toxic substances was written in 1980 and provides only the most fundamental information in comparison with the latest methods and technologies on the topic.

2.2.7 The ILO Today and Future Directions

In support of the original ILO goals to promote social justice and universal human and labor rights, the newest initiative is called "The Decent Work Agenda." Its focus is to advance economic and working conditions as a means to achieve peace, prosperity, and progress. It is based on the premise that the ability to have decent work is integral to peace and progress. Another initiative at the ILO involves efforts to prohibit the use of child labor. The ILO has supported child labor prohibition initiatives in more than 200 countries.

2.3 WORLD HEALTH ORGANIZATION

2.3.1 Background and History

The World Health Organization (WHO) was created on April 7, 1948. The primary role is to promote international health through the United Nations network. It is responsible for directing and coordinating international health initiatives by providing leadership and promoting collaboration and information exchange between the different Member States. The WHO sets norms and standards for promoting and monitoring health in the broadest of social terms. It shapes and influences the research agendas of Member States through ethical and evidence-based policies. It monitors global health trends and provides technical support and capacity building where needed.

Since the inception, the WHO has achieved numerous projects and landmarks highlighted here in the following list:

1950—Training and advice on the use of antibiotics

1952—57 Facilitation of mass global campaigns using newly invented poliovirus vaccine

1963—69 Facilitation and training in the use of new vaccines for measles, mumps, and rubella

1969—The first International Health Regulations are created by the Health Assembly that requires Member States to prevent the transmission of health risks across borders

1979—Following a 12-year campaign by the WHO, smallpox is eradicated
1983—HIV is discovered and WHO begins research and response programs
1987—The first antiretroviral medications for HIV infection are developed and become part of the WHO program combatting the disease and the AIDS
1988—The Global Polio Eradication Initiative is established with the use of vaccines and outreach
1999—A Global Alliance for Vaccines and Immunization is created to overcome barriers to vaccine distribution throughout the world
2003—The WHO adopts the Framework Convention on Tobacco Control aimed at reducing related disease and deaths worldwide
2004—The Strategic Health Operations Centre is created as a nerve center for global emergency response
2009—The H1N1 pandemic leads to collaboration with pharmaceutical industries to expand and hasten the development of vaccines
2014—WHO responds to the West Africa Ebola outbreak with a vast emergency response including the deployment of thousands of technical and medical experts, laboratory support, and mobile treatment centers
2016—The Zika virus is recognized as an international public health emergency, and the WHO begins education and outreach on the topic (WHO, 2016)

Today, the WHO employs more than 7,000 people in over 150 WHO country offices, six regional offices, the Global Service Center in Malaysia, and the headquarters in Geneva, Switzerland. Current regional offices are in Africa, the Americas, Southeast Asia, Europe, Eastern Mediterranean, and the Western Pacific. Employees have expertise and education in a broad range of technical fields including medicine, public health, epidemiology, statistics, administration, finance, economics, and emergency preparedness and response. Official languages of the organization are French and English, but many publications are available in other additional languages.

2.3.2 Administrative Format, Responsibilities, and Authorities

The work of the WHO is conducted by three different groups: the World Health Assembly, the Executive Board, and the Secretariat. The Health Assembly comprises up to three delegates from each Member State and meets in a regular annual session. Other special sessions may be convened at the request of the Board or a majority of members. During the annual session, various officers and a President are selected.

Some of the primary functions of the Health Assembly are to

- Set policies.
- Name the members entitled to designate a person to serve on the Board.
- Appoint the Director-General.
- Review and approve reports and activities of the Board and of the Director-General and to instruct the Board in regard to matters upon which action, study, investigation, or report may be considered desirable.

- Establish working committees.
- Supervise the financial policies of the organization and review and approve the budget.
- Instruct the Board and the Director-General to bring to the attention of members and of international organizations, governmental or nongovernmental, any matter with regard to health which the Health Assembly may consider appropriate.
- Invite any organization, international or national, governmental or nongovernmental, to participate in WHO activities.
- Consider recommendations bearing on health made by other branches of the United Nations, and to report to them on the steps taken by the WHO to respond to the recommendations.
- Promote and conduct research in the field of health.

During assembly meetings, various conventions or recommendations are proposed and debated. A two-thirds vote of the Health Assembly is required for the adoption of conventions or agreements. Each Member State has 18 months after the adoption of conventions to take action towards the acceptance of the convention or to notify the WHO with a statement of the reasons for nonacceptance. Each member must make an annual report to the WHO regarding their status with respect to the adoption of various conventions and agreements.

The Executive Board of the WHO consists of 34 persons selected by members of the Health Assembly. The terms are for 3 years and representatives may be reelected. The Board meets at least twice per year and elects its own Chairperson from among the members.

The primary purposes of the Executive Board are to provide oversight and administrative support to the Health Assembly. It advises and answers questions regarding conventions, agreements, and regulations. When necessary, the Executive Board may deal with emergency events and authorize the Director-General to take urgent actions to deal with the organization of relief to victims of a calamity or to combat epidemics.

The Director-General is the ex officio secretary of the organization and the chief technical and administrative officer. The Director-General maintains and prepares all financial statements and budget estimates of the organization. The Director-General appoints a staff to uphold the administration of the organization.

The Health Assembly has the power to create regional organizations in various geographical areas based on special needs or conditions. Each region has a committee and an office. Committees are made up of representatives from the Member States and Associate Members in the region. These regional groups govern themselves through their own rules of procedure and formulate policies and programs specifically designed to address matters of regional concern. They may conduct research, hold technical conferences, and coordinate directly with regional committees of the United Nations or other international organizations operating in the region. Other activities may be coordinated with or directed by the Health Assembly or the Director-General.

2.3.3 Funding Sources

The WHO's financing comes from assessed contributions from Member States. The amount paid by each Member State is calculated based on the country's wealth and population. In addition, a significant portion of funds come from voluntary contributions from Member States and various partners. The respective portions of funding and contributions are shown in Figure 2.1. A small portion of organizational funds come from private sources (WHO, 2016).

2.3.4 Programs Regarding OSH

Today, the WHO operates following six main leadership priorities that guide the direction of the organization. The priorities are oriented towards the concept of development and sustainability of health and well-being, for individuals of all ages internationally.

The first priority is the advancement of health coverage to ensure that all countries can provide and sustain access to health services for their populations. The goal is to provide health care to people that is within their financial capabilities and prevent financial catastrophe when illness occurs. WHO works with governments to provide practical advice on how health service systems can be best expanded within a particular government system.

The second priority is the continued battle against existing and historic challenges related to maternal and child health, HIV, malaria, tuberculosis, polio, and other tropical diseases. Newly discovered diseases also fall into this priority and are addressed as they become apparent.

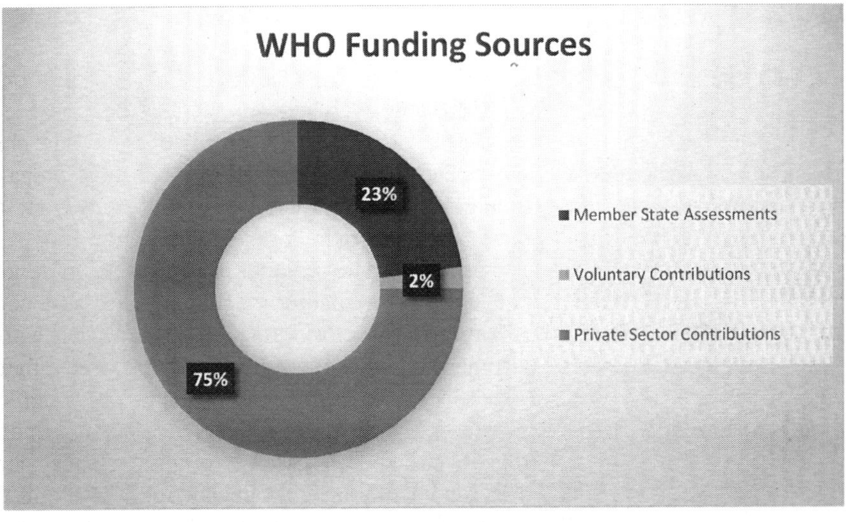

FIGURE 2.1 WHO funding sources. (Adapted from WHO, 2016.)

The third challenge to the WHO is the reduction of the impacts of noncommunicable diseases that result from tobacco and alcohol use, sedentary lifestyles, and unhealthy diets. WHO provides training and educational materials geared towards combatting these conditions and curbing policies that allow the conditions to prevail.

The WHO provides ongoing assistance for emergency preparedness and response to countries unable to adequately respond to disease outbreaks and other humanitarian crises. The WHO helps to plan for emergency response and ensure that national health-care systems can respond to and recover from a variety of public health stresses that are likely to occur from time to time.

Another goal of WHO is to improve access to medical products and technologies. WHO promotes national procurement of medicine and information necessary to support a modern health-care service structure. More broadly, the WHO evaluates a variety of social, economic, and environmental determinants of health in order to reduce health inequalities between countries.

As part of the overall plan for public health, the WHO has been committed to the improvement and protection of the health of workers. In 1999, a network of Collaborating Centers for Occupational Health was created to develop and strengthen institutional capacities to provide healthy workplaces in countries and regions. The mission of the collaboration is to provide technical expertise to economically developing countries to promote equity, justice, and fairness in OSH by strengthening the capabilities of national and regional systems. Activities and functions of the collaborating centers include the following:

- Collection, collation, and dissemination of information
- Standardization of terminology, technology, methods, and procedures
- Development of evidence-based technical guidance tools and resource materials
- Development and application of appropriate technology
- Development of collaborative research on occupational health topics
- Training
- Coordination of activities carried out by several institutions on a given subject
- Capacity-building work at a country level
- Monitoring, preparedness, and response services to deal with disease outbreaks and public health emergencies (WHO, 2014)

Collaborating centers and programs include support from the ILO OSH Branch and from other NGOs. Work is oriented according to a Global Master Plan (GMP), which describes project priorities, product goals, and activities necessary for achieving the desired objectives. The current 2012–2017 GMP lists seven major OSH priorities, which are as follows:

1. Regional and national programs on occupational noncommunicable diseases with focus on cancer, silica, and asbestos-related diseases
2. National programs and good practices for occupational health and safety of health-care workers

3. Tools, standards, and capacities for healthy workplaces
4. Strengthening health systems, governance, capacities, and service delivery for workers' health
5. Occupational health aspects of emerging technologies
6. Classification, diagnostics, and exposure criteria for occupational diseases
7. Occupational health and safety for vulnerable groups and high-risk sectors (WHO, 2017a)

Resolution WHA 60.26 "Workers' Health: Global Plan of Action" urges the creation and provision of occupational health services for all workers including those in the informal economy, migrants, contract workers, and those working in the agricultural sector. WHO is working to ensure that primary care centers are created to ensure preventive, curative, and rehabilitative services are available to all workers. It includes efforts to improve multidisciplinary capacity in primary care and occupational health specialties (WHO, 2017b).

2.3.5 WHO Publications

Over the many years, the WHO has created numerous documents regarding occupational health. Most are available with no fee. Some are published as periodicals with different frequencies, such as The African Newsletter on Occupational Health and Safety. It has been published since 1991 and targets 21 African countries with distributions to more than 100 countries. Reports of the Annual World Assemblies are routinely available from the WHO website. And many special reports are available, such as "Occupational Noise: Assessing the burden of disease from work-related hearing impairment at national and local levels" (WHO, 2004) and "Work Organization and Stress" (WHO, 2003).

2.3.6 International Agency for Research on Cancer

The International Agency for Research on Cancer (IARC) is a subdepartment of the WHO that performs specialized research and promotes international collaboration on cancer research. The agency uses an interdisciplinary approach with data and information from countries around the world using epidemiology, laboratory sciences, and biostatistics to identify the causes of cancer. The information from the research helps the IARC and the WHO prioritize medical research and prevention needed to reduce the global burden of disease from cancer. The IARC is a definitive source of information about cancer and is of great value to researchers and healthcare practitioners worldwide. Education and training is a priority of IARC with publications available online in a variety of formats, and through other fellowships and courses (IARC, 2016).

2.4 EUROPEAN UNION

In 1989, the European Union (EU) European Framework Directive on Safety and Health at Work (Directive 89/391 EEC) was adopted by the EU. It was a substantial milestone in improving safety and health at work and led to the creation of detailed directives that guarantee minimum safety and health requirements throughout Europe to which all Member governments must adhere to for worker safety (EU, 2016a). Separate from, but in addition to, directives, EU guidelines are nonbinding documents that aim to facilitate the implementation of European directives. Different types of guidelines include recommendations from various other organizations such as in EU social partners' agreements (EU, 2016b).

In 1994, the EU Council Regulation (EC) No. 2062/94 established a European Agency for Safety and Health at Work, EU-OSHA (EU-OSHA, 1994). EU-OSHA is allocated funds by the EU's budgetary authority, through the Treaty on the Functioning of the EU. EU-OSHA directives set by the European Parliament (directly elected Members of the European Parliament) and the Council of the EU (representatives of the 28 Member State governments) become laws which each of the Member States must transpose into national laws within set deadlines.

The goal of EU-OSHA is to promote improvements in working conditions for the health and safety of workers under existing Treaty conditions through the support of successive action programs (EU-OSHA, 1994). Some of the main goals of the regulation are to

- Collect and disseminate technical, scientific, and economic information in order to develop priorities.
- Perform OSH research and publish results.
- Promote cooperation and exchange of information.
- Organize conferences and seminars on occupational safety topics.
- Collect and disseminate information on OSH to developing countries.
- Promote OSH in small to mid-sized companies.
- Contribute to the development of community action programs that improve worker safety.

2.5 UNITED NATIONS ENVIRONMENTAL PROGRAM

The United Nations Environmental Program (UNEP) advocates environmental issues and environmental sustainability globally. They prioritize work on the topics of climate change, conflict, natural disasters, eco-management, chemical hazards, hazardous waste, and resource efficiency. Many UNEP projects and documents overlap into areas of occupational exposures and hazards in certain industries and even environmental and occupational exposures that occur in the informal work sector (UNEP, 2018).

2.6 WORLD BANK

The World Bank is supported by 189 member countries and provides funding and technical assistance to developing countries around the world as a means to reduce poverty globally. They support investments in education, administration, public health, private sector development, natural resource management, the environment, and agriculture. They work with governments, tripartite groups, and private enterprises. As they relate to social development and security, many funded World Bank projects include areas that overlap with occupational safety issues and concerns such as emergency and disaster preparedness, environmental hazards, and measures of identifying and minimizing industrial accidents (World Bank, 2018).

2.7 INTERNATIONAL MARITIME ORGANIZATION

The International Maritime Organization (IMO) is the UN special agency with responsibility for the safety of shipping (IMO, 2018a). Their main goal is to create performance standards and regulatory frameworks to be followed by member nations. As shipping is truly a global industry, it is an area where international occupational safety standards have the opportunity to be equal for all workers. IMO programs include training on occupational health and safety topics. IMO conventions overlapping with safety include the safety of fishing vessels and the carriage of hazardous and noxious substances by sea (IMO, 2018b).

2.8 INTERNATIONAL PROGRAM ON CHEMICAL SAFETY

The International Program on Chemical Safety (IPCS) was established in 1980 as a joint program between the three cooperating organizations: ILO, UNEP, and WHO. Standards of practice are developed to ensure the safety and health of workers, the public, and the environment. They cover a broad range of exposure scenarios in the presence of chemicals including industrial production, transport, and disposal. Programs include evaluation of chemical risks, hazard evaluation, and prevention of exposure to chemicals (IPCS, 2018).

2.9 CONCLUSIONS AND RECOMMENDATIONS

Countries that collect and transparently share objective data regarding OSH which work towards common goals and objectives can obtain mutual benefits for workers, societies, and economic stability. Guidance documents and standards not only can improve conditions within national boundaries but can also be used to minimize transboundary hazards created in one nation that have the potential of affecting neighboring nations. Standardized OSH programs can level the playing field in business and economic development of nations. These same standards can be used to ensure safe working conditions for all workers, regardless of their nationality or nation of origin. Decisions can be made collectively in guidance documents and treaties for assessing risks and setting priorities for appropriate controls. Balances can be struck within decision-making processes to reduce uncertainty and take a precautionary approach to potential hazards. Numerous OSH goals have been achieved globally through the international agreements between numerous international organizations that have been created over the past several decades.

There is always room for improvement and further collaborations. The bureaucratic monolithic structures of many government agencies, in addition to the IGOs created and supported by them, need to do more to include other stakeholders and community organizations who are willing to assist and provide resources in human capacity through volunteers with specialty expertise in OSH. NGOs have a significant capacity which is not being effectively or fully utilized by IGOs. Offers to assist with projects including document development, research, and capacity building are often ignored by large IGOs. Governments and IGOs would benefit greatly from increased awareness and collaboration with NGOs and other external groups that are doing great work.

REFERENCES

EU, European Union, European Directives (2016a) https://osha.europa.eu/en/safety-and-health-legislation/european-directives accessed June 17, 2016.

EU, European Guidelines (2016b) https://osha.europa.eu/en/safety-and-health-legislation/european-guidelines accessed June 17, 2016.

EU-OSHA, European Union, Council Regulation (EC) No. 2062/94, European Agency for Safety and Health at Work 8 July, 1994, Official Journal of the European Union (August 20, 1994) No. L. 216.

IARC, International Agency for Research on Cancer, IARC a unique agency—cancer research for cancer prevention, IARC, Lyon (2016) www.iarc.fr/en/about/iarc-brochure-web.pdf accessed June 22, 2016.

ILO, (June 18, 1998) *Declaration on Fundamental Principles and Rights at Work, International Labor Organization.* Geneva: International Labor Organization. ISBN 978-92-2-124804-0.

ILO (2003). *Safety in Numbers: Pointers for a Global Safety Culture at Work.* Geneva: International Labor Organization. ISBN: 92-2-113741-4.

ILO, (June 10, 2008) *Declaration on Social Justice for a Fair Globalization, International Labor Organization.* Geneva: International Labor Organization. ISBN 978-92-2-121617-9.

ILO, Origins and History, International Labor Organization (2017a) www.ilo.org/global/about-the-ilo/history/lang--en/index.htm accessed July 21, 2017.

ILO, Constitution, International Labor Organization, (2017b) www.ilo.org/dyn/normlex/en/f?p=1000:62:0::NO:62:P62_LIST_ENTRIE_ID:2453907:NO, accessed July 24, 2017.

IMO, Brief History of IMO, International Maritime Organization (2018a) www.imo.org/en/About/HistoryOfIMO/Pages/Default.aspx accessed July 20, 2018.

IMO, List of IMO conventions, International Maritime Organization (2018b) www.imo.org/en/About/Conventions/ListOfConventions/Pages/Default.aspx accessed July 20, 2018.

IPCS, International Program of Chemical Safety (2018) www.who.int/ipcs/en accessed July 20, 2018.

Kirchsteiger, C., Review of industrial safety management by international agreements and institutions, *Journal of Risk Research* (2005) Vol. 8, No. 1, pp. 31–51.

UNEP, Homepage, United Nations Environmental Programme (2018) www.unenvironment.org/ accessed July 20, 2018.

WHO (2003). Leka, S., Griffiths, A., Cox, T., Work *Organization and Stress: Systematic Problem Approaches for Employers, Managers and Trade Union Representatives,* protecting workers health series No. 3. Geneva: World Health Organization. ISBN 92-4-159047-5.

WHO (2004). Concha-Barrientos, M., Campbell-Lendrum, D., Steenland, K., *Occupational Noise—Assessing the Burden of Disease from Work-Related Hearing Impairment at National and Local Levels,* environmental burden of disease series, No. 9. Geneva: World Health Organization. ISBN 92-4-159192-7.

WHO, WHO Collaborating Centers for Occupational Health, World Health Organization, 2016 (2014) www.who.int/occupational_health/network/OH_CCs_WoW_20160502.pdf?ua=1 accessed July 27, 2017.

WHO (2016) The Guardian of Public Health, World Health Organization (2016) www.who.int/about/what-we-do/global-guardian-of-public-health.pdf?ua=1 accessed July 27, 2017.

WHO, Global Master Plan, World Health Organization (2017a) www.who.int/occupational_health/network/OH_GMP_2012-2017.pdf?ua=1 accessed July 27, 2017.

WHO, Universal Health Coverage of Workers, World Health Organization (2017b) www.who.int/occupational_health/activities/en/ accessed July 27, 2017.

World Bank, Homepage (2018) www.worldbank.org accessed July 20, 2018.

3 Nongovernmental International Occupational Safety and Health Professional Organizations

Thomas P. Fuller
Illinois State University

CONTENTS

3.1 INTRODUCTION

A number of international nongovernmental (NGOs) and professional organizations exist that cover a wide range of subject areas and topics. Many of the organizations' goals, objectives, and activities overlap substantially. And as a result, there exist many possible areas for collaboration and intersupport between the groups. The members of a national professional organization may automatically belong to an umbrella international organization due to existing arrangements or memorandum of understanding. Or members of one organization may join an international organization independently. Many professionals participate in several organizations and cross-collaborations at the same time.

This chapter briefly introduces each of the major organizations with activities associated with occupational safety and health (OSH) at the international level. Although many large national organizations have strong international outreach activities and objectives, and large international membership, this chapter only includes those that are truly "international" in nature, by either their name, membership, or charters.

3.2 INTERNATIONAL OCCUPATIONAL HYGIENE ASSOCIATION

The International Occupational Hygiene Association (IOHA) is a global community of occupational hygienists and professionals who are dedicated to the discipline and application of the inherent principles used to protect workers from hazards to reduce injury and illness (IOHA, 2018). There are 35 member organizations, in 32 countries, that represent more than 20,000 occupational hygiene professionals around the world. IOHA is recognized as an NGO by both the World Health Organization (WHO) and the International Labor Organization (ILO).

IOHA activities include the development of promotion of occupational hygiene as a professional field of practice and holding international conferences to support the exchange of information and ideas on occupational hygiene theory and practice. The IOHA National Accreditation Recognition (NAR) Committee reviews and approves occupational hygiene credentialing and certification schemes and designates those systems with comparable standards. There are currently 15 recognized certification schemes recognized by the IOHA NAR Committee. With more than 13,000 chartered safety and health practitioners under the NAR program, IOHA has more certified OSH professionals than any other organization (IOHA, 2018).

IOHA membership organizations around the world include the following:

- American Conference of Governmental Industrial Hygienists (ACGIH)
- American Industrial Hygiene Association (AIHA)
- Association of Hygienists of Argentina (AHRA)
- Australian Institute of Occupational Hygienists (AIOH)
- Belgian Society for Occupational Hygiene
- Brazilian Association of Occupational Hygienists (Associação Brasileira de Higienistas Ocupacionais [ABHO])
- British Occupational Hygiene Society (BOHS)
- Canadian Registration Board of Occupational Hygienists (CRBOH) (Conseil Canadien d'Agrément des Hygiénistes du Travail [CCAHT])
- Dutch Occupational Hygiene Society (Nederlandse Vereniging voor Arbeidshygiëne [NVvA])
- French Occupational Hygienists Society (SOciete Francaise des HYgienites du Travail or French Occupational Hygienists Society)
- Finnish Occupational Hygiene Society (Suomen Työhygienian Seura [STHS])
- German Society for Occupational Hygiene (Deutsche Gesellschaft für Arbeitshygiene [DGAH])

- Hong Kong Institute of Occupational and Environmental Hygiene
- Italian Industrial Hygiene Association
- Japan Occupational Hygiene Association
- Japan Association for Working Environment Measurement (JAWE)
- Korean Industrial Hygiene Association (KIHA)
- Macedonian Association of Industrial Hygiene and Occupational Health
- Malaysian Industrial Hygiene Association
- Mexican Industrial Hygiene Association (Asociación Mexicana de Higiene Industrial [AMHI])
- New Zealand Occupational Hygiene Society (NZOHS)
- Norwegian Occupational Hygiene Association (Norsk yrkeshygienisk forening [NYF])
- Occupational and Environmental Health Society of Singapore
- Occupational Hygiene Society of Ireland (OHSI)
- Polish Association of Industrial Hygienists (Polskie Towarzystwo Higienistów rzemysowych [PTHP])
- Spanish Association of Industrial Hygiene
- Southern African Institute for Occupational Hygiene—Certification Board (SAIOH-CB)
- Swedish Association of Occupational and Environmental Hygiene (Svensk Yrkes-och Miljöhygienisk Förening [SYMF])
- Swiss Society for Occupational Hygiene (Schweizerischen Gesellschaft für Arbeitshygiene [SGAH] and Société Suisse d' Hygiène du Travail [SSHT])
- Taiwan Occupational Hygiene Association (TOHA)
- Vietnamese Industrial Hygiene Association

3.3 WORKPLACE HEALTH WITHOUT BORDERS

Workplace Health Without Borders (WHWB) was founded in 2011 as a not-for-profit organization to address occupational health and safety issues in the developing world. WHWB engages volunteers with expertise and experience in occupational hygiene to donate time to support projects in economically developing countries to build capacity in OSH through training, research, and outreach projects. WHWB is built on a foundation of collaboration between numerous governmental, academic, and professional organizations (WHWB, 2018). WHWB International was originally founded in Canada. National branches of WHWB have since been created in the United States, the United Kingdom, and Australia.

In the past several years, WHWB has delivered OSH training in India, Tanzania, Vietnam, South Africa, Botswana, Mozambique, and Swaziland. The WHWB facilitated occupational hygiene instrument donations to Botswana, Uruguay, Pakistan, and Tanzania. In addition, personal protective equipment was donated through WHWB to fight the Ebola outbreak in Sierra Leone.

Research activities that have been funded by WHWB have included clay brick kiln exposure studies in Pakistan and Nepal. Figure 3.1 shows a typical brick kiln operation where a study has been completed by WHWB volunteers.

FIGURE 3.1 WHWB brick kiln study site in Nepal. (Photograph courtesy of Steve Thygerson.)

3.4 OCCUPATIONAL HYGIENE TRAINING ASSOCIATION

The Occupational Hygiene Training Association (OHTA) was formed to promote improved occupational hygiene practice and build professional occupational hygiene capacity throughout the world by developing and providing associated occupational hygiene training materials. The training materials are developed by volunteers with particular expertise on given subjects and are made available freely to students and training providers (OHTA, 2018a).

OHTA began as an informal collaboration but has grown into an international framework supported by professionals from numerous occupational hygiene professional organizations, including the IOHA. OHTA is registered as a charity in the

United Kingdom and operates through the support of volunteers who create training materials and support program activities. Association operating expenses are funded by small levies on training providers based on the number of students taking course examinations. Funds are also received from supporting professional associations, companies, and consultancies in the form of donations.

A main operating principle of OHTA is that materials are available to everyone for free, that even for-profit concerns can gain benefit from the use of the materials, and that they should not be excluded. It is also important that organizations with limited funds have access to the materials, and approved instructors who may also perform as volunteers.

Another goal of OHTA is to make access to the materials and associated training courses as easy as possible. Regional training providers are identified on the OHTA website, and, in addition, numerous national associations and other not-for-profit groups are available as approved trainers to provide courses. The use of volunteers keeps administrative costs to a minimum and increases the availability and accessibility of courses to those individuals and groups with limited financial resources.

OHTA works to be as transparent as possible in the implementation of courses and examinations. While maintaining high standards for course examination and confidentiality, the benefits of attaining the course certifications are held at high levels and are respected within the OH profession. Training quality is maintained for OHTA course by a thorough and rigid training provider review and approval process (OHTA, 2018b).

One of the most valuable aspects of the OHTA model is that it can be used by early occupational hygiene professionals, managers in other fields, and specialists in other related areas to build their knowledge base in occupational hygiene topics. Those in the occupational hygiene field can earn certification credentials that can lead to higher professional levels in occupational hygiene. Students can work on the certifications over time, at their own pace, to attain goals on their terms and timeframe. Courses offered at the "awareness" level introduce managers, supervisors, worker representatives, and employees to basic health risks and hazards typically associated with workplaces and identify ways to control the hazards to provide a safe work environment. Courses at the "foundation" level can be used to support knowledge development in a particular specialty area for professionals such as physicians or nurses. "Intermediate" courses build and expand upon basic concepts to improve technical understanding and provide hands-on and practical information about how to assess exposure, take measurements, and design and assess effectiveness of control measures. "Academic" courses are designed to assist those individuals with the responsibility for designing and delivering OH programs in the workplace. In many cases, these courses, in addition to other qualifications, can be used to support requirements for entry to professional qualification schemes. Lastly, "leadership" level courses provide senior occupational hygienists with information necessary to stay current in the field of occupational hygiene and grow into management and supervisory positions in the field of occupational hygiene (OHTA, 2018c). The OHTA occupational hygiene training and career ladder steps are shown in Figure 3.2.

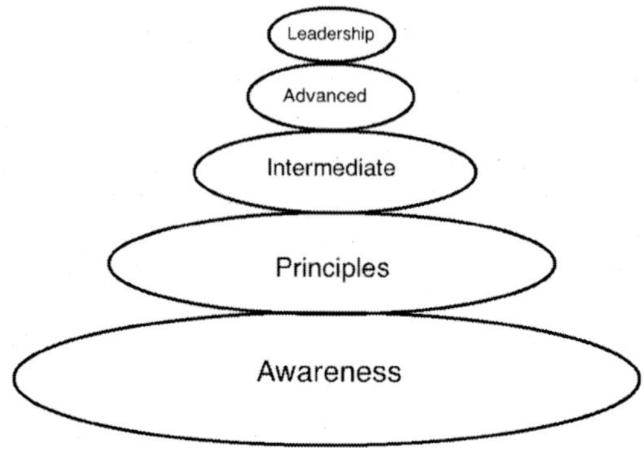

FIGURE 3.2 OHTA training and career ladder.

OHTA course materials that can be found online include slide presentations for topic lectures, course syllabi, in-class activities, homework assignments, student manuals, practice examinations, case studies, and laboratory activities. Current courses available through OHTA include the following:

- Basic principles in occupational hygiene
- Measurement of hazardous substances
- Thermal environment
- Noise—Measurement and its effects
- Asbestos and other fibers
- Control of hazardous substances
- Ergonomics essentials
- Health effects of hazardous substances

OHTA course examinations allow students to earn certifications in each of the topic subjects. Examinations are administered and graded under strict OHTA program requirements for qualified occupational health professionals. Recipients who have earned six of the above certificates and meet other basic educational and experience requirements can become qualified to sit for IOHA NAR scheme examinations and earn professional certifications, such as Certified Industrial Hygienist from the American Board of Industrial Hygiene.

3.5 INSTITUTION OF OCCUPATIONAL SAFETY AND HEALTH

The Institution of Occupational Safety and Health (IOSH) was founded in 1945 to promote health and safety at work and to support the development of safety professionals working in the field. With more than 47,000 members, this is one of the world's largest health and safety professional membership organizations (IOSH, 2018).

IOSH activities include ongoing training on health and safety topics by licensed instructors and development of books and e-resources on health and safety topics.

3.6 INTERNATIONAL COMMISSION ON OCCUPATIONAL HEALTH

The International Commission on Occupational Health (ICOH) is an international nongovernmental professional society founded in 1906 to promote occupational health and safety. There are currently more than 2,000 members from 93 countries. The organization is committed to fostering OSH understanding and capacity around the world. It works to disseminate scientific information through triennial conferences at varying locations around the world. Collaboration on technical projects is achieved through a broad variety of technical committees and working groups (ICOH, 2018). The ICOH is recognized by the United Nations as an NGO and has close working relationships with ILO and the WHO.

3.7 INTERNATIONAL NETWORK OF SAFETY AND HEALTH PROFESSIONAL ORGANIZATIONS

The International Network of Safety and Health Professional Organizations (INSHPO) is an international collaboration among professional organizations with the goal of improving safety and health at work (INSHPO, 2018). The INSHPO conducts meetings to share information and publishes OSH-related materials online as a means to spread awareness of workplace health and safety issues and controls. This organization is not-for-profit, and board members serve without compensation. Current projects include development and dissemination of core competencies for OSH professions, designation or moral codes of conduct, and educational equivalency comparisons.

3.8 EUROPEAN NETWORK EDUCATION AND TRAINING IN OCCUPATIONAL SAFETY AND HEALTH

The European Network Education and Training in Occupational Safety and Health (ENETOSH) was created in 2015 and is funded by the European Commission with the aim to mainstream occupational safety and health into education and training at all levels. It fosters the idea that health and safety are an integral part of life, and it should be incorporated in curricula at all levels of schooling and ongoing continuous professional development (ENTOSH, 2018).

A network of experts in various topical areas of health and safety is called upon to collaborate and create training and educational materials at all phases of an individual's education and career. Figure 3.3 shows a lifelong progression where OSH training can be mainstreamed into educational and training programs.

Through a variety of collaborations, the goals of the network are to create high-quality educational and training materials for all levels, and then find ways to mainstream these into various learning environments. Much of the work is done through open exchange of information and training materials between members. Examples of good practices are systematically collected and disseminated.

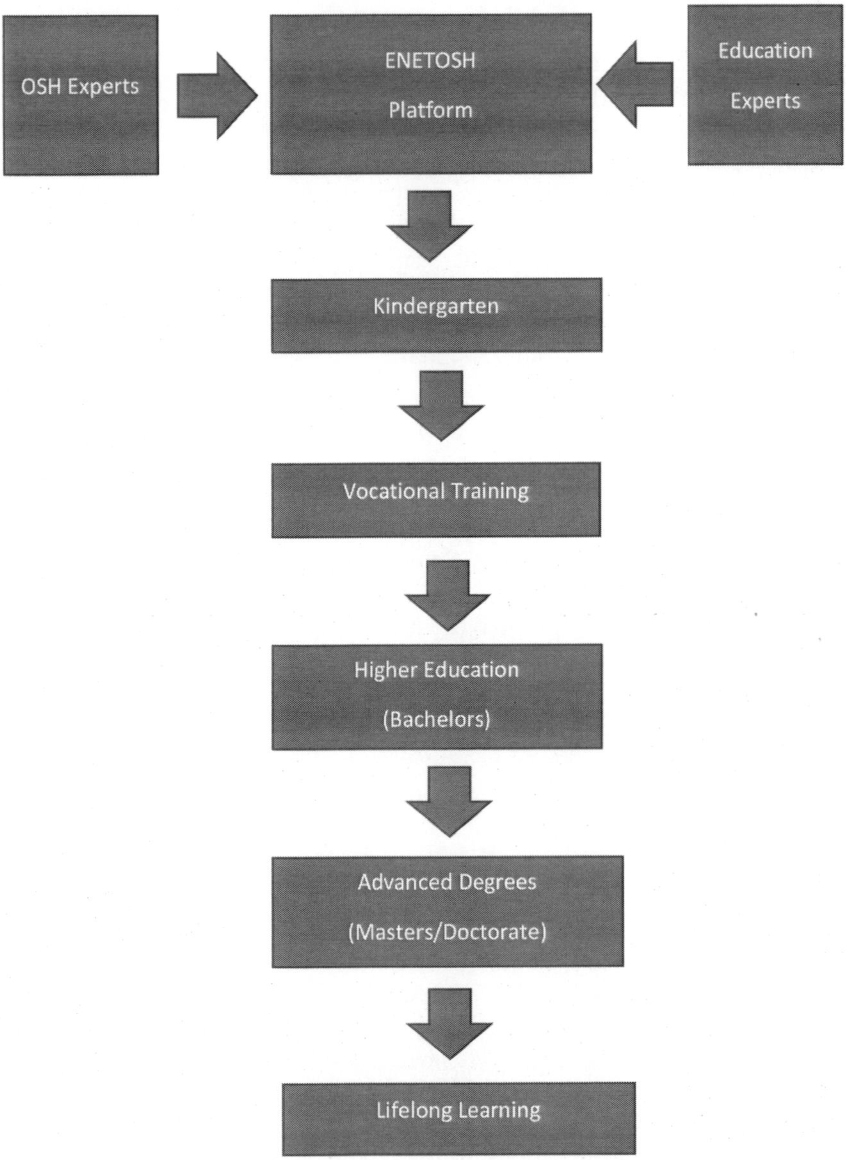

FIGURE 3.3 ENETOSH model of lifelong OSH learning. (Adapted from ENTOSH, 2018.)

3.9 EUROPEAN NETWORK OF SAFETY AND HEALTH PROFESSIONAL ORGANIZATIONS

The European Network of Safety and Health Professional Organizations (ENSHPO) was established in 2001. The objective of the organization is to bring safety and health professionals from different European organizations together to collaborate on projects and share information. The main objectives of ENSHPO are to create dialogue with national and international authorities, exchange opinions and viewpoints, identify and share good practices, and develop European-wide recognition of OSH professional qualifications and training (ENSHPO, 2018).

The current ENSHPO members are provided in the following list:

Spain—Asociacion de Especialistas en Prevencion y Salud Laboral
Italy—Associazionne Professionale Italiana Ambiente e Sicurezza
Denmark—Danish Association of Occupational Health and Safety Consultants
Romania—Romanian Association for Occupational Health and Safety
Czech Republic—Occupational Safety and Health and Fire Prevention Chamber of the Czech Republic
Cyprus—The Cyprus Safety and Health Association
Croatia—Croatian Institute for Health Protection and Safety at Work
United Kingdom—Institution of Occupational Safety and Health
Romania—Institute of Risk Management and Occupational Health and Safety
Malta—Malta Occupational Safety and Health Practitioners Association
Russia—National Association of the Centres for Occupational Safety and Health
Bulgaria—National Association for Health and Safety at Work
The Netherlands—Nederlandse Vereniging voor Veiligheidskunde
Belgium—Institute for Occupational Safety and Health
Denmark—The Danish Society of Occupational Safety and Health
Slovakia—Slovak Association of OSH and Fire Protection
Portugal—Sociedade Portuguesa de Seguranca e Higiene Ocupacionais
Finland—Tyoturvallisuuskeskus Centre for Occupational Safety
Germany—Verband für Sicherheit, Gesundheit und Umweltschutz bei der Arbeit

3.10 INTERNATIONAL COMMISSION ON RADIATION PROTECTION

The International Commission on Radiation Protection (ICRP) is an independent, international organization with more than 200 volunteer members from approximately 30 countries across six continents. These members represent the leading scientists and policy makers in the field of radiological protection. The main goal of the organization is to prevent cancer and other diseases caused by exposure to ionizing radiation and radioactive materials. It was created in 1928 to promote the development of international radiation protection standards.

ICRP is funded through ongoing contributions from organizations with an interest in radiological protection. It maintains formal relations and liaisons with several other international safety organizations including the ILO and the WHO (ICRP, 2018).

ICRP has published more than one hundred reports on all aspects of radiation protection. The International System of Radiological Protection has been developed by ICRP based on (1) the current understanding of the science of radiation exposures and effects, and (2) value judgments. These value judgments account for societal expectations, ethics, and experience gained in application of the system.

3.11 INTERNATIONAL RADIATION PROTECTION ASSOCIATION

The purpose of the International Radiation Protection Association (IRPA) is to provide a platform for professionals and experts to exchange information on radiation safety in science, engineering, medicine, and technology. Standards and guidelines are shared to assist in the creation of regulations to protect workers and the environment. IRPA also promotes education and training in the field of radiation safety. The organization maintains relations with several intergovernmental organizations including the WHO and the ILO (IRPA, 2018).

3.12 INSTITUTE OF ELECTRICAL AND ELECTRONICS ENGINEERS

The Institute of Electrical and Electronics Engineers (IEEE) is the world's largest technical professional organization with more than 430,000 members. The organization's primary goals are education and dissemination of technical information on electronics, computer engineering, and associated disciplines. The IEEE has a significant role in the development of various safety standards associated with electrical fields and equipment, and other technical subjects. IEEE produces technical publications and organizes professional conferences of which the proceedings are often published. Educational programs are a significant part of IEEE activities, and online courses are available for IEEE members (IEEE, 2018).

3.13 INTERNATIONAL STANDARDS ORGANIZATION

The International Standards Organization (ISO) is a not-for-profit international organization that was created in 1946, with a membership of 160 national standards bodies working together to share knowledge and create consensus-based voluntary standards of practice on a broad range of topics. ISO has published more than 22,230 International Standards covering numerous industries and technological disciplines. Standards are created to support important social or public health issues with such goals as ensuring safety of operations, products, and services. Standards may also be used to ensure system reliability and product quality. Strategic and management standards can be used to improve productivity, reduce errors, and minimize waste (ISO, 2018).

By developing and sharing good standards of practice, ISO believes it can benefit a wide range of global stakeholders, in addition to contributors and members. ISO products and services help level the playing field in global competition by expanding markets and improving the use of resources. The harmonization of standards of quality, service, and safety helps to ensure all countries are operating equally and transparently. Best practices are designed to ensure sustainability criteria are met to protect the environment, provide social justice, and ensure economic opportunity.

ISO standards are based on the consensus of more than 700 organizations and 100,000 experts. ISO office located in Geneva, Switzerland, is staffed by around 170 employees of the organization who are supported by contributions from members and fees for ISO products and documents. Much of the consensus development work is provided and funded by member organizations, and their members are voluntarily working through more than 230 different technical committees (ISO, 2016).

3.14 CONCLUSIONS

The number and scale of the activities in many of the described global OSH organizations is interesting to the extent that the groups work together, collaborate, and as a result seem to build synergy at the international level. Many of the organizations have collaborations with tripartite and government organizations in addition to those with other NGOs. Many of the organizations consciously and specifically have an interest in collaboration, exchange of ideas and information, and capacity building in OSH. Many of the NGOs have particular interests in building capacity and improving working conditions in economically developing countries. The use of volunteers to support research and training in developing countries is a good example; WHWB provides volunteers and network support to provide OHTA training courses at distant developing countries that would not receive such training otherwise. In addition, having the collaborations in place then offers structure and further funding support from national professional organizations and NGOs and the structured organizational ability to apply for other grants and funding from other tripartite and governmental sources.

With thousands of OSH professionals represented by the rosters of the organizations described above, it is difficult to estimate the potential for growth of support and activity in training and research in OSH capacity building. With improved communication and interface, the benefits and outreach could be substantially enhanced and increased. E-learning tools and an online OSH course using OHTA course materials are being developed as a special project of the American Industrial Hygiene Association (AIHA) through a memorandum of understanding with OHTA. This may be a first step with great potential for reaching larger audiences in the future.

REFERENCES

ENSHPO, ENSHPO Objectives, European Network of Safety and Health Professional Organisations (2018) www.enshpo.eu/home/enshpo-objectives accessed July 17, 2018.

ENTOSH, Who We Are, European Network Education and Training in Occupational Safety and Health (2018) www.enetosh.net/webcom/show_article.php/_c-221/_nr-1/_lkm-175/i.html accessed July 17, 2018.

ICOH, About ICOH, International Commission on Occupational Health (2018) www.icohweb.org/site/about-icoh.asp accessed July 17, 2018.

ICRP, Homepage, International Commission on Radiation Protection (2018) www.icrp.org accessed July 19, 2018.

IEEE, History of IEEE, Institute of Electrical and Electronics Engineers (2018) www.ieee.org/about/ieee-history.html accessed July 19, 2018.

INSHPO, International Network of Safety and Health Practitioner Organizations (2018) www.inshpo.org accessed July 18, 2018.

IRPA, Homepage, International Radiation Protection Association (2018) www.irpa.net accessed July 19, 2018.

IOHA, Welcome to IOHA, International Occupational Hygiene Association webpage (2018) https://ioha.net/ accessed July 17, 2018.

IOSH, About Us, Institute of Occupational Safety and Health webpage (2018) www.iosh.co.uk/About-us.aspx accessed July 17, 2018.

ISO, About ISP, International Organization for Standardization (2018) www.iso.org/about-us.html accessed July 19, 2018.

ISO, ISO In Brief, International Organization for Standardization (September, 2016) www.iso.org/files/live/sites/isoorg/files/archive/pdf/en/isoinbrief_2015.pdf accessed July 19, 2018.

OHTA, Occupational Hygiene Training Association (2018a) www.ohlearning.com/about-ohta/purpose-and-principles.aspx accessed July 18, 2018.

OHTA, OHLearning Approved Training Provider Scheme—Terms and Conditions for Training Providers (2018b) http://ohlearning.com/fileshare/Linked%20files/Terms%20and%20Conditions%20for%20Training%20Providers%20%20jan%202012.pdf accessed July 18, 2018.

OHTA, Training Levels (2018c) www.ohlearning.com/training/training-levels.aspx accessed July 18, 2018.

WHWB, Home, Workplace Health Without Borders (2018) www.whwb.org/home/ accessed July 17, 2018.

4 Cultural Issues in International Occupational Safety and Health

Thomas P. Fuller
Illinois State University

CONTENTS

4.1 INTRODUCTION

The globalization of business and industry in the past decades has led to increasingly complex supply chains that have introduced new and increased risks to workers and employers. Globalization has also caused large increases in the numbers of people traveling across international boundaries on a regular basis to perform their work. These workers then interact with a broad range of people from other cultures, often working side-by-side on projects. Managers of global companies and joint ventures may have large groups of new local employees that they will need to communicate with and motivate, to reach organizational objectives and goals. The managers of occupational safety and health (OSH) programs need to consider the culture of employees in order to fully develop a culture of safety within the regional operations. This chapter provides a discussion of what safety culture and climate are, and how national cultures might influence and shape the corporate safety cultures necessary to reach organizational goals.

4.2 SAFETY CULTURE DEFINED

Culture can be defined as similar ways of thinking and behaving demonstrated by members of a social group (Rousseau, 1988).

An organizational culture is the embodiment of certain values, beliefs, and underlying assumptions about how the organization operates (Fahlbruch, 1999). It encompasses a wide variety of phenomena to include behavior, values, norms, patterns of behavior, attitudes, and beliefs (Davies, 2000; Cox, 1998; Mearns, 1998).

Organizational safety culture is the assembly of characteristics and attitudes that establish the significance of safety shared by groups and individuals that comprise the institute (Cooper, 2000). It is a broad concept for explaining how the management of an organization shapes the safety beliefs and behaviors of the workers (Guldenmund, 2000). Safety culture is an enduring characteristic that is reflected in the consistent way that it deals with critical safety issues independent of temporary states (Wiegmann, 2004).

Poor corporate safety culture has been identified as a contributory factor in accidents by many industrial accident investigations, and it is now generally accepted that organizations with a strong and positive safety culture are more effective in preventing workplace accidents and injuries (Shiney, 2014). Organizational safety culture can be quantified through determinants such as clarity of responsibilities, levels of training, management selection criteria, safety audits and reviews, relations with regulators, and management actions and attitudes (IAEA, 1991). Other indicators of safety culture come from management style, commitment, and levels of communication. Five additional indicators of effective safety cultures include organizational commitment to safety, management involvement in safety, employee empowerment, reward systems that encourage safe behaviors, and systems to ensure problems are reported and resolved (Wiegmann, 2004). Production pressure perceived by workers, poor housekeeping, and lack of personal hygiene care facilities can also lead to poor safety culture in a work environment.

Organizations where management is perceived to be committed to safety tend to report fewer unsafe worker behaviors as a result (Mearns, 2003). Dedication to safety by management is also associated with the likelihood to report accidents and near misses, thus allowing for the evaluation of root causes and lowering the probability of recurrence (Mearns, 2001). So on a global scale, for countries with workplaces that tend to have authoritative management systems based on "command and control" methods that do not generally encourage or promote safety, it would follow that the culture of safety would not be as strong within its organizations or businesses, and accident and injury rates would be higher.

4.3 SAFETY CLIMATE DEFINED

Organizational climate is a descriptive measure that reflects the workforce's perceptions of the organizational atmosphere at a particular cross section of time (Fahlbruch, 1999). Safety climate is regarded as the surface features of the safety culture determined by visible parameters such as the attitudes, perceptions, and safety-related behaviors of the workforce at a given point in time (Gonzalez-Roma, 1999;

Mearns, 2001). Safety climate is like a snapshot of the safety culture that is subject to change and somewhat unstable. Safety climate is comprised of the attitudes and beliefs of the workers as primarily demonstrated by their workplace behaviors and perceptions (Bentley, 2010; Zohar, 1980). Behaviors might be demonstrated by the likelihood to wear the required personal protective equipment. Perceptions can be measured through surveys that ask such questions as "Do you believe your supervisor considers safety important?"

Interpretations of the collective climates obtained are an indication of the penetration of top managers' organizational view down to the other hierarchical levels (Flin, 2000). Management commitment to safety, and placing a high level of priority on safety, are closely related to levels of safety climate within organizations (Srinivasan, 2016). Safety management programs can also be structured in ways to maximize safety climate within specific organizations (Cheyne, 1998).

Measures of safety climate can be used as a predictive tool for the identification of potential safety problems (Clarke, 2006). Safety climate can be used to inform management of potential safety problems. As a result, effective use of accurate tools to measure safety climate can reduce overall accident and injury rates for organizations (Rodrigues, 2015). A measure of safety climate could be the likelihood of the use of safety eyewear as required by procedures and management oversight, as opposed to a strong safety culture in which the employee wears the safety eyewear because the worker understands the risks, and wants to protect their eyes. Culture represents more of a belief and a norm that a person would follow over time, even when they were working at home, for example. Safety climate is more a manifestation of the safety culture, expressed by the behavior and attitudes of the employees at a particular point in time (Cheyne, 2002).

Safety climate can be negatively influenced by organizational or environmental factors. Workplaces with a generally good safety climate and resulting perceptions and behaviors by the workers can develop poor safety climates when negative work environments of high levels of pressure or stress are introduced (Amponsah-Tawaih, 2016). Management systems can better ensure a climate of safety and related accident rate reductions, by ensuring appropriate worker training and a sound balance between safety and production goals (Kvalheim, 2016). Ultimately, a strong safety culture can be used to positively influence the temporary negative pressures that impact safety climate.

4.4 SAFETY CULTURE ASSESSMENT

Accurate evaluation of safety culture has been used increasingly by management systems to identify the problem areas in need of additional attention to reduce accidents (Carroll, 1998; Fuller, 2001). In addition, using safety culture assessment and information regarding opportunities for organizational learning, safety management and training can focus on corrective actions to target specific problem areas (Mearns, 2013; Lee, 2000).

It is important that international organizations have a clear picture of their safety culture and how it may change in different international settings. Many companies have begun to use safety culture assessment tools to profile the workforce of organizations in international operations. This is particularly true in high-risk industries such as nuclear power, aviation, and petrochemical industries (Reader, 2015; Taylor, 2010). These evaluations have become increasingly important with the realization that organizational safety cultures can be influenced by national cultural tendencies to practice safe behaviors and avoid risk (Hofstede, 2010).

Studies have shown that organizations which learn what and where their weaknesses are and then take actions to improve their safety cultures and lead to safer behaviors in employees have fewer accident-related injuries and illnesses (ILO, 2009). Although operational parameters affect the likelihood of accidents and incidents, the social environment of the workplace plays a significant role in the disposition of employees to follow safe practices (Brown, 2000). The implementation of safety culture enhancement programs has even been shown to significantly improve the benefits of technical advances and management systems (Yau, 2014).

4.5 CULTURAL DIFFERENCES AND ADAPTATION

In certain countries, the national norms influence such aspects of worker behaviors as innovation in problem solving, and dependence on static procedures to solve dynamic problems (Lee, 2013) (Helmreich, 1999). Different cultures have been shown to have different perceptions, attitudes, and behaviors towards risk (Weber, 1998). Studies have also demonstrated cultural differences in flexibility to act on emerging risks, such as reallocation of resources (Weiner, 2005). These national cultural tendencies then influence the safety cultures that are formed by the managers and workers in organizations (Gharpurea, 2018). These differences can impact the numbers and types of accidents that might occur in various operations, especially highly complex processes and industries (Strauch, 2010).

The influence of national culture on the status of organizational safety culture has only recently been evaluated scientifically. In a study by Reader (2015), he found that corporate safety culture can be measured and closely related to national cultural characteristics. By using known or measured national characteristics, it may be possible to identify areas of concern by relating these to what is known about how safety culture influences safety outcomes in the workplace. High power distance and authoritative cultures that emphasize downward communications generate unwillingness to challenge authority and communicate upward. Based on what is understood about corporate organizations, these characteristics tend to reduce safety culture and increase injury and illness outcomes. Poor communication leads to high uncertainty and poor group harmony, with high dependence on established practice, and avoidance of change and innovation (Reader, 2015).

The basis of a good safety culture is an environment where workers have high levels of trust in the organization. This culture of trust is based on open communication and free flow of information, the belief that reported information will be acted upon, the work environment is just and fair, the organization is flexible when necessary, and it is an organization that fosters learning and personal growth (Reason, 1997). It follows that in organizations that inhibit communication and information, the safety culture would suffer. Similarly, in rigid organizations that discourage change to correct reported problems, safety would be impacted overall. So it follows that where the social cultures of the country where the organization is located, and builds the workforce and management teams, the local culture could be expected to pervade and influence the organizational culture in significant ways.

In the global business setting, national cultural tendencies need to be considered and confronted by management systems directly to address cultural and communication differences between workers of different backgrounds and cultures (Manzey, 2009). For example, where national culture shapes behavior to avoid confrontation, special emphasis may be needed to encourage workers to report unsafe conditions or even exposures. In societies where women have been historically subservient to men, they may need special training or coaxing to become more involved in the safety process, in addition to the men, and management. Personal beliefs and attitudes about a certain gender or race of a co-worker can bias an individual's behavior in the workplace. A company operating a production facility overseas needs to have a good understanding of the societal culture to fabricate a culture of safety within the organization.

CASE

An industrial hygienist got a call from the university hospital occupational health physician requesting an evaluation of the endoscopy unit ventilation system because a worker had been seen for dizziness and shortness of breath after disinfecting endoscopes. After determining that the area ventilation system was operating properly the industrial hygienist interviewed the workers to try and determine how an exposure may have occurred. During the interviews it was determined that the worker had splashed glutaraldehyde on her forearm and it had dripped down into her glove. Rather than stop working, change the glove and wash her hand, she kept working because she did not want to report the incident to her supervisor. When asked why, she said in her Hispanic upbringing women were not supposed to confront men directly with personal problems, and she did not want to have to discuss the issue with her male supervisor. She thought she could wait until lunchtime to wash her hands. What are some things the industrial hygienist could do to ensure a similar event did not occur in the future with this worker or with others in the organization?

The organizational safety culture in globally operating corporations has been shown to vary according to local cultural norms (Kirkman, 2006; Schwartz, 1999). The national tendencies that exist within a society, and influence employee perceptions and beliefs, can play a significant role in the safety culture that forms within the local operation. Societal norms such as not contradicting superiors, admitting an error, or whistleblowing illegal activities of one's superiors or colleagues, can have drastic impacts on the overall safety of an operation. If an organization can become aware of the existing safety culture of the local society, it can play an important role in understanding their workers and what types of safety management or training would be effective.

National tendencies towards uncertainty avoidance (UA) have been shown to be negatively associated with the safety cultures that existed within organizations operating in those countries. Factors outside the control of management directly influence the safety culture within the organization through employee-related attitudes and practices (Noort, 2016). Uncertainty analysis is a cultural measure of society's tolerance for ambiguity, and measures member discontent with unstructured situations. It was one of the five cultural dimensions developed by Geert Hofstede to describe national cultural differences (Hofstede, 1983). Countries with high UA cultures tend to minimize ambiguity by creating strict rules and laws, minimizing social change, and strong singular religious or philosophical beliefs. Companies in countries with high UA scores will have poorer safety cultures than companies in countries with low UA scores. Some predictors of safety culture associated with organizational styles for high UA and low UA company scores are shown in Table 4.1.

The UA indices for a sample of countries are shown in Table 4.2.

Other cultural factors developed by Hofstede to describe different cultures include individualism, masculinity, long-term orientation, and Power Distance Index. Individualism refers to the degree to which individuals in a culture accept and pursue goals that are in their own best interests, rather than seeking those of the group to which they belong. Masculinity is a measure of the degree the people in the group demonstrate what are considered to be masculine traits such as being assertive, ambitious, and competitive as opposed to more caring and demure traits considered feminine. Long-term orientation reflects a person's values on characteristics such as thrift and perseverance and ordering relationships according to status. However, short-term orientation values tradition, fulfilling social obligations, and

TABLE 4.1
Organizational Styles for High and Low UA Companies for Various Safety Culture Predictors

Safety Culture Predictor	High UA	Low UA
Employee perceptions of management commitment to safety	Management discourages new ideas and open discussion	Managers open to new suggestions and approaches
Collaborating for safety	Workers are unlikely to vary from protocols, make suggestions for improvements, or report errors	Workers are open to new ideas, innovation, and working together to solve problems
Incident reporting	Tendency to avoid reporting deviations from norms, including injuries or illnesses	Reduced concern in reporting errors or problems, including accidents
Communication	Top-down following chain of command, information is not shared freely	Open and two-way, less constrained by protocol, information is easily accessible
Colleague commitment to safety	Workers do not feel that their coworkers adhere to the safety program	Workers feel that they can rely on coworkers to act safely
Management support of safety programs	Managers see safety as a necessary expense and give safety programs little autonomy	Managers demonstrate a commitment to safety by actions and funding throughout the organization

TABLE 4.2
UA Indices for a Sample of Countries

Country	UA Index
Greece	112
Poland	93
Japan	92
France	86
Mexico	82
Brazil	76
Egypt	68
Germany	65
Nigeria	54
Australia	51
The United States	46
China	40
India	40
Ireland	35
Denmark	23

Source: Adapted from Clearly Cultural (2018).

reciprocation of greetings, favors, and gifts. Power distance refers to the extent to which people accept inequality in the distribution of power and status, or the degree to which people depend on superiors in a group.

Although there is little research available on how these other indices might be used to evaluate safety culture or performance in various countries, they could be useful tools to consider in designing international safety programs and policies. One recent study used the five Hofstede cultural factors to evaluate how each of the tendencies would impact safety culture in the global oil and gas industry. They found that the cultural tendencies of masculinity, long-term orientation, and Power Distance Index each had a significant impact and direct correlation on organizational safety performance (Gharpurea, 2018). Yet despite the proven relationships in this study, the nuances of how and why the impacts occur within the organization were not identified, and this area remains one that is in need of more research in order to design interventions to better improve the management of international safety programs.

In a review of how culture can affect sociotechnical system operations, Strauch (2010) determined that cultural factors within workgroups can impact rates and degrees of human errors. In addition to language differences, cultural differences can lead to behaviors that are associated with increased numbers and more severe accidents in marine and aviation industries. In addition, culturally heterogeneous work groups had more accidents than homogeneous ones, presumably due to the differences in understanding different cultural signs.

4.6 GLOBAL DIFFERENCES IN SAFETY CULTURE

In 1986, two major industrial accidents occurred, which were both later attributed to poor safety cultures. The Space Shuttle Challenger exploded just after liftoff and killed seven astronauts, and the Chernobyl nuclear power plant accident eventually killed hundreds and contaminated large parts of the Ukrainian countryside with radioactive materials. Each of these accidents was shown to be the result of safety cultures that allowed a continuous accumulation of failures within the organizations that led to the accidents. In many ways, these incidents led to the advancement of safety culture study in the following decades (IAEA, 1986). Some of the common cultural features between National Aeronautics and Space Administration (NASA) and Chernobyl were as follows:

- Competition between departments for necessary information
- Lack of response or slow, incomplete response, to identified errors or problems
- Budget and program cuts to essential OSH operations
- Management incentives for workers to maintain the status quo
- Lack of job security and extensive use of temporary workers
- Bureaucratic safety management systems
- Performance reviews based on production without consideration of safety (ICSI, 2018)

In a detailed study of safety climate measures based on sociocultural and/or individual factors between the United States and Italy, Barbaranelli (2015) demonstrated that U.S. participants had significantly more positive safety climate perceptions overall than the Italian participants. These results could have useful implications for a global company with operations in both countries. The management approaches to creating a climate of safety in Italy may need to be strengthened in comparison with U.S. branches, in order to achieve equal levels of site safety and security. Although this study did not provide definitive reasons for the differences between the two countries, it provides strong evidence that safety climates would vary among workers from other countries.

In many developing economies, workplace health and safety regulations have taken a back seat to business development (Michaels, 1985). The false assumption that safety regulation and safety programs increase operating costs and reduce profits is still a broadly held belief by many developing country governments and businesses. Additionally, decades of downplaying the value of OSH have pervaded the workplace culture to impact how managers and workers think about OSH. Despite continuous advances in OSH program development and management in developed countries, occupational health continues to be a low national priority for many economically developing countries (EDCs) (Perrow, 1984).

There is a wide range of progress in the development and implementation of OSH regulations and programs internationally. In general, the value of OSH programs and policies has been realized by both businesses and governments in developed countries; however, there are few political models for the justification and creation of OSH regulation in EDCs (Jilcha, 2016). In many developing countries, the risks of occupational injury and illness are significantly higher than those in developed countries, largely because there are few requirements or standards for OSH to protect most of the workers (Tadesse, 2007; LaDou, 2003). Although more advanced countries have pockets of workers who are not well covered by OSH protections (e.g., informal workers, some small to midsized enterprises), and some EDCs have well-developed and enforced OSH regulations, the general status is that EDCs lag behind in OSH program development and implementation.

Significant percentages of the national production in EDCs are done within the informal economy; therefore, EDCs have less to gain directly from improved occupational health policies to protect informal workers who do not pay taxes and are not explicitly part of the economy. In these EDCs, the benefits of healthy workforces and populations need to be approached in a broader context of social and economic policy and equity (Swuste, 2002; Joubert, 2002).

As EDCs advance in OSH development, the early changes tend to occur on larger social issues such as national policies and regulations, social and economic structures, and human resource growth. Internal domains of organizations, including control of working conditions or hazards, tend to occur later in national development. Occupational health tends to be a low priority in developing countries partly due to limited resources and inadequate information. It is difficult to show government officials why OSH is significant to the country's economic growth, and therefore difficult to create an impetus for change (Jilcha, 2016; O'Neill, 2000).

Developed countries have political systems and finances to support scientific research that can drive policy and program development to support technical change (Verma, 2002). However, despite the evidence that having healthy workers in a society is closely related to a healthy economy, developing countries with poor credit, close to the poverty line, have few options for improving workplace OSH. The high number of workplace injuries, illnesses, and fatalities thus reduces the country's economic resources yet further, creating a spiraling decline (ILO, 2012).

Numerous examples exist to show how a country's political and social systems can have direct impacts on worker health and safety. There are as many scenarios for how culture affects OSH as there are countries. Companies with international operations not only need to be aware of regulations in those regions, but they must understand the social, economic, and cultural environments that will impact worker health and safety.

Case: Safety Culture in France

Food production is the leading manufacturing sector in France, with the highest number of employees at approximately 600,000 (Caroli, 2009). It provides an excellent example of how market and political forces act together to impact occupational health and safety. About 40% of French food-manufacturing jobs are classified as unskilled, and about two thirds are blue-collar. Many of these workers work at or slightly above minimum wage. But despite low wages, French workers are protected by strict Employment Protection Legislation that sets such operational parameters as working hours, holiday and vacation time off, and job security. Although only about 8% of the workers in France fall into unions, 90% of employees are covered under some sort of public service regulated status.

With increasing global competition in the manufacturing sector, French firms have had to intensify work in order to take advantage of capital investments and maximize profits. Since overtime is strictly regulated, more employees must be hired and as a result, the employers must incur all of the additional overhead costs typically associated. In order to take advantage of expensive manufacturing equipment, many employers have begun operating on two and three shifts, which requires more workers yet. So despite shorter working hours, the low level employees are subjected to harsher working conditions, like shiftwork and nighttime hours. In some cases the intensification of work has led to increased absenteeism (possibly due to increased injury?) Which then requires firms to hire yet more replacement staff.

(Caroli, 2009).

4.7 CONCLUSIONS/RECOMMENDATIONS

Strong safety cultures influence the abilities of effective safety climates and greatly improve organizational safety performance and operational efficiencies. Organizations should strive to induce a culture of safety into their organizations in any country and should be especially aware of the cultural variations in workers going abroad or being hired in foreign nations for international operations.

Weak safety cultures in developed and sophisticated organizations have been shown to lead to accidents with devastating consequences. When hazardous industries are transferred to developing nations with limited resources and experience, emphasis on the development of a strong safety culture can play a significant role in reducing the inherent risks.

In developing countries, the large number of workers in the informal industry warrants enhanced development of an occupational safety culture at a national level (Kim, 2016). In addition, worker health and safety in the informal sector is generally unregulated (Rantanen, 2009; Naidoo, 2009). Additional research needs to be conducted on the associations between informal work and OSH working conditions. National OSH reporting systems need to be expanded to include this sector and these workers to better understand and address OSH issues and problems. Once the hazards to informal workers are identified, and controls are developed, workers need to receive associated necessary training to minimize the risks.

Technologically advanced countries with sophisticated OSH regulatory systems are also more likely to have a national culture of workplace safety in the working population. Advances in management systems and further growth in the development of workplace safety culture will continue to reduce the rate of incidents moving forward. Developing counties not only need support in the creation of technical and policy advances in OSH, but, in parallel, need training and development in safety culture in order to fully minimize workplace injuries, illnesses, and fatalities. Cultures of injury and illness prevention need to be germinated and nurtured at the national levels of developing countries to be sure to include all workers and organizational activities, including the informal sector (Kim, 2016; Ametepeh, 2013).

Management systems of organizations can use the available information regarding national culture and the tendencies in workers to target specific areas for specialized attention to reduce possible negative outcomes. Cultural norms are influential in creating a corporate safety culture, but they can be influenced positively by targeted management actions (Mearns, 2009). Although research has suggested that training can moderate the impact of cultural tendencies of workers on an organization in certain cases (Klein, 2007; Sutton, 2006), numerous questions remain with regard to overall performance improvement and how to avoid culturally induced safety problems in complex operations (Ahasan, 2001).

REFERENCES

Ahasan, M. R., Partanen, T., Occupational health and safety in the least developed countries: A simple case of neglect, *Journal of Epidemiology* (2001) Vol. 11, pp. 74–80.

Amponsah-Tawaih, K., Adu, M., Work pressure and safety behaviors among health workers in Ghana: The moderating role of management commitment to safety, *Safety and Health at Work* (2016) Vol. 7, pp. 340–346. doi:10.1016/j.shaw.2016.05.001.

Ametepeh, R., Adei, D., Arhin, A., Occupational health hazards and safety of the informal sector in the Sekondi-Takoradi metropolitan area of Ghana, *Research on Humanities and Social Sciences* (2013) Vol. 3, No. 20, ISSN 2222-1719 https://pdfs.semanticscholar.org/f176/e1be5a59c0d60057de76179866007dabab55.pdf accessed June 29, 2018.

Bentley, T., Tappin, D., Incorporating organisational safety culture within ergonomics practice, *Ergonomics* (2010) Vol. 53, No. 10, pp. 1167–1174. doi:10.1080/00140139.2010.512981.

Barbaranelli, C., Petitta, L., Probst, T., Does safety climate predict safety performance in Italy and the USA? Cross-cultural validation of a theoretical model of safety climate, *Accident Analysis and Prevention* (2015) Vol. 77, pp. 35–44.

Brown, K., Willis, P., Prussia, G., Predicting safe employee behavior in the steel industry: Development and test of a sociotechnical model, *Journal of Operations Management* (2000) Vol. 18, pp. 445–465.

Carroll, J.S., Safety culture as an ongoing process: Culture surveys as opportunities for enquiry and change, *Work and Stress* (1998) Vol. 12, pp. 272–284. doi:10.1080/02678379808256866.

Caroli, E., Gautie, J., Lamanthe, A., The French food-processing model: High relative wages and high work intensity, *International Labor Review* (2009) Vol. 148, No. 4, pp. 375–394.

Cheyne, A., Cox, S., Oliver, A., Tomas, J., Modelling safety climate in the prediction of levels of safety activity, *Work & Stress* (1998) Vol. 12, No. 3, pp. 255–271.

Cheyne, A., Oliver, A., Tomas, J., Cox, S., The Architecture of employee attitudes to safety in the manufacturing sector, *Personnel Review* (2002), Vol. 31, pp. 649–670.

Clearly Cultural, Uncertainty Avoidance (2018) www.clearlycultural.com/geert-hofstede-cultural-dimensions/uncertainty-avoidance-index/ accessed June 20, 2018.

Clarke, S., Contrasting perceptual, attitudinal and dispositional approaches to accident involvement in the workplace, *Safety Science* (2006) Vol. 44, No. 6, pp. 537–550.

Cooper, M., Towards a model of safety culture, *Safety Science* (2000) Vol. 36, pp. 111–136. doi:10.1016/S0925-7535(00)00035-7.

Cox, S., Flin, R., Safety culture: Philosopher's stone or man of straw? *Work & Stress* (1998) Vol. 12, pp. 189–201.

Davies, H. T. O., Nutley, S. M., Mannion, R., Organisational culture and quality of health care, *Quality in Health Care* (2000) Vol. 9, pp. 111–119.

Fåhlbruch, B., Wilpert, B., (1999). System safety—an emerging field for I/O psychology. In C. L. Cooper, I. T. Robertson (Eds). *International Review of Industrial and Organizational Psychology*, (Vol. 14, pp. 55–93). New York: John Wiley & Sons Ltd.

Flin, R., Mearns, K., O'connor, P., Bryden, R., Measuring safety climate: Identifying the common features, *Safety Science* (2000) Vol. 34, pp. 177–192.

Fuller, C. W., Vassie, L. H., Benchmarking the safety climates of employees and contractors working within a partnership arrangement: A case study in the offshore oil industry, *Benchmarking: An International Journal* (2001) Vol. 8, pp. 413–430. doi:10.1108/EUM0000000006386.

Gharpurea, S., Roya, S., Purang, P., Bhattacharyya, S., Role of cultural dimensions in safety performance of global oil and gas industry, *Recent Advances in Petrochemical Science* (2018) Vol. 5, No. 1. ISSN: 2575-8578.

Gonzalez-Roma, V., Peiro, J., Lloret, S., Zornoza, A., The validity of collective climates, *Journal of Occupational and Organizational Psychology* (1999) Vol. 72, pp. 25–40.

Guldenmund, F., The nature of safety culture: A review of theory and research, *Safety Science* (2000) Vol. 34, pp. 215–257.

Helmreich, R. (1999). Building safety on the three cultures of aviation. Paper Presented at the *Proceedings of the IATA Human Factors Seminar.* Bangkok, Thailand, August 12, 1998, pp. 39–43.

Hofstede, G., The cultural relativity of organizational practices and theories, *Journal of International Business Studies* (1983) Vol. 14, pp. 75–89.

Hofstede, G., Hofstede, G. J., Minkov, M. (2010). *Cultures and Organizations: Software of the Mind* (3rd ed.). New York: McGraw-Hill.

IAEA, International Atomic Energy Agency, International Nuclear Safety Advisory Group (HSC) (1986). *Summary Report on the Post-Accident Review Meeting on the Chernobyl Accident, Safety Series No. 75-INSAG-1.* Vienna: International Atomic Energy Agency.

IAEA, International Atomic Energy Agency (1991). *Safety Culture, Safety Series No. 75-INSAG-4.* Vienna: IAEA. IBSN 92-0-123091-5.

ICSI, Institut pour une culture de securite industrielle, Safety Culture: from Understanding to Action (2018) www.icsi-eu.org/documents/88/csi_1801-_safety_culture_from_understanding_to_action.pdf accessed June 28, 2018. ISSN 2100-3874.

International Labour Organization (ILO), Information on Decent Work and a Health and Safety Culture [Internet]. Geneva (Switzerland): Office; 2009 [cited 2015 Sep 15] (2009) www.ilocarib.org.tt/portal/index.php?option.com_content&task.view&id.1138&Itemid.1141.

International Labour Organization (2012). *Estimating the Economic Costs of Occupational Injuries and Illnesses in Developing Countries: Essential Information for Decision-Makers.* Switzerland: International Labour Organization www.ilo.org/publns.

Kim, Y., Park, J., Park, M., Creating a culture of prevention in occupational safety and health practice, *Safety and Health at Work* (2016) Vol. 7, pp. 89–96.

Kirkman, B. L., Lowe, K. B., Gibson, C. B., A quarter century of culture's consequences: A review of empirical research incorporating Hofstede's cultural values framework, *Journal of International Business Studies* (2006) Vol. 37, pp. 285–320. doi:10.1057/palgrave.jibs.8400202.

Klein, H. A., Steele-Johnson, D. (2007). Training for multinational teamwork. In R. R. Hoffman (Ed). *Expertise Out of Context: Proceedings of the Sixth International Conference on Naturalistic Decision Making* (pp. 473–506). New York: Lawrence Erlbaum.

Kvalheim, S., Dahl, O., Safety compliance and safety climate: A repeated cross-sectional study in the oil and gas industry, *Journal of Safety Research* (2016) Vol. 59, pp. 33–41.

LaDou, J., International occupational health, *International Journal of Hygiene and Environmental Health* (2003) Vol. 206, pp. 303–313.

Lee, T., Harrison, K., Assessing safety culture in nuclear power stations, *Safety Science* (2000) Vol. 34, pp. 61 97. doi:10.1016/S0925 7535(00)00007 2.

Lee, S., Trimi, S., Kim, C., The impact of cultural differences on technology adoption, *Journal of World Business* (2013) Vol. 48, No. 1, pp. 20–29.

Jilcha, K., Kitaw, D., A literature review on global occupational safety and health practice & accidents severity, *International Journal for Quality Research* (2016) Vol. 10, No. 2, pp. 279–310.

Joubert, D. M., Occupational health challenges and success in developing countries: A South African perspective, *International Journal of Occupational Environmental Health* (2002) Vol. 8, pp. 119–124.

Manzey, D., Marold, J., Occupational accidents and safety: The challenge of globalization, *Safety Science* (2009) Vol. 47, pp. 723–726.

Mearns, J., Havold, J., Occupational health and safety and the balanced scorecard, *The TQM Magazine* (2003) Vol. 15, No. 6, pp. 408–423.

Mearns K, Yule, S., The role of national culture in determining safety performance: Challenges for the global oil and gas industry, *Safety Science* (2009) Vol. 47, pp. 777–785.

Mearns, K., Flin, R., Gordon, R., Fleming, M., Measuring safety climate on offshore installations, *Work & Stress* (1998) Vol. 12, pp. 238–254.

Mearns, K., Whitaker, S., Flin, R., Benchmarking safety climate in hazardous environments: A longitudinal, interorganizational approach, *Risk Analysis* (2001) Vol. 21, No. 4, pp. 771–786.

Mearns, K., Kirwan, B., Reader, T. W., Jackson, J., Kennedy, R., Gordon, R., Development of a methodology for understanding and enhancing safety culture in air traffic management, *Safety Science* (2013) Vol. 53, pp. 123–133.

Michaels, D., Barrera, C., Gacharna, M.G., Economic development and occupational health in Latin America: New directions for public health in less developed countries. *American Journal of Public Health* (1985) Vol. 75, pp. 536–542.

Naidoo, R., Kessy, F., Mlingi, L., Petersson, N., Mirembo, J., Occupational Health and Safety in the Informal Sector in Southern Africa—The WAHSA Project in Tanzania and Mozambique, WAHSA Special Issue (2009) www.occhealth.co.za accessed June 29, 2018.

Noort, M., Reader, T., Shorrock, S., Kirwan, B., The relationship between national culture and safety culture: Implications for international safety culture assessments. (2016) *Journal Occupational and Organizational Psychology* (2016) Vol. 89, pp. 515–538.

O'Neill, D., Ergonomics in industrially developing countries: Does its application differ from that in industrially advanced countries? *Applied Ergonomics* (2000) Vol. 31, No. 6, pp. 631–640. doi:10.1016/s0003-6870(00)00033-8.

Perrow, C. (1984). *Normal Accidents: Living with High-Risk Technologies, Basic Books.* New York. ISBN: 978-0-691-00412-9.

Rantanen, J., Occupational Health Services for the Informal Sector. Africa Newsletter on Occupational Health and Safety, No. 2 [Internet]. 2009 [cited 2015 Sep 17] (2009) www.ttl.fi/en/publications/electronic_journals/african_newsletter/african_archives/Documents/african_newsletter2_2009.pdf.

Reader, T. W., Noort, M. C., Shorrock, S., Kirwan, B., Safety sans frontieres: An international safety culture model, *Risk Analysis* (2015) Vol. 35, pp. 770–789. doi:10.1111/risa.12327.

Reason, J. T. (1997). *Managing the Risks of Organizational Accidents.* Aldershot: Ashgate. ISBN: 978-1-84014-105-4.

Rodrigues, M., Arezes, P., Leao, C., Multilevel model of safety climate for furniture industries, *Work* (2015) Vol. 51, pp. 557–570.

Rousseau, D. (1988). Quantitative assessment of organizational culture: the case for multiple measures. In L. C. Cooper, I. Robertson (Eds). *International Review of Industrial Organizational Psychology.* Chichester: Wiley.

Schwartz, S. H., A theory of cultural values and some implications for work, *Applied Psychology* (1999) Vol. 48, pp. 23–47. doi:10.1111/j.1464-0597.1999.tb00047.x.

Shiney, C., Medha, K., Safety culture: The buzzword to ensure occupational safety and health, *Procedia Economics and Finance* (2014) Vol. 11, pp. 130–136.

Srinivasan, S., Hughes, L., Shakouri, M., Nahmens, I., Harvey, C., 5S impact on safety climate of manufacturing workers, *Journal of Manufacturing Technology Management* (2016) Vol. 27, No. 3, pp. 364–378.

Strauch, B., Can cultural differences lead to accidents? Team cultural differences and sociotechnical system operations, *Human Factors: The Journal of the Human Factors and Ergonomics Society* (2010) Vol. 52, pp. 246–263.

Sutton, J. L., Pierce, L. G., Burke, C. S., Salas, E. (2006). Cultural adaptability. In C. S. Burke, L. G. Pierce, E. Salas (Eds.). *Understanding Adaptability: A Prerequisite for Effective Performance within Complex Systems.* Advances in human performance and cognitive engineering research (Vol. 6, pp. 143–173). Oxford, UK: Elsevier.

Swuste, P., Eijkemans, G., Occupational safety, health, and hygiene in the urban informal sector of sub-saharan Africa: An application of the prevention and control exchange (pace) program to the informal-sector workers in healthy city projects, *International Journal of Occupational and Environmental Health* (2002) Vol. 8, No. 2, pp. 113–118. doi:10.1179/107735202800338920.

Tadesse, T., Kumie, A., Prevalence and factors affecting work-related injury among workers engaged in small and medium-scale industries in Gondar woreda, North Gondor zone, Amhara Regional State, Ethiopia, *Ethiopian Journal Of Health Development* (2007) Vol. 21, No. 1, pp. 25–34. doi:10.4314/ejhd.v21i1.10028.

Taylor, J. B. (2010). *Safety Culture: Assessing and Changing the Behaviour of Organisations.* Farnham, UK: Gower.

Verma, D., Translating evidence about occupational conditions into strategies for prevention, *Occupational and Environmental Medicine* (2002) Vol. 59, No. 3, pp. 205–214. doi:10.1136/oem.59.3.205.

Weiner, B. J., Lewis, M. A., Linnan, L. A., Waarts, E., Van Everdingen, Y., The influence of national culture on the adoption status of innovations: An empirical study of firms across Europe, *European Management Journal* (2005) Vol. 23, pp. 601–610. doi:10.1016/j.emj.2005.10.007.

Weber, E. U., Hsee, C., Cross-cultural differences in risk perception, but cross-cultural similarities in attitudes towards perceived risk, *Management Science* (1998) Vol. 44, pp. 1205–1217.

Wiegmann, D., Zhang, H., von Thaden, T., Gibbons, A., Sharma, G., Safety culture: An integrative review, *International Journal of Aviation Psychology* (2004) Vol. 14, No. 2, pp. 117–134.

Yau, B. (2014). Occupational Safety Culture Index E Measuring the Community and Employees Awareness, Attitude and Knowledge Towards Workplace Safety and Health in Hong Kong [power point slides]. XX World Congress on Safety and Health at Work 2014: Global Forum for Prevention. Germany: Frankfurt.

Zohar, D., Safety climate in industrial organizations: Theoretical and applied implications, *Journal of Applied Psychology* (1980) Vol. 65, No. 1, pp. 96–102.

5 Risk Assessment
Global Perspectives

James F. Whiting
Risk@workplaces Pty Ltd

CONTENTS

5.1 INTRODUCTION

In these days of business globalization, professional safety and health engineering (SH&E) practitioners often need to be aware of what are the international practices and standards around the world. Global organizations need assurance of the uniformity, consistency, and harmonization of their risk management standards, policies, and processes, in general, and their SH&E standards, in particular.

In 2009, the International Standards Organization (ISO), an international body charged with achieving global standardization, finalized a standard risk management system to achieve consistency and reliability in risk management by creating ISO 31000, a standard that is applicable to all forms of risk. It should be noted that in 2011, ISO 31000:2009 was nationally adopted in the United States as an American National Standard, ANSI/ASSE Z590.2-2011, "Risk Management Principles and Guidelines," and is identical to the ISO standard. The standard includes principles, a framework, and one common overarching RM process shown in Figure 5.1.

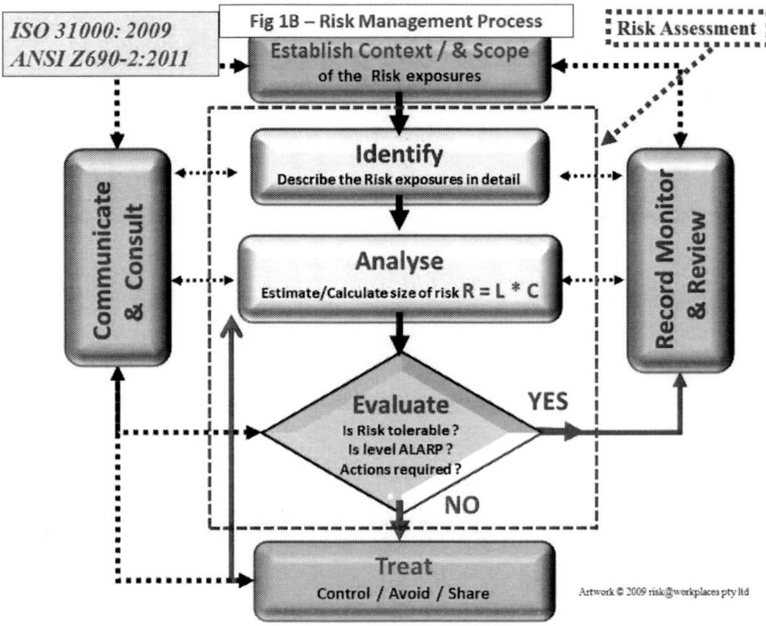

FIGURE 5.1 The ISO 31000/ANSI Z690.2 RM process.

The ISO 31000 Standard and a Conformity Assessment tool (Whiting, 2012) can be applied throughout the life of an organization or a project, and to a wide range of activities, including strategies, decision-making, operations, processes, functions, projects, products, services, and assets. As a benchmark measurement at any given time, they provide an effective means of following change management and continuous improvement. They can be used by any public, private or community enterprise, association, group, or individual. Therefore, this standard is not specific to any industry or sector. In addition, this standard and conformity assessment tool can be applied to any type of risk domain, whatever its nature, and whether positive (upside risk) or negative (downside risk) consequences are being considered. Establishing internal and external benchmarking at a single organization, national or international level, is possible.

5.2 GLOBAL UPTAKE OF ISO 31000—INTERNATIONAL RISK MANAGEMENT STANDARD

Since 2009, there has been widespread acceptance and adoption by many countries of the ISO 31000 Standard as their national RM standards. ISO 31000 is not used for certification but does provide freedom for organizations to decide how deeply and fully they establish and adhere to its principles.

The real strength of ISO 31000 is that it encourages risk managers and their organizations to understand and make use of the relationships, commonalities, and differences between various risk management methods, standards, and best practices across all risk domains, not just SH&E.

It is designed to be used to harmonize risk management processes in existing and future management system standards such as ISO 9001:2015 QMS, ISO 14001:2015 EMS, and ISO/DIS 45001:2017 OHSMS. It provides a common approach in support of standards dealing with specific risks and/or sectors, but does not replace those standards. A partial list of countries that have adopted ISO 31000 includes Canada, the United States (ANSI/ASSE Z590.2), Brazil, Russia, China, India, and Australia.

5.3 GLOBAL COMPARISON OF RISK TOLERABILITY CRITERIA

A critical stage in the RM process is *evaluation*. At that stage in the process, the risk manager needs to evaluate the sizes of the risks calculated or estimated in the preceding *analysis* stage by comparing them with predetermined criteria developed in the initial *establish context and scope* stages. The criteria of most importance are the prior agreed risk tolerability levels and whether the risk level is being continually managed down so far as is reasonably practicable (SOFAIRP).

5.4 INDIVIDUAL RISK AND SOCIETAL RISK

Risks to people can be represented in two ways. Both are a combination of the likelihood of an event happening (e.g., an accident at a major hazard installation) and the possible consequences—in terms of harm to people.

5.4.1 Individual Risk

Individual risk is the likelihood or probability or chance that a particular individual at a particular location under specific exposure circumstances will be harmed. It is usually described in numerical terms such as "a 1 in 20,000,000 chance of being killed by lightning per annum (p.a.)." But the assessment of individual risk does not take account of the total number of people at risk from a particular event. Individual risk is usually expressed as the probability of fatality of an individual per year such as

$$1 \text{ in } 1{,}000 \text{ p.a.}, 1 \text{ chance in } 1{,}000 \text{ p.a.}, 10^{-3} \text{ p.a.}, 1\text{E}-03 \text{ p.a.}$$

As an example, if the fatality rate for a traffic risk is estimated as 10,000 driver fatalities p.a. in 100,000,000 drivers, then the individual risk is expressed as 10,000/100,000,000 p.a. = 1 in 10,000 p.a. or 10^{-4} p.a., or 1E–04 p.a.

If the risk of work fatalities for a generic or specific risk exposure is estimated for an organization of 20,000 exposed employees to be 4 per annum, then the individual risk is expressed as 4/20,000 p.a. = 1 in 5,000 p.a. = 2 in 10,000 p.a. = 2 chances in 10,000 p.a. = 2×10^{-4} p.a. = 2E–04 p.a.

In reverse, if the individual risk for a given exposure is estimated as 2 chances in 10,000 p.a., then

- For an organization of 20,000 exposed employees, there is a risk of 20,000 × 2/10,000 p.a. = 4 fatalities p.a.
- For an organization of 5,000 employees (all other risk factors the same), there is a risk of 5,000 × 2/10,000 = 1 fatality p.a.
- For an organization of 1,000 employees (all other risk factors the same), there is a risk of 1,000 × 2/10,000 = 0.2 fatality p.a. OR 1 fatality in 5 years.

5.4.2 Societal Risk

Societal risk is a way to estimate the chances of numbers of people being harmed from an incident. The likelihood of the primary event (an accident at a major hazard plant) is still a factor, but the consequences are assessed in terms of level of harm and numbers affected, to provide an idea of the scale of an accident in terms of numbers killed or harmed.

Societal risk can also be expressed as a *potential loss of life* (PLL), which is the number of fatalities that may be expected to occur each year, averaged over a long period. The number should be small: if 100 people are each exposed to a risk level of 10 in a million per year, the PLL is 0.001.

The PLL is a useful basis for cost–benefit analyses (CBAs) of risk reduction measures, via the *implied cost of averting fatality* (ICAF): ICAF = cost of measure/(initial PLL – reduced PLL). Such calculations are often controversial as they appear to require a value to be placed on life, but these calculations are commonly used internationally and may aid decision-making in regard to adopting control measures for major hazards. For example, a *low ICAF* for a proposed risk reduction/treatment measure implies that the measure is *highly effective* because the cost is low compared to the risk reduction achieved. Conversely, a high ICAF implies a relatively ineffective risk reduction measure, indicating that the money should be diverted to an alternative.

5.5 TOLERABILITY CRITERION FOR INDIVIDUAL RISK

In the United Kingdom and the Netherlands, many decades ago, public safety risk criteria for new industrial hazardous activities have been set by government regulation. The starting point for determining these criteria was the statement "that any additional risk from exposure to a new hazardous activity to a member of the public should not be significant when compared with risk in everyday life." Following this approach, for new major hazard installations (say, a petrochemical plant or nuclear reactor), the maximum acceptable level for individual risk for a nearby resident has been taken arbitrarily as the risk level that increases the risk of death by *a maximum of 1%* compared with all other causes. For developed countries, the individual "natural death" risk for the population group of 10- to 14-year-olds is close to 10^{-4} p.a. or 1 in 10,000 p.a. and has been taken as the basic "background" fatality risk. The maximum tolerable *individual risk* was hence established as follows:

$$1\% \times 1 \text{ chance in } 10{,}000 \text{ p.a.} = 1/100 \times 1 \text{ in } 10{,}000 \text{ p.a.} = 1 \text{ chance in } 1{,}000{,}000 \text{ p.a.} = 10^{-6} \text{p.a.}$$

In other words, the risk of a fatal accident to which an individual anonymous member of the public is exposed because of his/her continuous presence (365 days per year) in the neighborhood of a new hazardous activity shall be less than one in a million years. Risk exposure levels of less than 10^{-8} per year, or less than once in 100 million years, are considered to be very low/almost negligible* (see note). This individual risk level is consistent with low/almost negligible risk in other areas.

Note: *The authors recommend that a very low risk NOT be classified as "negligible" or "insignificant" as that implies that they can be neglected. No risk can ever be entirely neglected. Rather a more appropriate management approach is to recognize that even a very low risk can increase and become of concern if the risk factors change unfavourably. Hence it is recommended to use "very low" risk as the descriptor only after considering if the risk factors would not change significantly and/or quickly then the monitoring frequency to revisit a risk assessment can be set as very low as well.*

In the United States, the concept of 10^{-6} p.a. was originally an arbitrary number, finalized by the U.S. Food and Drug Administration around 1977 as a screening level of *essentially zero* or *de minimis risk*. This concept was traced back to a 1961 proposal by two scientists from the National Cancer Institute regarding methods to determine "safety" levels in carcinogenicity testing.

The proposal for *de minimis risk* was contained in a 1973 notice and eventually adopted in 1977 in the Federal Register entitled "Compounds Used in Food-Processing Animals: Procedures for Determining Acceptability of Assay Methods Used for Assuring the Absence of Residues in Edible Products of Such Animals," commonly called the "Sensitivity of Method" regulations. The term *de minimis* is an abbreviation of the legal concept, *de minimis non curat lex*—translated as *the law does not concern itself with trifles*. In other words, 10^{-6} p.a. was developed as a level of risk below which was considered a "trifle" and not of regulatory concern.

A survey of worldwide risk tolerability criteria (AIChE/CCPS, 2009; Appendix B) shows similarities among the criteria around the world. The data have been extracted from various publications and provide a benchmarking perspective. The author's consideration of numerous worldwide, reasonably well-established, and widely accepted criteria for individual fatality risk tolerability criteria leads to the indicative levels in Table 5.1.

TABLE 5.1
Generally Accepted Global Individual Fatality Risk Tolerability Levels

Individual Fatality Risk Tolerability Criteria	Exposed Worker	Exposed Member of the Public
Max tolerable threshold	1 in 1,000 per annum 1×10^{-3} per annum	ln 10,000 per annum 1×10^{-4} per annum
Broadly tolerable or acceptable levels	1 in 1,000,000 per annum 1×10^{-6} per annum	1 in 1,000,000 per annum 1×10^{-6} per annum

Caution: *It must be emphasized that these data can be used for internal assurance but cannot be used for regulatory compliance without checking with the appropriate regulatory authority in the home country.*

5.6 TOLERABILITY CRITERIA FOR PLANNING NEW OPERATIONS

The foundation for choosing quantitative risk tolerability criteria considers the principles

1. That the exposed persons such as nearby residents should not be involuntarily subject to a risk from a new exposure that is significant compared to the "background" risk associated with existing hazards.
2. That individual and societal risk should be considered separately.

Land use planning departments and regulators of major hazard facilities in different countries have established quantitative risk criteria for new land use developments adjacent to existing land users according to specific sensitivities of exposed public persons, which are shown in Table 5.2. They represent very low risks compared to other everyday risks associated with their existing land uses.

5.7 INVESTMENT TO PREVENT A FATALITY

An interesting global perspective in risk management is how different countries use measures related to human life values when calculating *CBA* for deciding to commit to spending on proposed new or changed risk controls for a particular risk, for example, *should all school buses be fitted with seat belts?*

Extract Quote: *Washington Post*, August 25, 2011 (www.washingtonpost.com/local/feds-reject-request-to-require-seat-belts-on-school-buses/2011/08/25/gIQATJhseJ_story.html).

TABLE 5.2
Fatality Risk Criteria for Land Use Planning and Locations of New Exposures

Location of Exposed Persons	Fatality Risk Criteria (per million pa)[a]
Sensitive, for example, in hospitals, schools, child care, aged care	0.5
Residential including hotels, motels, resorts	1
Commercial	5
Sporting including open space, parks	10
Industrial	50

[a] NSW Department of Planning 2011.

> In Thursday's Federal Register, NHTSA cited its 2002 report to Congress, which said that shoulder lap belts are effective in reducing school bus fatalities, but the addition of the belts "*would increase capital costs.*" NHTSA estimated equipping each bench-style seat would cost between \$375 and \$600, a total of between \$5,485 and \$7,346 for each large bus.
>
> "The benefits would be achieved at a cost of *between \$23 and \$36 million per equivalent life saved*," NHTSA said. Rather than face a federal mandate, NHTSA said state and local governments should be left to decide whether to spend the money. Texas and California require school bus belts.

The standard approach to CBA of risks to life is to convert them into equivalent costs. The monetary valuation of risks to life is often described as a "*value of life*." This phrase is convenient but inaccurate and also evokes a strong emotional response. CBA evaluates small changes in risks for many people and does not attempt to value individual lives. The accumulation of risk to many people, which can be expected on average to result in the saving of one fatality, is better described as a "statistical fatality." For example, *a reduction in risk of* 10^{-3} *per year for each of 100 individuals over a period of 10 years would amount to a saving of one statistical fatality.* This distinction is important because it is much more reasonable to place a value on small changes in statistical risk than on individually identifiable lives. Presentation of this difficult and often emotive concept can be improved by using the term *value of preventing a statistical fatality* (*VPF*). This emphasizes that what is being valued is the reduction in risk to many lives, rather than the actual lives that are at risk of being lost.

There are many different ways of considering \$ values of saving lives at work and in societies generally. Some expressions used are as follows:

- Value of a statistical life (VSL)
- VPF, or statistical economic value for preventing a fatality
- Investment per equivalent life saved
- Investment to prevent a fatality
- Cost of a year of life saved
- Willingness to pay
- PLL
- ICAF (= cost of measure/(initial PLL – reduced PLL).

Such calculations are often controversial as they appear to require a value to be placed on life, but these calculations are commonly used internationally and may aid decision-making in regard to adopting control measures for major hazards. For example, a low ICAF for a proposed risk reduction measure implies that the measure is highly effective because the cost is low compared to the risk reduction achieved. Conversely, a high ICAF implies a relatively ineffective risk reduction measure, indicating that the money should be diverted to an alternative.

A VSL is the \$ amount that a group of people, say a government, is willing to pay for a fatal risk reduction in the expectation of saving one anonymous life. These values can be estimated from what governments reflecting social values in

TABLE 5.3
Variations between Countries in VoLS—VSL (Thousands of 1995 U.S. Dollars)

Country	Mean Value
Japan	8,280
Switzerland	7,525
The United States	3,472
France	4,435
New Zealand	1,625
Taiwan	956
South Korea	620

Source: Adapted from Miller (2000).

TABLE 5.4
Examples of International VPFs/VoLS

Context	VPF or VoLS	Source
U.S. school bus seat belts	\$23 → \$36 million per equivalent life saved	NHTSA August 2011
Road fatalities on U.S. roads	\$5.8 million as the statistical economic value for preventing a human fatality.	U.S. DOT February 5, 2008
New Zealand VPF accident compensation system	\$3,4 million	Wren (2011)
OECD and EU countries	\$1.5 → \$5.4 million	OECD (2012)

how much is invested in coronary care units, road safety engineering measures, vaccinations, etc.

These values are determined by the risk tolerance of the group [not always consciously] and are strongly influenced by social, cultural, and economic factors. Tables 5.3 and 5.4 show variations among countries according to these factors in the 1990s but would be quite different now as countries' economic circumstances have changed.

5.8 SHIFTING THE PARADIGM FROM ABSOLUTE SAFETY TO RISK MANAGEMENT

Internationally, more and more countries have shifted, and are shifting their safety philosophies and practices from unrealistic *absolute safety* or even *zero risk* beliefs to more appropriate models of managing risk to as low as reasonably practicable (ALARP). The safety regulators in many countries have been traditionally *prescriptive* in describing absolute obligations to detail how safety risks were to be eliminated or prevented or stopped. In recent decades, legal and regulatory safety obligations are being described in *ALARP/performance-based frameworks* rather than prescriptive, unconditional *ensure without risk* models. National and international safety standards are now more and more written in terms of ALARP or the equivalent.

5.9 WHAT IS REASONABLY PRACTICABLE?

A risk-informed performance-based regulatory approach to safety laws defines *reasonably practicable*, in relation to a duty to ensure health and safety, as meaning that what was reasonably able to be done in relation to ensuring health and safety, taking into account and weighing up all relevant matters including the following:

a. The *likelihood* of the hazard or the risk concerned occurring
b. The *degree of harm* that might result from the hazard or the risk
c. What the person concerned *knows*, or ought reasonably to know, about—
    i. The hazard or the risk
    ii. Ways of eliminating or minimizing the risk
d. The *availability and suitability* of ways to eliminate or minimize the risk
e. *After* assessing the extent of the risk and the available ways of eliminating or minimizing the risk, the *cost* associated with available ways of eliminating or minimizing the risk, including whether the cost is *grossly disproportionate* to the risk

It is important to note some misunderstandings of CBA and ALARP. The concept of ALARP or SOFAIRP does NOT imply that *cost* alone nor *capacity to pay* for a reasonably practicable risk control, mitigation, or treatment measure can ever be legal justifications or defenses for not implementing the measure. Moral and legal factors will always emphasize the expectation that even if a risk is tolerable the closer the risk level is to an agreed intolerable level, then the more investment in risk management is required to drive the risk lower. Hence, the $ width of the triangle widens in Figures 5.2 and 5.3. In terms of *individual risk*, ALARP or SOFAIRP tolerability principles are described in Figures 5.2 and 5.3 with an emphasis on "grossly disproportionate."

Of *societal or group risk*, ALARP or SOFAIRP tolerability principle is shown in Figure 5.4 with a common expression of there being "no tolerable" frequency of incidents involving more than 1,000 public fatalities per incident (e.g., Bhopal explosion/gas release or 2 × 500 passenger planes colliding in midair).

Note: *As with individual risk, it is recommended that risk zones such as "negligible" and "insignificant" or "broadly tolerable" are not used. If a risk is to be tolerable, it needs to be below the intolerable threshold AND must be able to be shown to be ALARP regardless of how much it is below the threshold.*

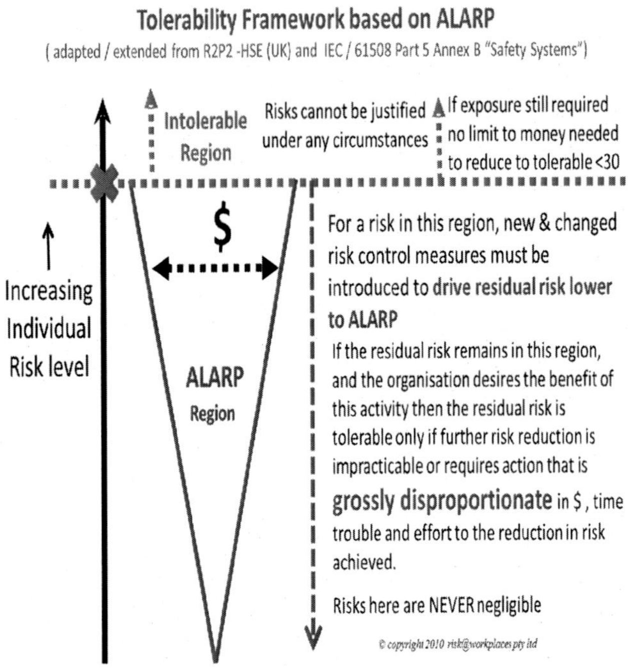

FIGURE 5.2 ALARP risk tolerability framework for individual risk.

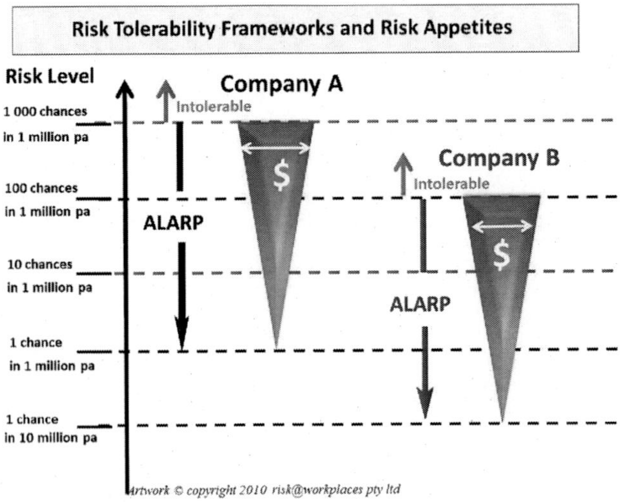

FIGURE 5.3 Examples of risk tolerability and appetites for individual risk.

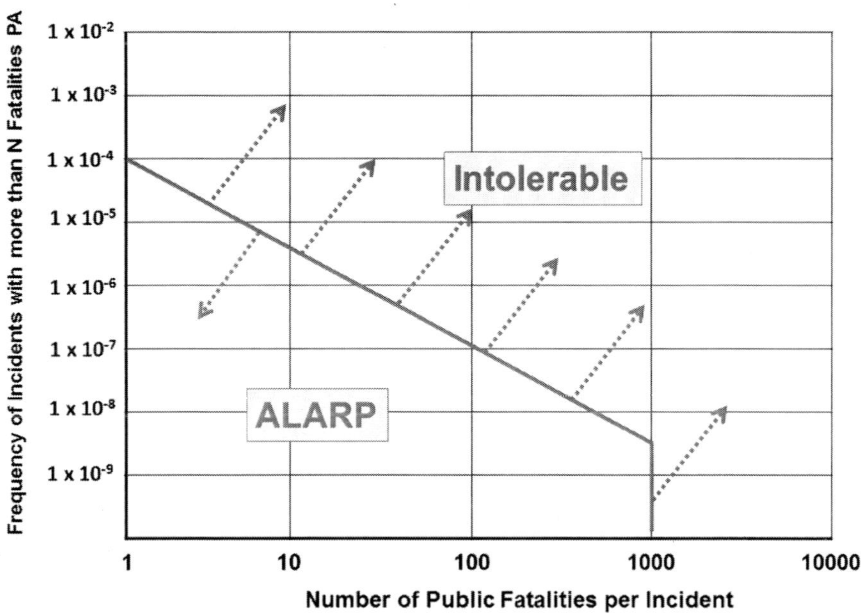

FIGURE 5.4 Indicative societal or group risk tolerability framework (generic example for all fatality risks to members of the public).

5.10 MOVING TOWARDS RISK-BASED LANGUAGE FOR MORE EFFECTIVE RISK CONVERSATIONS

If an SH&E practitioner wants to converse with others, internationally as well as nationally, then safety risk-based language can improve the quality and effectiveness of the conversation. Risk-based conversations can make safety discussions more realistic, objective, and solution-focused as well as less argumentative. If risk-based conversations are to achieve this purpose, they need a common agreed language to express sometimes complex safety concepts.

The transition from compliance-based safety to risk-based safety in large part has occurred in many parts of the world and is beginning to take place in the United States. This requires risk-centric organizations and progressive risk professionals to use better terminology and language for risk-based conversations. For example, outdated terms such as loss control and loss prevention are now being replaced with safety risk management and risk control.

Appendix A shows more commonly used risk-based language that is becoming more universal as preferred and recommended so as to clarify and reduce biases, misunderstandings, and misperceptions of the group and its individual members during safety-related discussions.

APPENDIX A
Better Terminology and Language for Risk-Based Conversations

Traditional Safety Terminology	Preferred and Recommended Risk-Based Language
Loss control/loss prevention	**Safety risk management and risk control**—profits as well as losses—enabling positive outcomes as well as preventing negatives—maximizing the chances of gains, profits, and benefits—safety is about a focus on *maximizing chances of gains* NOT *minimizing chances of losses* Risk management includes maximizing and exploiting *opportunities*
Safety—as the absence of harm, double negative	**Safety**—as the presence of well-being, double positive
Safe acts/conditions	**Standard, agreed acts/behaviors/conditions**
Unsafe acts	**Nonstandard, non-agreed behaviors/conditions**
At-risk behaviors/ at-risk conditions	
To **accept** a risk or **acceptable risk**	To **tolerate** a risk—working with, never passively accepting always uncomfortable—chronic unease—looking for how to make the risk ALARP or **tolerable risk**
Safe	When risk is managed **ALARP**
Safer/safest	**Lower risk level/lowest risk level**
Event/scenario	If used interchangeably creates confusion, for example, the expression: The same *event* can lead to different consequences—is a valid statement but the same *scenario* can lead to different consequences—is NOT valid Reserve the term ***event*** for each discrete happening/action. Reserve the term ***scenario*** for the whole sequence of all the events and circumstances needed to describe "How"/"When"/"Where"/"Who"/"What" an incident DID occur or a risk COULD occur
If safety is involved, money does not count!	Sounds like a good caring philosophy but it is an untrue, unbelievable statement which corrodes credibility, trust, and respect. Better to use expressions such as "WHEN a risk exceeds our agreed defined intolerable threshold level, and IF continued exposure to the risk is needed or desirable for legal, moral, or commercial reasons, there is no limit to time money effort needed to introduce measures that reduce the risk below the threshold." The reduced risk then also needs to be shown as always being managed to ALARP—not just at one point in time. *Tolerable* means BOTH below intolerable and ALARP
Alertness, vigilance	**Situational awareness and mindfulness**
Violation, breach, failure, negligent, reckless	Use nonjudgmental terms *such as* ***variation, alternative, deviation, and work-around*** so you will look for deeper underlying root causes of the variations
Shortcut	Smarter way of doing a job which can be an approved variation but only after a formal authorization/approval process that must involve qualitative or semiquantitative risk assessments (Whiting, 2012) Always distinguish between • Finding a shortcut (*smart*) and • Taking a shortcut without risk assessment (*dumb*)

(*Continued*)

APPENDIX A (*Continued*)
Better Terminology and Language for Risk-Based Conversations

Traditional Safety Terminology	Preferred and Recommended Risk-Based Language
Safety measures, preventative measures, safeguards, barriers, layers of protection, mitigating factors, corrective actions	Use the single term **risk controls** for all of them
Causes of incidents and risks	All causes are missing or ineffective risk controls due to deeper underlying root causes based on systemic, physical, and work environment factors
Behavioral causes	Behaviors are consequences of deeper underlying root causes NOT seen as causes in themselves
Human error	Use term **human factor** in preference to **human error** to emphasize that error is not a cause of an incident or a risk of an incident. It is a consequence of the underlying human factors/mismatches between a job's requirements and the person's capabilities and limitations. The mismatches are usually due to systemic, physical, and work environmental factors
Possible, probable, potential used interchangeably and hence confusingly	**Possible** = absolute, YES/NO, black/white—it is or it is not—has no range of values—cannot be used to express a level of likelihood—cannot use meaningless terms quite possible or *remotely possible* **Probable** = relative not absolute—can use *likely, chances, odds*—always has a range of values—used to express a level of likelihood **Potential** = confusing—it can be used to express either *possible* or *probable*
Probability	**Likelihood, chances, and odds** are risk terms preferred for nonquantitative users
Likelihood can be expressed as either a **frequency** or a **probability**	**Frequency** can be used *retrospectively* to indicate how often an actual incident has been occurring in the past AND ALSO It can be used *prospectively* to predict how often the risk of an incident may occur in the future **Likelihood, chances, and odds** can be used ONLY prospectively to express predictive estimate of how likely the risk will occur Often better to use the terms "chance" or "odds" NOT decimal 0.001 or unfamiliar exponential 1E-03 notation, for example, 1 chance in 100 ladder climbs 1 chance in 10,000 valve operations The odds are 1 in 1,000 holes drilled Avoid using fractions of %—hard to interpret For example, use 1 chance in 1,000 rather than 0.1% Always question any assessor's perception that 1% or 1 chance in 100 is a small likelihood. It is a large likelihood
Exposure	How often and how long exposed (in financial RM, it is $ quantum)
Frequency of exposure	How often, for example, exposed to noise daily (or yearly or every shift)
Duration of exposure	How long, for example, exposed to asbestos 3 h/shift (or 100 h p.a.)

5.11 A CAUTIONARY CONCLUDING NOTE

Risk management is fundamentally about how to make risk-informed decisions when life and business circumstances are uncertain. RM is not about more *risk taking*, rather better *risk understanding* of the exposures we are currently not managing as well as we need to, or could. Safety is *managing risk to ALARP*, not *zero risk*. The *risk makers* and the *risk takers* need to be the *risk managers*.

RM is needed when VUCA circumstances exist (i.e., when volatility/ Uncertainty/ complexity/ambiguity need to be managed with vision/understanding/clarity/agility).

REFERENCES

AIChE/CCPS (2009). Appendix B—survey of worldwide risk criteria In *Guidelines for Developing Quantitative Safety Risk Criteria*. Center for Chemical Process Safety (New York) and American Institute of Chemical Engineers, Inc. (New York) https://onlinelibrary.wiley.com/doi/pdf/10.1002/9780470552940.app2 accessed January 5, 2019.

ANSI/ASSE/AIHA Z10-2012 (2012). *American National Standard—Occupational Health and Safety Management Systems*. Fairfax, VA: American Industrial Hygiene Association.

ANSI/ASSE Z590.3-2011 (2011). *Prevention through Design: Guidelines for Addressing Occupational Hazards and Risks in Design and Redesign Processes*. Des Plaines, IL: American Society of Safety Engineers.

Eddowes, M., Page, N. (1999), A Review of International Risk Criteria and Related Environmental Cleanup Standards by AEA Technology for US DOE Center for Risk Excellence European Maritime Safety Agency EMSA, 2013, Risk Level and Acceptance Criteria for Passenger Ships. First Interim Report, Part 2: Risk Acceptance Criteria, Report No.: PP092663/1-1/2, Rev. 2 Document No.: 18KJ9LI-6 Date: 2014-04-28.

Kelly, K. E. (1991), The Myth of 10-6 As A Definition Of Acceptable Risk, Paper Originally Presented at the *84th Annual Meeting Air & Waste Management Association Vancouver*, B.C., Canada 16–21 June 1991.

Miller, T. R., Variations between countries in VoLS—Value of a statistical life, *Journal of transport, Economics and Policy* (2000), Vol. 34, No. 2, pp. 169–188. http://scholar.google.com.au/scholar_url?url=http://www.researchgate.net/profile/Ted_Miller/publication/238369152_Variations_Between_Countries_in_Values_of_Statistical_Life/links/0c96052cde36d95d2c000000.pdf&hl=en&sa=X&scisig=AAGBfm3WreOfiy-2C_5vSKZv89c3cmI9IfQ&nossl=1&oi=scholarr&ei=zvcxVa3JMMemmAXNm4D-gAg&ved=0CB0QgAMoATAA.

NSW Department of Planning, Australia (2011), Hazardous industry planning advisory paper No.4 – Risk Criteria for Land Use Safety Planning (HIPAP 4) (January 2011).

OECD, Recommended Value of a Statistical Life Numbers for Policy Analysis (2012), pp. 125–136 www.oecd-ilibrary.org/environment/mortality-risk-valuation-in-environment-health-and-transport-policies/recommended-value-of-a-statistical-life-numbers-for-policy-analysis_9789264130807-9-en.

SafetyNet, Cost-Benefit Analysis (2009) http://ec.europa.eu/transport/road_safety/specialist/knowledge/pdf/cost_benefit_analysis.pdf accessed April 18, 2015.

UK HSE (1989). Risk Criteria for Land Use Planning in the Vicinity of Major Industrial Hazards.

UK HSE, The Tolerability of Risk from Nuclear Power Stations, HMSO (1992) www.onr.org.uk/documents/tolerability.pdf.

UK HSE (2001). Reducing Risks, Protecting People R2P2—HSE's Decision-Making Process.

UK HSE (2003). Good Practice and Pitfalls in Risk Assessment—HSL Research Report 151.

UK HSE (2006). Offshore Installations (Safety Case) regulations 2005 Regulation 12—Demonstrating Compliance with the Relevant Statutory Provisions, HSE Offshore Information Sheet No. 2/2006.

UKOOA (1999). Industry Guidelines on a Framework for Risk Related Decision Support, Issue 1.

U.S. Food and Drug Administration (USFDA), Compounds used in food-producing animals. Procedures for determining acceptability of assay methods used for assuring the absence of residues in edible products of such animals. Proposed rule, *Federal Register* (1973), Vol. 38, 19226–19230.

Whiting, J. F., (2012), Using the New RM Standard ANSI Z690.2 to assess the maturity of your risk management system ASSE PDC Safety 2012—Denver, Session 662 Tuesday 5th June 2012 jim@workplaces.com.au.

Wren, J. (2011), The Value of Preventable Injury Fatality (VPF) in New Zealand, *Presentation for New Zealand Association of Economists Conference*, June 30, 2011.

6 Occupational Health and Safety Management Systems

Charles Redinger
Institute for Advanced Risk Management

CONTENTS

Managing occupational health and safety (OH&S) hazards and risk has traditionally been driven by governmental regulations, often referred to as command-and-control approaches. Over time, and with advances in organizational science, nonregulatory approaches for OH&S management have proliferated. This is seen in nongovernmental consensus standards, as well as innovative approaches developed by professional organizations and companies. A theme in these newer approaches is the application of system-based principles and methods to OH&S management. This is observed with the evolution of formal OH&S management system (OHSMS) standards and guidelines.

This chapter provides an introduction to OHSMS concepts, principles, and frameworks. An overview of the International Organization for Standardization's (ISO) OHSMS, ISO 45001:2018, is presented along with several other legacy approaches. Closely linked to the use of OHSMS standards is what is referred to as conformity assessment, which deals with numerous important issues, namely, third-party certification. It is important to be clear that an OHSMS standard, such as ISO 45001:2018, can be implemented in an organization, but that third-party certification is not required.

6.1 APPROACHES TO OH&S MANAGEMENT

The reduction of workplace risk and meeting regulatory compliance are two dominant drivers that have shaped the historic development of OH&S management practices. A common term used to describe OH&S management practices up to the 1980s has been "traditional compliance-based" management approaches. This reflects that regulations imposed by national, regional, and local governmental agencies were a large driver that impacted how OH&S professional defined their duties and organizations arranged their OH&S structures and practices.

Two central purposes for governmental regulations are as follows: (1) to control externalities (costs) that organizations do not or are not willing to adsorb, whereby the costs are passed on to the public, which is also referred to as "risk transfer," and (2) to provide warnings and protection in instances where activities are hazardous.

Organizations have found over time that compliance with governmental regulations does not necessarily provide the protections thought or desired, thereby leading to lower risks, increasing risk resilience, or helping overall OH&S performance. The challenge for many organizations has been to maintain regulatory compliance while introducing new beyond-compliance risk management and performance measurement practices as embodied in OHSMS approaches.

6.2 BASIC SYSTEMS CONCEPTS

Most organizations have some sort of OHSMS in place. That is, there is an identifiable OH&S process, there are inputs to the process, there are probably some sort of feedback loops, and there are outputs. The advent of formal management system standards (MSSs) provides robust means for characterizing and defining an OHSMS.

Research on the breadth and similarities of OHSMSs has been conducted (IOHA, 1998). This work identified more than 30 OHSMSs in use or being developed in the late 1990s. With the trend towards the development of international standards, there was a trend of national standards development efforts to defer to the international efforts. The International Occupational Hygiene Association (IOHA)-sponsored research found that many of the existing management system approaches at that time had many common elements.

Existing management system concepts can be traced to numerous sources. The most common source is plan–do–check–act model. While there are numerous definitions provided for a system, the common four elements presented are input,

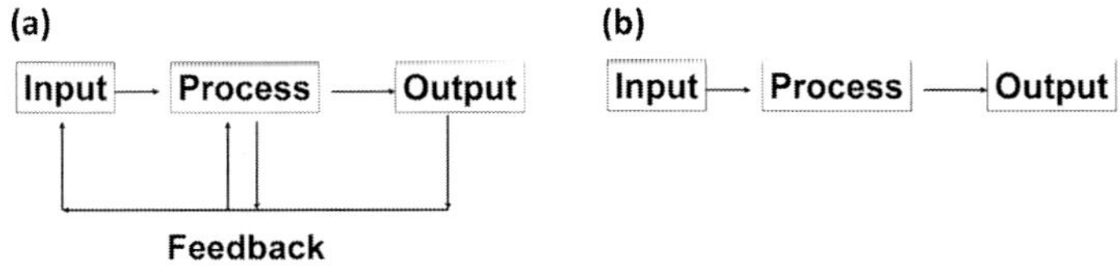

FIGURE 6.1 (a) Simple system diagram and (b) feedback.

process, output, and feedback. The relationship between these four elements is depicted in Figure 6.1a.

A question that often comes up when talking about an OHSMS is "what is the difference between a system and program?" One way to describe this difference is in terms of an information feedback loop. That is, feedback in a system is essential and an integral component of the system. Conversely, this is not the case with programmatic approaches where feedback is not necessarily part of a structural design. This is depicted in Figure 6.1b.

A system can be further characterized as being either open or closed. In the case of open systems, there are identifiable pathways whereby the system interacts—exchanging information with and gaining energy—from its external environment. This phenomenon is readily observed in biological systems. Conversely, closed systems do not have such pathways and thus limit their ability to adapt or respond to changing external conditions.

In traditional OH&S management approaches, the focus has been on trailing indicators, such as illness, injury, and fatality statistics. In a systems approach, regulatory compliance and trailing indicators are not neglected; however, commonly, there is a shift in focus towards performance variables and measurements from the input and process components of the system. These components can be thought of as being "upstream" from the system output, or leading indicators.

6.2.1 Programs vs. Systems

The distinction is made here between traditional programmatic approaches and newer systems approaches to OH&S management. A program is operationally defined as singular, vertical, and based on traditional command-control regulations. The focus is on compliance with the program standard/regulation, not the broader impact on OH&S performance in the organization. In this conceptualization, programs do not have strong, if any, feedback or evaluation mechanisms whereby the program is adjusted or modified.

Conversely, a systems approach—while not losing sight of programmatic requirements and opportunities for improvement—broadens in perspective to address the manner in which the program affects other programs, and the extent to which the program may or may not improve worker health and safety. Further, a systems approach is driven by OH&S improvement, more so than by programmatic regulatory compliance. A key distinction of a systems approach is that there are clear feedback and evaluation mechanisms whereby the system responds to both internal and external events. A systems approach integrates individual programs within the business operations and the external environment and is thus more comprehensive than any single program.

6.3 OHSMS STANDARDS

Systems concepts have been used in managing OH&S arrangements for decades. A new era was entered with ISO's entry into the management system codification arena with 9001 in the late 1980s. At that time, there were few formal OHSMS approaches throughout the world. In the early 1990s, OH&S and environmental management professionals and standards developers began to consider how the ISO 9001:1987 principles could be applied to environmental and OH&S arrangements. ISO 9001 was updated in 2000 (ISO, 2000).

In 1994, an ISO Technical Committee devoted to "materials, equipment, and offshore structures for petroleum and natural gas industries" began to develop an integrated health, safety, and environment management system. This effort produced a draft standard but was not continued when ISO 14001:1996 was published. Around that time, several OHSMSs were also published (e.g., BSI 8800, Australia's SafetyMAP). By the late 1990s, numerous nation-states, along with professional organizations (e.g., the Japan Industrial Safety and Health Association, the American Industrial Hygiene Association (AIHA), the Chemical Manufacturers Association), had started to develop OHSMS standards and guidelines.

In the mid-1990s, OSHA in the United States began to consider rulemaking for a comprehensive OH&S program standard. Activities on this effort continued through the early 2000s. Over time, the priority of these efforts diminished and was off OSHA's agenda by 2003.

Researchers at the University of Michigan (UM) developed an ISO 9001-based OHSMS that was published by the AIHA in 1996 When it was published, the UM/AIHA OHSMS received a significant attention from various stakeholders and standards-making organizations (Mansdorf, 1996). After this, the UM group developed and published a universal OHSMS assessment instrument in 1999 (Redinger, 1999a, b). As part of the development of the assessment instrument, they developed a generic OHSMS model. This model has since been used widely throughout the world by standards-making bodies to assist their development efforts and by private companies (IOHA, 1998; European Union, 2002; ANSI, 2005).

In 1996, ISO considered the development of an OHSMS standard. It elected at that time to not proceed. It was during those deliberations that standards-making experts put forth the idea, and recommendation, that the International Labour Office (ILO) would be a more suitable international organization to develop standards and guidelines in this area. With this mandate, in 1997, the ILO began to conduct background research on management systems as a precursor to forming the tripartite group of experts that developed ILO-OSH 2001.

While the ILO was performing these background efforts, two developments occurred. First, ISO elected for a second time to not develop an ISO OHSMS. Second, in Britain, the British Standards Institute (BSI) published OHSAS 18001:1999, which followed the structure of ISO 14001:1996. This document was published specifically for use as an auditable standard. In its introduction, OHSAS 18001:1999 stated that the document was developed "in response to urgent customer demand for a recognizable occupational health and safety management system standard against which their management systems can be assessed and certified."

In 2000 in the United States, the AIHA solicited the American National Standards Institute (ANSI) to form a committee to develop an ANSI standard in this area. The committee (Z10) held its first meeting in 2002 and issued a standard in 2005. The second edition of ANSI Z10 was published in 2012.

In its fourth attempt, ISO was successful in starting the process to develop an ISO OHSMS. In 2013, an ISO Project Committee (PC 283) was formed and subsequently published ISO 45001:2018.

Dating back to the 1980s and 1990s, when management system approaches were generally evolving (e.g., ISO 9001:1987 and ISO 14001:1996) and popular in some areas, there was not universal acceptance of them. This general skepticism spilled into OHSMS development efforts and was based in concerns about (1) costs to develop, implement, and maintain a management system, and (2) costs and complexities to seek and maintain certification by an external third party. These concerns diminished in the early 2000s. By the time that ISO 45001:2018 was published, the rationale and acceptance of OHSMS was well established.

When ISO 45001:2018 was published in 2018, the British-based OHSMS, OHSAS 18001:2007, was generally considered as the dominant internationally used OHSMS, even though strictly speaking it was not an international standard; it had been formally adopted by numerous countries (e.g., Singapore and Korea) and was considered the *de facto* standard used in certification schemes.

While there are numerous OHSMS standards developed within specific countries, the focus here is on ISO 45001:2018 and briefly on the ILO OHSMS guidelines. But first, a brief background on the development of ISO's "high-level" MSS framework.

6.3.1 ISO's Drive to a Unified MSS Framework

In an effort to bring uniformity to its management system efforts, ISO's Technical Management Board (TMB) formed the ISO Ad Hoc Group on Management System Standards shortly after ISO 14001:1996 was published. This group published "Guidelines for the Justification and Development of Management System Standards" (ISO Guide 72) in 2001. This guide presented common MS elements as follows:

a. Policy
b. Planning
c. Implementation and operation
d. Performance assessment
e. Improvement and
f. Management review

These elements followed the structure of ISO 14001:1996 and were found many nation-specific approaches at that time.

In the early 2000s, the Ad Hoc Group on Management System Standards recommended the formation of the Joint Technical Coordination Group (JTCG) to work on establishing consistency between ISO's various MSSs; the TMB subsequently formed the JTCG on MSS. This group developed ISO Guide 83, "High Level Structure, Identical Core Text and Common Terms and Core Definitions for use in Management Systems Standards." This document was never formally adopted, but was issued in December 2011. In it was the recommendation to establish what is often referred to as ISO's "high-level MSS structure." These recommendations were subsequently adopted, and published in 2013, in Annex SL of ISO's *Directives* (also referred to as the *ISO Supplement*). Annex SL formally presented the new high-level and generic MSS that all future ISO MSSs were required to follow. This high-level MSS structure has ten sections, which are as follows:

1. Scope
2. Normative references
3. Terms and definitions
4. Context of the organization
5. Leadership
6. Planning
7. Support
8. Operation
9. Performance evaluation
10. Improvement

6.3.2 ISO 45001:2018—OHSMS—Requirements with Guidance for Use

The group (PC 283) that developed ISO 45001:2018 was required to follow the MSS structure presented in Annex SL. A brief description of the ten sections follows. For brevity, the term "ISO 45001:2108" will be simply stated as 45001 here. However, take note that it is important to be rigorous in indicating the year a standard was adapted or published, when referring it.

6.3.2.1 Scope

This section covers the areas addressed in 45001. It does not contain any auditable requirements. 45001's scope states, "This document helps an organization to achieve the intended outcomes of its OH&S management system." Identified outcomes include continual improvement of performance, fulfillment of legal and other requirements, and achievement of OH&S objectives. An important point in the scope is that the standard "does not state specific criteria for OH&S performance, nor is it prescriptive about the design of an OH&S management system." The importance of this is the intent for the use of the standard as risk management tool rather than as prescriptive requirements, as seen in regulations.

6.3.2.2 Normative References

This section is required by ISO's MSS development criteria. No normative references are indicated for 45001.

6.3.2.3 Terms and Definitions

Thirty-seven terms and definitions are addressed in 45001. Several are highlighted here.

Worker (3.3) is defined as "person performing work or work-related activities that are under the control of the organization." This definition is different than in earlier OHSMSs that focused on employees. Of particular interest are three "notes" to the definition that indicate inclusion of contract and temporary workers, as well as workers provided by agencies.

Injury and ill health (3.18) is defined as "adverse effect on the physical, mental or cognitive condition of a person." New here is the inclusion of "cognitive condition." Earlier OHSMSs included mental (18001) and psychosocial factors (ILO).

Risk (3.20) is defined as "effect of uncertainty." This is a generic definition used throughout ISO's MSS. To OH&S professionals, this at first seems like an odd definition since the common definition in OH&S is "combination of the likelihood of an occurrence of a hazardous event or exposure(s) and the severity of injury or ill health that can be caused by the event or exposure." This more common definition is included in 45001 as "OH&S risk" (3.21). One way to think of the singular term "risk" is to think of it as indirectly related to OH&S risks such as the risk (or the "effect of uncertainty") related to the provision of resources.

OH&S opportunity (3.22) is defined as "circumstance or set of circumstances that can lead to improvement of OH&S performance." The notion of "opportunities" is new in 45001 and can be viewed as a more positive aspect towards improving performance versus how risk is traditionally viewed.

6.3.2.4 Context of the Organization

Examining, clarifying, and understanding organizational context establishes the foundation upon which the OHSMS is developed. A first step is to "determine external and internal issues that are relevant to its [the organization's] purpose and that affect its ability to achieve the intended outcome(s) of its OH&S management system" (4.1). This includes gaining an understanding of the needs and expectations of workers and interested parties; new to 45001 from previous OHSMSs is "interested parties." As indicated above related to OHSMS's scope, requirements are included in this section related to formally establishing and stating the OHSMS's scope.

Consideration of an OHSMS's scope is a critical first step in development and implementation in an organization. These considerations include clarity on operations covered by the OHSMS, such as a single plant, multiple ones, or corporate-wide, and what activities are covered, such as possibly environmental, sustainability, or product safety.

6.3.2.5 Leadership and Worker Participation

These two pieces—leadership and worker participation—are historically considered linchpins of the OHSMS approach. Detailed requirements are included for each in

45001. Dating back to the earliest OHSMS approaches (e.g., BS 8800, OSHA guidelines, SafetyMap, 18001:1999), there has been ongoing refinement and clarification of these pieces, as seen in the ILO's OHSMS, 18001:2007, and ANSI Z10, to name a few.

In 45001, worker participation requirements are nested throughout the standard. In this section, these requirements are broadly summed up as "the organization shall establish, implement, and maintain a process(es) for consultation and participation of workers at all applicable levels and functions, and, where they exist, workers' representatives, in the development, planning, implementation, performance evaluation and actions for improvement of the OH&S management system" (5.4). Key here is providing time, training, and resources, as well as removing barriers for effective participation. Clause 5.4.d suggests emphasizing the consultation of nonmanagerial workers in a range of activities, including OH&S policy (d.3); assigning organizational roles, responsibilities, and authorities as applicable (d.4); planning, establishing, implementing, and maintaining an audit program (d.7); and others.

All OHSMS approaches include top management leadership and commitment requirements; 45001 continues this. In section 5.1 (Leadership and commitment), 13 clauses contain auditable elements. Two clauses of interest are 5.1.j and 5.1.m. Clause j requires that "top management shall demonstrate leadership and commitment with respect to the OH&S management system by developing, leading, and promoting a culture in the organization that supports the intended outcomes of the OH&S management system." The requirement related to "culture" is new in 45001, and while its importance cannot be understated, the practicality of demonstrating this is not trivial from an audit perspective. Clause m requires that "top management shall demonstrate leadership and commitment with respect to the OH&S management system by supporting the establishment and functioning of health and safety committees."

Establishing, implementing, and maintaining an OH&S policy are required in this section (5.2). All earlier OHSMS approaches contained this requirement with varying degrees of specificity. Of interest in 45001 is a requirement that the policy contains "a commitment to eliminate hazards and reduce OH&S risks" (5.2.d) and is expanded on in section 8.1.2 "Eliminating hazards and reducing OH&S risks." This idea is noble and resonates philosophically. However, practically, some experts argue that it is impossible to eliminate all hazards and that it is sounder to think in terms of the reduction or elimination of risks.

6.3.2.6 Planning

Requirements in this section have increased from earlier OHSMS approaches, and focus on actions to address risks and opportunities (6.1) and establishing OH&S objectives and plans to achieve them. New in 45001 is the requirement to consider OH&S opportunities as well as OH&S risks. On the surface, this new consideration is straightforward.

Robust requirements are included related to hazard identification (6.1.2.1), "the organization shall establish, implement, and maintain a process(es) for hazard identification that is ongoing and proactive." From an audit perspective, consideration needs to be given to what constitutes "ongoing." In well-functioning OHSMSs, this issue points to establishing feedback channels for hazard identification-related data that arise from any number of activities, such as audits, accident reports, or worker complaints. The requirement to be proactive, while not absent in early OHSMS approaches, is clearly stated here. Of particular interest is the requirement that the hazard identification "process(es) shall take into account how work is organized, social factors (including workload, work hours, victimization, harassment, and bullying), leadership and the culture in the organization" (6.1.2.1.a). Language in this section makes clear that hazard identification extends into "locations not under the direct control of the organization" (6.1.2.1.e.3). As well, to multiemployer work locations, "situations not controlled by the organization and occurring in the vicinity of the workplace that can cause injury and ill health to persons in the workplace" (6.1.2.1.f.3).

A distinction is made between OH&S risks and "other risks to the OH&S management system" (6.1.2.2). OH&S risks refer to what could be considered tradition risks, such as a chemical exposure, slips, trips, and falls. Risks to the OHSMS refer to things that can affect OH&S performance, such as day-to-day operations and decision-making, regulatory changes, the organizational culture, and changes in resources, to name a few. A methodology for assessing OH&S risks is required, and this needs to "be defined with respect to their scope, nature, and timing to ensure they are proactive rather than reactive" (6.1.2.2).

Being proactive in identifying OH&S opportunities is an undercurrent in 45001. This is specifically seen in requirements for "assessment of OH&S opportunities and other opportunities to the OH&S management system" (6.1.2.13). Subclause "a" states that "The organization shall establish, implement, and maintain process(es) to assess OH&S opportunities to enhance OH&S performance, while taking into account planned changes to the organization, its policies, processes or its activities and (1) opportunities to adapt work, work organization and work environment to workers; and opportunities to eliminate hazards and reduce OH&S risks."

The standard requires that action plan(s) be developed and that they address: identifying risks and opportunities; legal and other requirements; and preparation and response to emergency situations (6.1.4). A significant evolution here from earlier OHSMS approaches is that "the organization shall plan how to integrate and implement the actions into its OH&S management system processes or other business processes; and evaluate the effectiveness of these actions." Key here is the inclusion of "business processes."

6.3.2.7 Support

Activities associated with supporting an OHSMS include resources (7.1), worker competency (7.2), worker awareness (7.3), communication process(es) (7.4), and documented information (7.5).

While requirements for worker awareness on a number of items are found in earlier OHSMS approaches, 45001 has them bundled in one section (7.3). Included in them is a unique clause (7.3.f) that states "workers shall be made aware of the ability to remove themselves from work situations that they consider present an imminent and serious danger to their life or health, as well as the arrangements for protecting them from undue consequences for doing so."

As with earlier OHSMS approaches, 45001 contains robust communication requirements. Section 7.4.1 states that "The organization shall establish, implement, and maintain the process(es) needed for the internal and external communications relevant to the OH&S management system." This includes determining what will be communicated; identifying the parties involved in communications, including contractors and visitors; how to communicate; and to take into account language issues. More rigorous from earlier OHSMS standards is a requirement that "the organization shall ensure that the views of external interested parties are considered in establishing its communication process(es)" (7.4.1). Unique here is considering external parties views in establishing the process(es).

6.3.2.8 Operation

The OH&S profession has done operational planning and control (8.1) since its earliest days. Requirements in this section are familiar with the profession. Two items clearly stated here are the requirement to "adapt work to workers," and "at multiemployer workplaces...to coordinate relevant parts of the OH&S management system with the other organizations."

As seen in other OHSMS approaches, use of the hierarchy of controls to eliminate hazards and control OH&S risks is highlighted (8.1.2). It is noted that "in many countries, legal requirements and other requirements include the requirement that personal protective equipment (PPE) is provided at no cost to workers."

Management of change (8.1.3) has been central from the earliest days of the OHSMS approach. It is often cited as being one of the most crucial sections in an OHSMS. While all requirements of an OHSMS are important, advocates suggest that this piece is one to highlight in training and activities associated with increasing awareness (7.3). The standard states that

> The organization shall establish a process(es) for the implementation and control of planned temporary and permanent changes that impact OH&S performance, including:
>
> a. new products, services, services, and processes, or changes to existing products, services and processes, including:
>   - work locations an surroundings;
>   - work organization;
>   - working conditions;
>   - equipment;
>   - work force;
>
> b. changes to legal requirements and other requirements;
> c. changes in knowledge or information about hazards and OH&S risks;
> d. developments in knowledge and technology.
>
> The organization shall review the consequences of unintended changes, taking action to mitigate any adverse effects, as necessary.

Requirements are included in 45001 for procurement (8.1.4), outsourcing (8.1.4.3), and emergency preparedness and response (8.2).

6.3.2.9 Performance Evaluation

Many requirements in this section are familiar and aligned with earlier-generation OHSMSs. Clause 9.1.1 states that "The organization shall establish, implement, and maintain a process(es) for monitoring, measurement, analysis and performance measurement." This includes determining what needs to be monitored and measured, as well as the methods and their validity. Requirements for the "evaluation of compliance" are clearly stated in Clause 9.1.2, where it states "the organization shall: determine the frequency and method(s) for the evaluation of compliance [with legal and other requirements]; evaluate compliance and take action if needed; and maintain knowledge and understanding of its compliance status with legal requirements and other requirements."

Internal audits of an OHSMS are required "at planned intervals to provide information" related to conformance with 45001, as well as with the organizations OH&S policy and objectives (9.2.10). Internal audit criteria need to be established, along with ensuring auditors selected to do audits are objective and impartial (9.2.2). Results need to be reported to "relevant managers" and workers, and where deficiencies (nonconformities) are found, they need to be addressed.

A critical legacy element in the OHSMS approach is management review (9.3). This standard contains robust requirements for this. It states "Top management shall review the organization's OH&S management system, at planned intervals, to ensure its continuing suitability, adequacy, and effectiveness." Key here is the evaluation of the OHSMS's suitability, adequacy, and effectiveness. Top management needs to determine whether maintaining the OHSMS is aligned with strategic objectives, and if not, should maintaining conformance to it be continued, and if it is, to ensure that proper resources and support are being given to it.

6.3.2.10 Improvement

The term "improvement" finds its way into 45001 from ISO's MSS requirements. A number of "improvement"-related activities are bundled here: incident, nonconformity, and corrective action responses; and continual improvement. The standard states that "The organization shall establish, implement, and maintain, a process(es) including reporting, investigating and taking action to determine and manage incidents and nonconformities" (10.2). Specific requirements include timely response; conducting root cause analysis, with worker involvement; assessing potential historical trends; and ensuring that findings are feed back into the planning process.

A hallmark of the OHSMS approach is continual improvement. 45001 continues this trajectory, stating in 10.3 "The organization shall continually improve the suitability, adequacy and effectiveness of the OH&S management system: by enhancing OH&S performance; promoting a culture that supports an OH&S management system; promoting the participation of workers in implementing actions for the continual improvement of the system; and, communicating the relevant results of continual improvement to workers, and where they exist, workers' representatives." When implementing a management system, early in the process, ways to demonstrate conformity to this requirement should be given.

6.3.3 ILO OHSMS:2001—Guidelines on Occupational Safety and Health Management Systems

The ILO began its OHSMS development activities shortly after ISO elected not to pursue development activities in this area in late 1996. The ILO's SafeWork division leads the agencies effort with the initial step of contracting a research study with IOHA on OH&S and environmental management systems in use or development in the late 1990s. With the study findings in hand, the ILO formed a 21-person working group of experts to develop the standard that followed its tripartite structure with seven experts each from government, labor, and industry. The working group efforts culminated with a 2-week meeting in 2001 in Geneva where the bulk of the standard was written.

As with OHSAS 18001:1999 and ISO 14001:1996, the ILO OHSMS has front-end sections that address scope and context issues, with the meet of the standard contained in section three, organized within five sections:

- Policy
- Organizing
- Planning and implementation
- Evaluation and
- Action for improvement

A unique feature of the ILO's OHSMS front end is Section Two that contains a model that governments or nation-state standards developers can follow in the development of OHSMS standards unique to individual countries or industries. This section is titled "A national framework for occupational safety and health management systems." A schematic of this structure is shown in Figure 6.2.

Strictly speaking, the ILO OHSMS is not a standard, but rather it is a guideline with recommendations. It contains a mixture of "should"- and "shall"-based clauses that makes auditing against it difficult. Use in third-party certification schemes is not precluded, but its introduction states that its "application" does not require certification.

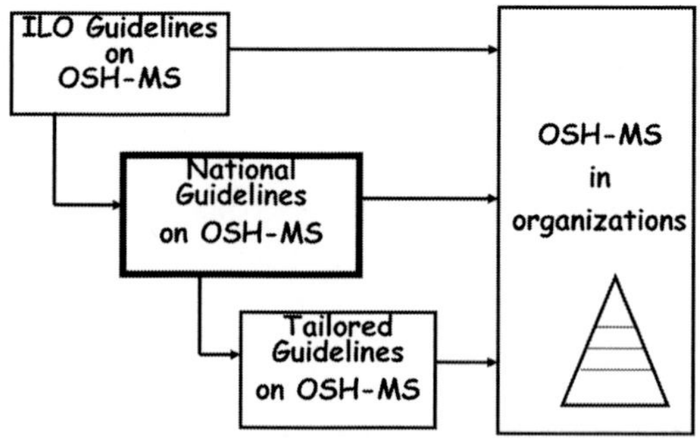

FIGURE 6.2 ILO framework schematic.

6.4 CONFORMITY ASSESSMENT

Conformity assessment refers to the activities associated with determining whether an implemented management system conforms with a formal OHSMS standard, such as ISO 45001. Activities and distinctions associated with conformity assessment are certification; auditing; first, second, and third parties; registrar; and accredited. It is important to understand and consider conformity assessment issues and how they relate to management systems because they are central to strategic considerations regarding the rationale for implementing and measuring the performance of a management system (NRC, 1995).

It is common to think of the term "certification" when discussing OHSMSs. A common misconception is that pursuing or implementing a management system necessarily means that certification must be pursued. This is not the case. Whereas some organizations do pursue certification of their management systems by a third-party registrar, many do not. Making the distinction between a management system and conformity assessment activities can reduce unnecessary confusion. Conformity assessment deals with the activities associated with determining how well a given system approach (e.g., 45001) has been implemented in an organization.

Conformity assessment frameworks commonly have three levels:

- Primary level—assessment
- Secondary level—accreditation
- Tertiary level—recognition

The primary level represents measurement and auditing activities. Workplace air sampling or safety surveys are examples of assessment activities, as are management system audits. The secondary level addresses the formal qualifications of the entities performing primary-level activities and the bodies that provide confirmation of the qualifications. An example is with Certified Safety Professionals (CSP) or Certified Industrial Hygienists (CIH) who perform workplace assessments. The CSP and CIH designations are given, respectively, by the Board of CSP (BCSP) and the American Board of Industrial Hygiene (ABIH). The certification function performed by the BCSP and ABIH represents secondary-level activities. With management system certification, registrars perform audits, a primary-level activity, and accreditation agencies, accredit them to perform the registration audits. Finally, an example of tertiary-level recognition is found in regulations that require certain activities be performed by CSPs or CIHs. With management systems, recognition is given by regulatory agencies who might give organizations with a certified OHSMS some sort of regulatory relief. And possibly more important here is recognition by the marketplace.

Each of the conformity assessment levels can be performed by first, second, or third parties. The designation first party refers to activities performed internal to an organization. An example is with self or internal audits. The second party refers to activities done at a given level by a customer or entity that may not be completely independent. An example is when a customer audits a supplier. Finally, the third party refers to activities performed by an independent entity such as an accredited registrar who performs a certification audit of an OHSMS.

It is important to understand that management system auditing systems are framed within a broader conformity assessment system that ensures the audit, inspection, certification, registration, accreditation, and/or workplace compliance inspection system works. Critical features of a robust conformity assessment structure include the following:

1. Having a standard (e.g., 45001) against which assessments are made
2. A way to perform the assessment that there is an agreed upon measurement method (e.g., validated tools and protocols)
3. A strong accreditation mechanism whereby first, second, and third parties can be certified to perform assessments and
4. A Quality Assurance/Quality Control mechanism whereby assessor performance is evaluated and modified as needed—a means where affected parties can register complaints.

6.5 THE MANAGEMENT SYSTEM FRAMEWORK PYRAMID

This framework provides a powerful tool to help develop and implement an OHSMS. It is commonly presented in terms of four levels, or tiers, where Level 1 addresses policies, Level 2 addresses procedures and process descriptions, Level 3 addresses detailed work instructions and best practices, and Level 4 addresses records and documents.

It is common to create a matrix based on this framework when performing an initial gap analysis to see what OHSMS pieces are in place, and then when implementing it. The matrix contains four rows, one for each level/tier, and then columns for each OHSMS element, such as OH&S policy, internal audit, and management review. Each cell of the matrix is then examined to see whether documents exist for that cell. For instance, with OH&S policy, Level 1, does a policy exist? Does it need to be modified to conform with the OHSMS requirements? At Level 2, is there a procedure that addresses how the OH&S policy statement is created, maintained, revised? At Level 3, this might not be applicable in small organizations, but in larger ones, there may be a need to have specific policy development instructions in business units or plants. At Level 4, are there documents and records related to the policy development process that can be used to demonstrate conformance?

6.6 FUTURE TRENDS

The publication of ISO 45001 represents a significant milestone in the evolution of OHSMS standards development. As the first truly international OHSMS standard, and as an ISO standard, it promises to have a significant impact on worker health and safety throughout the world.

It is anticipated that there will be ongoing efforts to integrate OH&S performance improvement with other operational risk management activities, such as environmental and sustainability efforts. Integration efforts are noble and make sense; however, OH&S professionals need to be vigilant to ensure that worker health and safety issues are not diminished through integration efforts. Another caution when developing and implementing an OHSMS in an organization is to not fall into a "checklist" mentality, or to get so focused on the OHSMS, and potentially lose sight of fundamentals of anticipation, recognition, evaluation, and control of occupational hazards and risks.

REFERENCES

American Industrial Hygiene Association. (1996). *Occupational Health and Safety Management System: An AIHA Guidance Document.* Fairfax, VA: American Industrial Hygiene Association.

American National Standards Institute. (2005). *American National Standard – Occupational Health and Safety Management Systems.* Fairfax, VA: ANSI/AIHA Z10-2005 and ANSI/AIHA Z10-2012.

British Standards Institute. (1996). *Guide to Health and Safety Management Systems.* London, England: British National Standard, BS 8800:1996.

British Standards Institute. (1999) *Occupational Health and Safety Management Systems – Specification.* London, England: BSI OHSAS 18001:1999 and OHSAS 18001:2007.

European Union, European Agency for Safety and Health at Work. (2002). *OSH Systems and Programmes: The Use of Occupational Safety and Health Management Systems in the Member States of the European Union.* Luxembourg: Office for Official Publications of the European Communities.

International Occupational Hygiene Association (IOHA). (1998). *Occupational Health and Safety Management Systems: Review and Analysis of International, National, and Regional Systems; and, Proposals for a New International Document.* Geneva: IOHA.

International Organization for Standardization. (1987). *Quality Systems – Model for Quality Assurance in Design/Development, Production, Installation and Servicing.* Geneva: International Standard ISO 9001:1987.

International Organization for Standardization. (1996). *Environmental Management Systems-Specifications with Guidance for Use.* Geneva: International Standard ISO 14001:1996(E).

International Organization for Standardization. (2000). *Quality Systems - Model for Quality Assurance in Design, Development, Production, Installation and Servicing.* Geneva: International Standard ISO 9001:2000(E).

International Organization for Standardization. (2013). *Directives, Part 1, Consolidated ISO Supplement, Procedures Specific to ISO* (4th ed.). Geneva: ISO.

International Organization for Standardization. (2018). *Occupational Health and Safety Management Systems – Requirements with Guidance for Use.* Geneva: ISO 45001:2018.

Mansdorf, Z., Mirer, F., Wright, M., Presentations given at the American National Standards Institute's Workshop on International Standardization of Occupational Health and Safety Management Systems: Is there a need? *Workshop Proceedings*, Rosemont, IL, May 7 (1996).

NRC, National Research Council. (1995). *Standards, Conformity Assessment, and Trade: Into the 21st Century.* Washington, DC: National Academy Press.

Redinger, C.F., Levine, S.P. (1999a). Occupational Health and Safety Management System Performance Measurement: A Universal Assessment Instrument, AIHA, Falls Church, VA, ISBN-13: 978-0932627926.

Redinger, C., Dalrymple, H., Dyjack, D., Levine, S., Mansdorf, Z. (1999b) *Occupational Health and Safety Management Systems: Review and Analysis of International, National and Regional Systems; and, Proposals for a New International Document.* Geneva: The International Labour Office.

7 Benchmarking in International Safety and Health

Charles Redinger
Institute for Advanced Risk Management

CONTENTS

7.1 INTRODUCTION

Benchmarking is a popular management and organizational learning tool. It is a multifaceted technique that can be used to identify operational and strategic gaps, and in the search for best practices to close these gaps. Benchmarking use and popularity has increased in global efforts to improve occupational health and safety (OH&S) performance since the 1990s. The European Union Agency for Occupational Safety and Health EU-OSHA) defines benchmarking as follows:

> a planned process by which an organisation compares health safety processes and performance with others to learn how to reduce accidents and ill health, improve compliance with health and safety law and/or cut compliance costs.
>
> *(EU-OSHA, 2015)*

Xerox Corporation's process improvement efforts in the late 1970s and early 1980s are often cited as the beginning of structured organizational benchmarking. Xerox investigated warehouse operations of L. L. Bean to gain insights on how Xerox's operations could be improved. Another early benchmarking example is seen in Nissan/Infiniti's benchmarking efforts to improve its strategic planning. In this effort, Nissan/Infiniti examined practices at a number of companies, including Walt Disney and McDonald's. Quality improvement efforts, such as seen with the Malcolm Baldrige Quality Award, are attributed to the proliferation of benchmarking in global companies. Since the late 1980s, there has been an increase in companies sharing quality and process improvement information (Yasin, 2002).

While there are numerous ways to characterize benchmarking, the organizational benchmarking literature commonly identifies three types of organizational benchmarking (Drew, 1997):

- *Process benchmarking*—used to compare operations, work practices, and business processes
- *Product/service benchmarking*—used to compare product and/or service offerings and
- *Strategic benchmarking*—used to compare organizational structures, management practices, and business strategies

OH&S management involves a wide range of activities that include identifying hazards in the workplace; determining their level of risk; taking actions to reduce hazards and associated risks; and performing a bundle of measurement activities related to these activities. A common way of describing this sequence of activities is anticipation, recognition, evaluation, and control. Activities traditionally included in the evaluation phase are collecting air samples, performing safety audits, and conducting job-specific self-assessment, to name a few. With the increased use of OH&S management systems (OHSMSs), the evaluation phase has expanded to include management reviews and global assessment of a company's OH&S programs and systems.

OH&S performance assessment includes a range of metrics, which can also be referred to as indicators. Common terms used are leading, trailing (also called outcome), and process (also called activities) indicators or metrics. All three of these can be the focus of an OH&S benchmarking endeavor; however, the most common OH&S benchmarking activity looks at leading and OH&S process indicators, which include, for instance, the organization's overall OHSMS, or a specific program such as training, respirator fit testing, confined space, ergonomics, and lock-out-tag-out programs, for example. Companies also benchmark trailing indicators, such as injury incident rates or agency citations, against industry or governmental databases; while there is value in this, it is difficult to affect change without looking at leading and process indicators/metrics.

Another form of OH&S benchmarking includes internal benchmarking, where the metric or indicator of interest is benchmarked between business units, or longitudinally over time, within a specific business unit or operation.

Reasons to perform OH&S benchmarking identified in a robust document prepared by WorkSafe Australia (1996) include the following:

- Recognition from governmental entities and certification bodies
- Better understanding of the workings of effective OHSMSs
- Overcome complacency about the role of OH&S in the enterprise.
- Build and reinforce broad commitment to change in OH&S management practices.
- Achieve quantum leaps in improvement in OH&S performance.
- Identify, understand, and implement the international best practice.
- Develop a shared vision for OH&S in the enterprise.

Global OH&S benchmarking also has historical links to standards and regulations in the United Kingdom, the United States, the International Organization for Standardization (ISO), and the International Labor Organization (ILO) (Idoro, 2011).

While there is value in performing benchmarking and it is widely done, it does have detractors and caution is voiced. Two primary weaknesses are commonly identified. The first relates to mission and vision. That is, critics caution that an entity that is benchmarking itself to another entity, or bundle of entities, might lose sight of aspects of its mission and vision, that is, benchmarking against different goals or ideals. The second relates to broader validity, reliability, and data quality issues; that is, the benchmarked metrics or indicators might not queue well between the benchmarking entities, and the quality of the benchmarking data might not be known.

Some important distinctions to keep straight when engaging in benchmarking endeavors are as follows:

- *Benchmarking process versus improvement from benchmarking* (Moriarty, 2011): This is a subtle but valuable distinction made by Moriarty. The point here is to be clear about the purpose and intended outcome from a benchmarking endeavor.
- *Benchmarks versus benchmarking*: This points to the role that performance goals (e.g., benchmarks) play in benchmarking. While benchmarking is a process, the activity of establishing meaning from it requires attention. A qualitative example here is whether a goal is to be "best in class" versus "good in class." Further, caution is needed when establishing benchmarks and potential validity issues when benchmarking entities are not similar, and indicators have different levels of maturity.

7.2 PERFORMANCE MEASUREMENT

Benchmarking is a measurement endeavor (Fuller, 1997). It is important to understand and consider basic measurement concepts when conducting benchmarking. Key here is the validity and reliability of performance indicators and metrics at the core of a benchmarking effort. In measurement theory, it is emphasized that prior to identifying any given variable or measurement for a variable, it is first necessary to identify the concepts and indicators with which they are associated. Thus, in order to make reliable and valid performance measurements, the indicators, variables, measurement units, and their logical relationships must be established. This measurement hierarchy is summarized in Figure 7.1.

7.2.1 Measurement Hierarchy

When considering measurement issues, it is important to understand a basic measurement hierarchy. This measurement hierarchy describes the measurement process from the conceptual to the operational levels. When issues of reliability and validity are considered (discussed below), it is critical to understand the relationship between concepts, indicators, variables, and operational definitions associated with a given measurement question. This hierarchy can be summarized as follows and in Figure 7.1.

1. Determine the construct or concept to be measured (e.g., OH&S performance).
2. Identify the indicator(s) associated with the construct(s) (e.g., employee participation).
3. Identify the variable(s) associated with the construct(s) or indicator(s) (e.g., continuous improvement meetings).
4. Develop operational definitions for each variable (e.g., the number of team meetings per year).

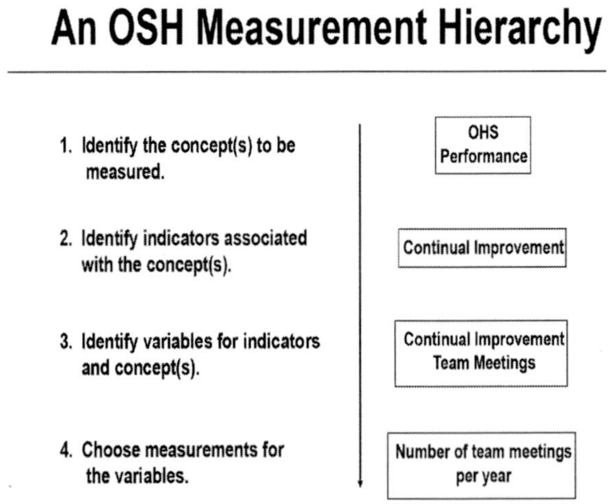

FIGURE 7.1 An OSH Measurement Hierarchy.

In Figure 7.1, a simple example is given showing how this hierarchy can be used in metrics development. In this example, the construct is OH&S performance. An indicator of OH&S performance is continual improvement. A variable associated with continual improvement could be meetings held by a continual improvement team. The number of team meetings held, or other aspects of the meetings, could be described in terms of an operational definition.

7.2.2 Measurement Levels

There is a range of measurement activities conducted within OH&S management. Following consideration of measurement hierarchy issues, attention needs to be given to the different measurement levels. These levels refer to the different levels found when developing operational definitions.

There are four measurement levels: nominal, ordinal, interval, and ratio. It is important to identify in what measurement level a particular metric resides as this impacts the type of data that can be collected and how it will be collected. Traditional OH&S metrics such as effluent discharge quantities or lost workdays are commonly interval or ratio level measurements.

Nominal measurements tell only what class a unit falls with respect to the property (e.g., male/female; yes/no). These are commonly used in the initial development of global benchmarking efforts, such as asking, "is a lock-out-tag-out" program present? Or, are OH&S risk assessments performed?

Ordinal measurements determine when one unit has more of a property than does another. This allows for ranking. Ordinal measures do not indicate how much more of the property is present, in terms of a linear scale, simply that more or less is present. With the lock-out-tag-out programs or OH&S risk assessments, beyond seeing if they are done, questions would look at how well they are done, or at their quality. Likert-like scales are often used and are an example of an ordinal measurement.

Interval measurements identify when one unit differs by a certain amount of the property from another. Air sampling and noise dosimetry are examples of interval measures.

Ratio measurements are the same as interval, except that there is a zero point. Temperature is a good example of a ratio measure.

7.2.3 Types of Metrics

The types of metrics usually mentioned in OH&S performance measurement are as follows:

Trailing versus leading: The distinction between trailing and leading indicators refers to the order in which events take place. A leading indicator happens before a trailing indicator. The goal is to understand the causal relationship between leading and trailing indicators such that leading indicators provide valid predictions of trailing indicators. An important concept here is causation.

Process versus outcome: The distinction between process and outcome indicators has been made for many years in quality assurance circles and provides a good example of leading and trailing indicators. Process indicators are considered leading indicators and are part of an actual manufacturing process or the system of interest. Outcome indicators can be considered trailing indicators since they measure the outcomes of the process or system.

Output versus outcome: Output indicators and measurement address the implementation of a system, process, or procedure. They address the result generated by the implemented system, process, or procedure. For instance, the implementation of an OHSMS is an output. The difference that the system makes (e.g., reduced injury rates) is an outcome.

7.2.4 Reliability and Validity

Reliability and validity issues are central to any measurement endeavor. These are two of the most important issues to consider when conducting global benchmarking and developing indicators/metrics that will be used. Concerns here are whether a given metric, or measurement effort, is measuring what it is intended to measure (validity) and whether successive measurements provide the same result (reliability). These terms can be used when considering individual metrics or an entire assessment endeavor.

Understanding the reliability of a measurement is a necessary precursor to establishing its validity. When the reliability is high, there is enhanced confidence that changes in individual measurements are due to identified interventions, and not just a random fluctuation in the measurement process. Reliability is commonly assessed in three forms: test–retest, alternate form, and internal consistency. Intra-observer and interobserver reliabilities are important considerations.

Validity refers to the degree to which a metric, or assessment activity, measures what it claims to be measuring. There are four types of validity: face, content, criterion, and construct.

Face validity and content validity are most applicable to global OH&S benchmarking at this point in the technique's evolution. *Face validity* is the least scientific measure of all the validity measures. It is based on a cursory review of items by untrained judges to determine whether a metric makes sense from a common sense perspective. Decisions made during the initial stages of a metrics development effort are based on face validity considerations. *Content validity* is a subjective measure of how appropriate the items seem to a set of reviewers who have some knowledge of the subject matter. The assessment of content validity typically involves an organized review of the metric (or instrument) to ensure that it includes everything it should and does not include anything it should not. Content validity and face validity are very similar and can be confused. In many efforts, face validity considerations are performed in conjunction with content validity considerations.

7.3 CONDUCTING A BENCHMARKING STUDY—STEPS AND GUIDANCE

A wide range of benchmarking approaches are found in the benchmarking literature (Drew, 1997; Sanders, 2016; Fuller 1997, 1999; Evans, 2012; Moriarty, 2011). A bundle of common steps are found in this literature and articulated well in an OH&S guidance document developed by WorkSafe Australia (1996), which are as follows:

- Step 1—Establish benchmarking project
- Step 2—Select teams and train them
- Step 3—Identify processes to benchmark
- Step 4—Analyze your own process
- Step 5—Select benchmark partners
- Step 6—Build relationships and conduct visits
- Step 7—Analyze performance gaps and develop improvement strategies

These seven steps are reflected in many of the studies and approaches presented in this chapter. They also reinforce performance measurement issues presented here.

In Step 1, a key point is to develop clarity on why a benchmarking project is being conducted, its scope, and intended outcome. That is, how are findings expected to be used? Step 6 reflects that many OH&S benchmarking endeavors are performed against external entities, that is, other enterprises; however, it is not uncommon to use steps such as these in internal efforts, that is, for instance, benchmarking different business units within a single enterprise. Addressed in the WorkSafe approach, but not explicitly seen in the step descriptors, is the importance of data collection and its impact on OH&S benchmarking validity and reliability.

WorkSafe's OH&S benchmarking tool kit provides enterprises and OH&S practitioners with an excellent bundle of guidance and tools on how to conduct an enterprise benchmarking study. It does not contain specific benchmarking indicators, as other studies in this chapter; rather, it provides guidance and criteria for an enterprise to use in developing indicators most appropriate to them.

7.4 BENCHMARKING APPROACHES AND FRAMEWORKS

There is a range of OH&S benchmarking approaches and frameworks used by enterprises, nongovernmental organizations, and governmental agencies. Some of these focus on individual performance indicators, whereas others look more broadly larger program elements and entire system/programs; depending upon one's perspective, overlap can be seen between these. OHSMS frameworks (e.g., OHSAS 18001, ISO 45001, ILO OSH-MS) have evolved into popular OH&S benchmarking frameworks. Scorecard frameworks are also vogue in numerous sectors; these are commonly based on the Balanced Scorecard (BSC) method developed at Harvard University in the early 1990s. Finally, the social responsibility and sustainability arena provides basic bundles of OH&S performance indicators that can be used in benchmarking.

7.4.1 Individual Performance Indicators

There are numerous ways to identify key occupational health and safety (OHS) performance indicators in an organization, and with which to use in benchmarking endeavors. In 1971, the Organization for Economic Cooperation and Development (OECD) established a guideline for developing safety performance indicators (SPI). Jennings and Schulberg (2009) observe that developing an SPI program affords a company the ability to

- Assess whether it is implementing appropriate chemical safety programs and policies.
- Evaluate whether these programs and policies are achieving their desired objectives.
- Help determine the extent to which such programs and policies are making a difference.

These three points are examples of what a company may use as the foundation of an OHS benchmarking endeavor.

The OECD guideline presents a robust seven-step process for establishing an SPI program (OECD, 2008), which are as follows.

1. Establish an SPI team.
2. Identify key issues of concern.
3. Define outcome indicator(s) and related metrics.
4. Define activities indicator(s) and related metrics.
5. Collect the data and report indicator results.
6. Act on findings from SPIs.
7. Evaluate and refine SPIs.

The OECD SPI program approach provides a valuable framework companies can use for international OHS benchmarking. As an international framework with a well-defined process, issues related to validity are minimized since benchmarking entities would be following the same process. Within this framework, individual performance indicators/metrics (outcome or activity) can be benchmarked, as well as the overall SPI program.

7.4.2 Program Elements

Beyond individual performance indicators, larger pieces of OHS management can be the focus of benchmarking, and these are referred to as program elements. An example of this type of benchmarking activity as seen in a safety climate assessment across nine North Sea oil and gas operations that identified five best practices in a benchmarking study between British Petroleum, Conoco, and

Royal/Dutch Shell Group (Mearns, 2001). The list of best practices that evolved from this is as follows:

1. *Health and Safety Executive (HSE) policy documents*: Top management commits to HSE goals; "the policy is strong, concise, and visionary"; the policy refers to striving towards zero accidents; performance is monitored and made public.
2. *Assurance of policy compliance*: Annual self-assessments and reports.
3. *Operation and governance*: One managing director has board-level responsibility for health and safety; a corporate health and safety advisor recommends policy and chairs a committee comprising senior business managers.
4. *Joint ventures/subsidiary policy*: The parent company health and safety policy applies to joint ventures under the parent company; external health and safety reports are made for joint ventures.
5. *Linkage of health and safety into the business*: Health and safety is a core value and part of company culture; risks are assessed, targets are set, and performance is monitored.

Establishing a set of best practices such as those above can be an OH&S stopping point. It can also be the starting point for ongoing longitudinal activities where measure scales are created to determine the strength and maturity of each of these practices.

Another example of a program element approach done internally in a UK water utility was based on the HSE's six key elements for effective OH&S management (Fuller, 1999), which are as follows: policy, organization, planning and implementation, performance, auditing, and review. These elements formed the basis of an OH&S audit program used for internal benchmarking. Three dimensions of OH&S performance were assessed in this effort: how well OH&S management was understood, how well OH&S procedures were implemented, and accident frequency rates (Fuller, 1999).

7.4.3 Management Systems

The use of formal OHSMS has proliferated in the 2000s with the advent of the ILO's OHSMS (ILO, 2001), the British Standards Institutes (BSI) 18001, numerous country-specific approaches, and recently the publication of ISO 45001:2018. OHSMS standards provide a robust framework for OH&S benchmarking. A universal OHSMS model was developed in the 1990s at the University of Michigan and has been used in numerous benchmarking studies (Redinger, 1998; EU-OSHA, 2002). This chapter provides background information on various OHSMS approaches.

A key distinction in OHSMS approaches is continual improvement. Benchmarking techniques provide a way to demonstrate the ongoing improvement of specific OHSMS elements and the entire system. And when improvement is not found, identifying such performance gaps is valuable. Another key distinction in this approach is performance measurement. Specific performance measurement requirements are found in all OHSMS approaches, such as auditing, management review, and accident investigations. While benchmarking *per se* is not specifically required or mentioned in OHSMSs, it is not uncommon for organizations to include it in its OHSMS policies and procedures.

A challenge with using OHSMS elements in an external benchmarking effort is that there is often variation in implementation and maturity. That is, there are numerous ways elements can be implemented and change and mature over time.

7.4.4 Use of the "BSC"

The "BSC" is a popular business performance measurement framework developed in the early 1990s (Kaplan, 1992, 1996). Since BSC's introduction by Kaplan, numerous iterations of the concept have been developed. The BSC approach has been important in highlighting the difference between leading and trailing indicators and in turning increased attention to the measurement of leading indicators. The "scorecard" concept that this approach initiated is now a common concept in business and OH&S management, and within the concept, there are key performance indicators that populate a scorecard. Kaplan's BSC contains a diverse set of performance measures that are divided into four groups: financial performance, customer relations, internal business processes, and learning and growth (Mearns, 2003).

The UK's HSE supports the multifaceted BSC approach: "Organisations need to recognize that there is no single reliable measure of health and safety performance." "What is required is a 'basket' of measures or a 'balanced scorecard' providing information on a range of health and safety activities" (HSE, 2001). Numerous examples of applying Kaplan and Norton's BSC framework to OH&S management and performance measurement are found in the literature and in companies (Koper, 2009; Paranjape, 2006; Mearns, 2003; Redinger and Birkner, 2003). Beyond providing a robust internal OH&S benchmarking framework for an organization, it is not clear that the BSC approach is used in larger intraorganizational OH&S benchmarking efforts.

7.4.5 Role in Social Responsibility/Sustainability

Social responsibility and sustainability have been gaining increased prominence in the development of governmental and nongovernmental standards and guidelines. Approaches developed by these entities provide bundles of indicators that can be used in benchmarking endeavors. While OH&S is not the focus of social responsibility and/or sustainability standards and guidelines, OH&S criteria and indicators can be found in them.

ISO's Guidance on Social Responsibility, ISO 26000, contains some guidance on OH&S issues (ISO, 2010). While not robust, these do provide OH&S with visibility within the social responsibility space.

Since its inception in the late 1990s, the Global Reporting Initiative (GRI) has provided a sustainability benchmarking platform for enterprises. Within a general rubric of "labor practices and decent work," GRI's framework includes criteria and performance indicators for OH&S. Compared to criteria and performance indicators found in the OH&S space, these are not robust. The Center for Occupational Health and Safety was formed in 2012, in part, to increase OH&S impact within GRI's structure and in sustainability reporting in general. As social responsibility and sustainability standards and approaches continue to evolve, OH&S professionals should monitor them for performance measurement guidance.

7.5 INTERNATIONAL BENCHMARKING—STANDARDS, GUIDELINES, AND EXAMPLES

7.5.1 International Association of Labour Inspection—International Benchmarking OSH Regulation

On behalf of the International Association of Labour Inspection (IALI), Singapore's Ministry of Manpower and Britain's Health and Safety Executive have developed an International Benchmarking on OSH Regulation (IALI, 2017). This document provides nations with a guideline upon which they can assess and organize regulatory activities; its focus is macro; that is, it addresses high-level governmental activities related to OH&S. Its aim is

- To help countries develop effective labor inspectorates systematically.
- To provide a structured framework for labor inspectorates to monitor progress.
- To develop a global network of OH&S collaborating centers.

The document is an application of IALI's International Common Principles for Labour Inspection to OH&S, and incorporates the aspects of the following:

- BS EN ISO/IEC 17020:2012—Conformity assessment—Requirements for the operation of various types of bodies performing inspection
- IALI's Global Code of Integrity for Labour Inspection
- ILO Labour Inspection Convention, 1947 (No. 81)
- ILO Occupational Safety and Health Convention 1981 (No. 155)

An inspection assessment protocol is included for use in determining a nation's current OH&S inspection status. Competency criteria are included for personnel who perform the assessment. Tools are also provided to assist in performing assessments, which include a questionnaire and an assessment-scoring checklist.

7.5.2 Association of Southeast Asian Nations—Survey of Good OH&S Practices

The Association of Southeast Asian Nations (ASEAN) OH&S network, with support from the ILO, conducted an OH&S best practices survey of Cambodia, Indonesia, Lao PDR, Malaysia, Myanmar, the Philippines, Singapore, Thailand, Vietnam, Japan, and Republic of Korea (ILO/ASEAN, 2012). Each country that participated reported on the status of their OH&S regulatory structures, national performance statistics, and strategies for OH&S performance improvement. Highlights from four country reports are as follows:

Lao PDR reported efforts to address worker safety and health expanded after the nation embarked on a new and more open economy in the 1980s. Into the early 2000s, there were challenges with many workers lacking experience in factories, and their awareness about safety and health risks was low. Efforts in 2005 introduced a comprehensive OH&S program that increased training of inspectors, enhanced existing labor laws, and increased collaboration with ASEAN and ILO. It is reported that OH&S awareness has improved along with strengthening training program requirements and delivery.

Malaysia reported a gap in safety cultures between workers directly engaged by multinational companies and those working for subcontractors (supply chain issue). The Safety Passport program was designed and implemented to address this gap. In this program, workers receive training on workplace safety and receive a "passport" after they successfully pass an assessment administered by the agency. Malaysia reported that it was in the process of benchmarking this program against similar systems in Europe.

Cambodia reported that its national OH&S efforts have been historically challenged by the lack of political stability and slow economic development. With support from the ILO, trade unions, and employers' organizations, an OSH Master Plan was developed in 2009. The plan identified six action areas: (1) strengthen the national OH&S system, (2) improve safety and health inspections and compliance, (3) promote OH&S activities within employers and workers' organizations, (4) implement a special program for hazardous occupations, (5) extend OH&S protection to small enterprises, and (6) build OH&S protection mechanisms for the rural sector and information economy.

Thailand. In the ASEAN study, Thailand highlighted safety officer training. Employment of safety officers has been mandated in some high-risk industries and expanded in 1997 to include mining, quarries, petrochemical, manufacturing, construction, and transportation. The range of the mandate was expanded further in 2006 to include hotel, department store, health care, financial, physical testing unit, sports entertainment, and chemical or biological laboratory. In 2006, the Ministry of Labour implemented an eight-step reform of the safety officer system. Following the report, the ministry reported that the quality of training was strengthened. In concert with the "zero accident" campaign, improvement of the safety officer training is credited with a reduction in occupational accidents and disease in Thailand.

7.5.3 EU-OSHA—Review of Successful OH&S Initiatives

As part of its Healthy Workplace Campaigns, the EU-OSHA commissioned a comprehensive study of 24 OH&S benchmarking schemes (EU-OSHA, 2015). Detailed summaries of 13 schemes are found in the study's report. Numerous examples of benefits to participant organizations in benchmarking schemes are reported. Drivers for participating in a benchmarking scheme were identified as calibration of an organization's performance against the market and achievement of improvements in accident and incident rates. Regarding improvements, the Finnish Zero Accident Forum members' accident rates fell by 46% between 2008 and 2012, whereas the national accident rate did not significantly alter over the period.

There was a variation between the schemes—information-sharing requirements varied significantly and included both quantitative (e.g., OH&S outcome) data and qualitative good practice (e.g., process) data. A distinction was found between schemes that require reporting quantitative performance data, versus ones with a lower participation threshold with requirements to share qualitative good or best OH&S practices.

Participation in the study/program was limited in some cases by both poor and good OH&S-performing organizations. Poorly performing organizations were found to be reluctant to participate because of potential scrutiny from regulators or insurers. Good performers were reluctant to share information believed to provide a competitive advantage.

7.5.4 European Agency for Safety and Health at Work—OHSMS

A universal OHSMS framework developed at the University of Michigan (Redinger, 1998) was used by EU-OSHA to conduct an OHSMS best practices benchmarking study in the European Union (EU-OSHA, 2002). The framework's 32 primary content variables (e.g., continual improvement, employee participation, performance measures) were used to assess and compare the extent to which the content variables were present in the OSH approaches in 11 companies. The study was conducted at a time (2000–2001) when the OHSMS approach was relatively new, and as such, many of the universal OHSMS elements were not found in the case companies. Content variables that were identified in the case companies tended to be ones associated with traditional programmatic approaches to OH&S management, such as training programs, OH&S policy, and auditing, to name a few.

7.5.5 An International Benchmarking Culture Survey

In 2015, the University of Western Australia's Centre for Safety reported on safety culture benchmarking survey that included eight countries and 879 respondents (Andrei, 2015). Respondents were primarily from mining interests (69.5%); other industries represented included construction (19.3%), oil and gas (7.2%), manufacturing (2.5), and power utility (1.5%). Respondent countries included the economically developing countries: Gambia ($n=80$), Indonesia (100), Ivory Coast (136), Kenya (84), Mongolia (164), Nigeria (106), the Philippines (97), and Zambia (106). The aim of the survey was to gather information on OHS needs in developing countries and to learn about improvement needs.

Factors have been identified that have been historically known to impact safety behavior and cultures. These factors were grouped into three main areas: risks, organizational, and individual, and are discussed as follows (Andrei, 2015):

Risks: Risks present in the workplace; the perceived likelihood of risks occurring; the perceived severity of risks.

Organizational: Perceptions of safety climate within the organization (e.g., management values, communication, training, and safety systems); how safety is rewarded by the organization; how the organization overall learns from mistakes; how the organization overall monitors and is alert to risk; the perceived safety of the physical work environment of the organization; whether high-performance safety systems were in place in the organization.

Individual: Individual beliefs about the importance of safety in the workplace; how individuals think about mistakes in the workplace; individual compliance and participation in safety procedures; what motivates peoples' behaviors in the workplace.

The survey found that risks commonly identified in developed economies were also identified in the participant countries. Risks unique in the participant countries included poor working conditions, hygiene and sanitation, and concerns about safety equipment quality. Gaps were identified between safety vision and physical workplace safety; the importance of embedding a safety vision was identified, as was actual support of employees to act safely. Workplace safety activities were generally characterized as being reactive as opposed to proactive. That is, regulatory compliance was the primary safety driver, as opposed to proactive employee participation.

7.6 CORPORATE BENCHMARKING

7.6.1 The Conference Board 2003 Survey—Striving for Best Practices

The Conference Board is a business membership and research organization in the United States. This entity conducted a benchmarking project to identify OH&S best practices used in global companies (Conference Board, 2003). A benchmarking survey of 58 companies was done first, followed by a detailed examination of Alcoa, Baxter International, Eastman Kodak, and Motorola. The survey questions were grouped into four categories: practices and programs, management support, supervisory procedures, and employee involvement.

The findings of the study reported that not only companies striving for outstanding safety and health records are ensuring strict regulatory compliance, but also they developed their own best practices to enhance their performance. The primary drivers for a beyond-compliance orientation are reported as a strong conviction that accidents and injuries are unacceptable in their operations and firm belief that business benefits—directly through reduced costs and indirectly through improved morale and increased productivity.

The survey and detailed examination for the companies that the core elements—leading indicators—of successful safety and health strategies are as follows:

- *Leadership at the top*: If the top executive believes in the work strategies, sets expectations for other managers, follows through on those expectations, and commits appropriate resources, shared beliefs, norms, and practices will evolve.

- *Confidence on the part of all employees* that the company values safety and health comparable with other values, and an understanding by all employees of how to achieve the expected performance. Everyone must be committed and engaged.
- *Creating and implementing a safety and health management system* that works for the individual company.
- *Monitoring performance regularly*: Companies must continually assess their norms and provide frequent feedback to all employees and to external stakeholders.

Themes that stand out in the Conference Board study are clear management visibility and leadership; ownership of safety and health by all employees, moving from "involvement" to "empowerment"; accountability at all levels of an organization, including positive and negative performance feedback; open sharing of knowledge and information throughout the organization; and incorporate safety into the business process and an operational strategy.

7.6.2 Organization Resource Counselors and Avery Dennison

In the early 2000s, the Organization Resources Counselors, Inc. (ORC) in the United States began to offer its member organizations an OH&S benchmarking platform. The platform's development began by including performance indicators and metrics associated with OH&S department structure, leadership practices, OHSMSs, and trailing indicators such as injury and illness rates (Redinger, 2001). Participant organizations include many Global 500 companies, representing a diverse set of industries.

This OH&S benchmarking platform contains a robust bundle of performance indicators that are organized within four categories. The organizing categories with example performance indicators follow. Performance indicators are measured (quantified) with a series of questions that are scored on a 5-point Likert scale.

Leadership commitment and support: (1) *Written* safety and health standards of performance are in the form of *measurable short- and long-term goals and* objectives. (2) Safety and health *roles and responsibilities*, based on goals and objectives, are established in writing for each major function and department. (3) Safety and health goals and objectives are updated annually.

Employee involvement: (1) Workers participate in the development of safety and health programs and policies, conducting training and education, and safety and health program audits and reviews. (2) Management encourages and authorizes employees to stop activities that present potentially serious safety or health hazards. (3) A system is in place to assess employee perceptions about workplace safety and health.

Risk identification, elimination, and safe practices: (1) A routine hazard identification process is in place. (2) A hazard control process is in place. (3) A medical surveillance program is in place.

Accountability: (1) Safety *and health criteria* are included in *performance standards* for employees, supervisors, and managers throughout the organization. (2) Safety and health performance against goals is included in regular, periodic, quantitative performance reviews with feedback for all managers, supervisors, and employees. (3) Objectives of all levels of management include *incident prevention activities* such as hazard elimination, safety, and health performance improvement.

Participant organizations report an improvement in OH&S improvement through participating in the ORC's OH&S benchmarking platform. A common attribute for performance improvement is the ability to make connections between leading and lagging performance indicators, and the ability to share best practices.

7.7 SUMMARY AND LOOKING TO THE FUTURE

OH&S benchmarking has evolved into an important component of OH&S performance measurement, both at the macro-national level and at the micro-organizational level. There has been a trend in OH&S benchmarking to use standardized frameworks, such as those contained in OHSMSs (e.g., OHSAS 1800 and ISO 45001). Developing economies find value in benchmarking with evolved economies such as ones in the European Union and Asia. The International Benchmarking on OSH Regulation published by Singapore's Ministry of Manpower (IALI, 2017) provides a robust model for developing economies to use in evolving their OH&S strategy and structure. In like fashion at the enterprise level, WorkSafe Australia's (1996) OH&S benchmarking tool kit provides an excellent guide on how to create and execute an OH&S benchmarking initiative.

OH&S benchmarking practices are relatively mature. However, there is always room to evolve these practices in order to improve validity and reliability, and also increase the predictive value of benchmarking.

There will always be new OH&S performance indicators coming on to the scene, and used in benchmarking endeavors. This is seen with increased focus and measurement of psychosocial and general well-being factors in the workplace. Evolution in the exposure and risk assessment arena, such as in the use of control banding, is one OH&S professionals should keep tabs on, as well as developments in the "total worker health" and social responsibility and sustainability arenas.

Advances in organizational risk management are anticipated to impact OH&S management. The increased movement towards integrated systems, as reflected in ISO's development of a generic management system framework, is one that OH&S professionals should keep tabs on as it can impact performance measurement activities. The close links between OH&S benchmarking and corporate social responsibility reporting should also be followed and strengthened.

In organizational risk management, there is also increased attention on what can be characterized as "soft issues" or culture-related issues that possibly have a greater impact on performance outcomes than do tradition focus areas, such as the systems and auditing. A promising area of research is awareness-based risk management and the potential of OH&S professionals driving organizational improvements

(Redinger, 2017). This evolving research area may have an impact on OH&S benchmarking as it links OH&S's historical strengths with transforming organizational risk management and organizational performance.

REFERENCES

Andrei, D., Griffin, M., Wang, L., Choe, W. K. (2015). *Benchmarking the Status of Safety: Safety Practices, Risks and Beliefs in Developing Countries.* Perth: The University of Western Australia, Centre for Safety.

Conference Board. (2003). *Driving Toward "0": Best Practices in Corporate Safety and Health, How Leading Companies Develop Safety Cultures.* Research Report R-1334-03-RR. New York: Conference Board.

Drew, S., From knowledge to action: The impact of benchmarking on organizational performance, *Long Range Planning* (1997) Vol. 30, No. 3, pp. 427–441.

European Agency for Safety and Health at Work, EU-OSHA. (2002). *The Use of Occupational Health and Safety Management Systems in the Member States of the European Union.* Luxembourg: Publications Office of the European Union.

European Agency for Safety and Health at Work, EU-OSHA. (2015). *Review of Successful Occupational Safety and Health Benchmarking Initiatives.* Luxembourg: Publications Office of the European Union.

Evans, M. T., Twenty-first century benchmarking: Searching for the next generation, *Benchmarking: An International Journal* (2012) Vol. 19, No. 6, pp. 760–780.

Fuller, C., Key performance indicators for benchmarking health and safety management in intra- and inter-company comparisons, *Benchmarking for Quality Management & Technology* (1997) Vol. 4, No. 3, pp. 165–174.

Fuller, C., Benchmarking health and safety performance through company safety competitions, *Benchmarking: An International Journal* (1999) Vol. 6, No. 4, pp. 325–337.

Health and Safety Executive, HSE. (2001). *A Guide to Measuring Health and Safety Performance.* London, England: Health and Safety Executive.

Idoro, G., Comparing Occupational Health and Safety (OHS) management efforts and performance of Nigerian construction contractors, *Journal of Construction in Developing Countries* (2011) Vol. 16, No. 2, pp. 151–173.

ILO, (2001). *Guidelines on Occupational Safety and Health Management Systems.* Geneva: International Labor Organization. ISBN 92-2-111634-4 www.ilo.org/wcmsp5/groups/public/---dgreports/---dcomm/---publ/documents/publication/wcms_publ_9221116344_en.pdf accessed December 13, 2018.

ILO/ASEAN. (2012). *Good Occupational Safety and Health Practices 2008/2009.* Lao PDR: ASEAN-OSHNET, Ministry of Labour and Social Welfare.

International Association of Labour Inspection, IALI. (2017). *International Benchmarking for OSH Regulation.* Singapore: IALI Secretariat, Ministry of Manpower.

International Organization for Standardization, ISO. (2010). *Guidance on Social Responsibility.* Geneva: ISO 26000.

Jennings, K., Schulberg, F., Guidance on developing safety performance indicators, *Process Safety Management* (2009) Vol. 28, No. 4, pp. 362–366.

Kaplan, R., Norton, D., The balanced scorecard—Measures that drive performance, *Harvard Business Review* (1992) Vol. 70, No. 1, pp. 71–79.

Kaplan, R., Norton, D. (1996). *The Balanced Scorecard: Translating Strategy into Action.* Boston, MA: Harvard Business School Press.

Koper, B., The occupational safety and health scorecard—A business case example for strategic management, *Scandinavian Journal of Work, Environment & Health* (2009) Vol. 35, No. 6, pp. 413–420.

Mearns, K., Håvold, J. I., Occupational health and safety and the balanced scorecard, *The TQM Magazine* (2003) Vol. 15, No. 6, pp. 408–423.

Mearns, K., Whitaker, S. M., Flin, R., Benchmarking safety climate in hazardous environments: A longitudinal, Interorganizational approach, *Risk Analysis* (2001), Vol. 21, No. 4, pp. 771–786.

Moriarty, J., A theory of benchmarking, *Benchmarking: An International Journal* (2011) Vol. 18, No. 4, pp. 588–612.

OECD. (2008). *Guidance on Developing Safety Performance Indicators*, Paris, France: OECD Environmental, Health, and Safety Publications, Series on Accidents, No. 18. Organisation for Economic Cooperation and Development.

Paranjape, B., Insights from the balances scorecard performance measurement systems: Successes, failures and future: A review, *Measuring Business Excellence* (2006) Vol. 10, No. 3, pp. 4–14.

Redinger, C., Development and evaluation of the Michigan occupational health and safety management system: A universal OHSMS performance measurement tool, *American Industrial Hygiene Association Journal* (1998) Vol. 59, pp. 572–581.

Redinger, C. (2017). Awareness-based risk management: seeing, transforming, and unleashing organizational capacity. Poster presented at the *Society for Risk Analysis Conference*, December 2017. The poster won best poster award.

Redinger, C., Birkner, L. (2001). *Avery Dennison, Environmental Health and Safety Benchmarking Project.* San Rafael, CA: Redinger & Associates, Inc.

Redinger, C., Birkner, L. (2003). *Developing a Balanced Scorecard for Occupational Health and Safety.* Dallas: Professional Development Course, American Industrial Hygiene Conference and Exposition.

Sanders, L., International benchmarking for performance improvement in construction safety and health, *Benchmarking: An International Journal* (2016) Vol. 23, No. 4, pp. 916–936.

Worksafe Australia. (1996). *Benchmarking Occupational Health and Safety.* Canberra: Australian Government Publishing Service.

Yasin, M., The theory and practice of benchmarking: Then and now, *Benchmarking: An International Journal* (2002) Vol. 9, No. 3, pp. 217–243.

8 International Reporting of Occupational Injuries, Illnesses, and Fatalities

Thomas P. Fuller
Illinois State University

CONTENTS

8.1 INTRODUCTION

Perhaps one of the most important aspects of a successful occupational safety and health program, whether for an organization or a nation, is a robust and accurate means to measure and record occupational injury and illness statistics. Without an accurate assessment of where injuries and illnesses, and fatalities, are occurring, it is difficult to accurately identify the causes and take corrective action. Without sophisticated means to analyze complex data and associations, it is easy to be led off track from the true causes of poor worker health, and away from more effective improvements to occupational safety and health programs.

Accurate global estimates of occupational injury and disease are needed to understand where priorities for control and prevention should be placed by governments and businesses, and as a means to motivate and educate governments and enterprises to take corrective and proactive actions. Accurately reporting injuries and illnesses in order to make meaningful comparisons between nations is a big challenge. Recent studies have shown that fatal and nonfatal injuries in developing countries are greatly underestimated (Hämäläinen, 2006). It has been shown that between 29% and 81% of occupational injuries go unreported (Tucker, 2014; Pransky, 2010; Shannon, 2002; Rosenman, 2006; Erickson, 2000; Moll Van Charante, 1998). In one study of rural South African workers, it was found that only 5% of workplace fatalities of women were reported (Schierhout, 1997). In the United States, where there are very sophisticated and regulated reporting requirements, between 33% and 69% of occupational injuries go unreported. Up to 88.3% of agricultural injuries go unreported (Leigh, 2004). In developing countries with poor communication infrastructure, and unsophisticated data collection methods or systems, the accurate reporting of work-related injuries and illnesses can be nearly nonexistent.

The starting point for the creation, development, and advancement of any safety and health program must be a clear understanding of occupational risks. The probability of an accident or illness based on a workplace exposure, multiplied by the severity associated with the injury or illness, will provide the level of risk, which will guide further research into the causes of the injuries, where and how workers are exposed, the progression of disease or injury, and the best means to control, minimize, and if possible eliminate the risk. It is widely recognized that transparent data regarding workplace safety can be a key driver of social change (Brown, 2005).

Estimates of workplace risks can guide private or public policy with regard to the use of limited resources to address the most significant and costly hazards, in terms of both money and suffering to workers and their families. Social benefits in terms of both are self-evident. According to the International Labor Organization (ILO), workplace injuries and illnesses worldwide cost the global gross domestic product (GDP) of about 4%, with losses in some particular countries as high as 10% of their GDP (ILO, 2012a).

The true value of an injury, illness, and fatality reporting system for a nation is the identification of programs and means that will reduce the economic burden on society by showing which systems, industries, and regions need the most improvement and assistance. That is, which workers are being affected most, and where will measures to minimize exposures to the hazards be most effective and efficient. Many successful and sophisticated companies already do this in order to maximize profits, but countries need to use the same techniques to improve GDP and protect citizens (ILO, 2002). A visual depiction of how occupational injury, illness, and fatality data can be used is provided in Figure 8.1 (ILO, 2013).

In the past few decades, there has been a significant expansion in the use of comparative risk assessment (CRA) to use available data and sophisticated mathematical methods to identify and prioritize risks to environmental factors (Embleton, 1996).

Despite powerful analytical and mathematical means to estimate the health outcomes associated with workplace exposures to hazardous agents and chemicals, most inaccuracies in assessing the risks associated with the hazards are the consequence of insufficient input data. Problems in defining what constitutes an illness or injury, what is considered work or a workplace, which workers are included in the data pool, and the accuracy of reporting are just some of the problems in creating accurate estimates. Additionally, accurate measures of exposure are extremely difficult to obtain and document in many parts of the world as well as for many different

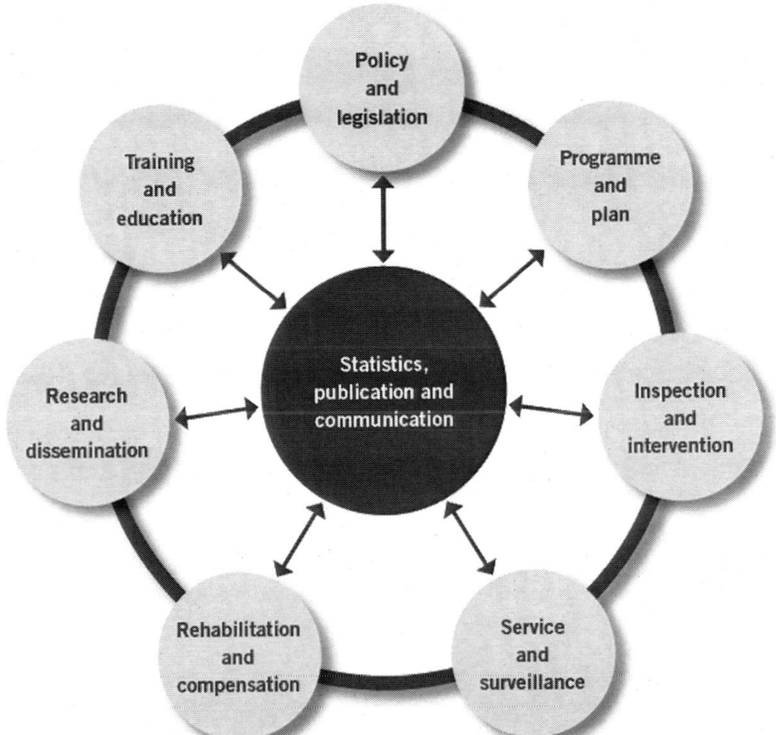

FIGURE 8.1 Use and application of occupational disease data. (Courtesy of ILO, 2013. With permission.)

jobs or tasks. Even within a single national reporting system, collection methods and data can vary between different regions or states.

8.2 VALUE TO ORGANIZATIONS

In developed countries (countries with GDP per capita exceeding $12,000), where in general, workers are considered valuable and limited resources, national governments and enterprises typically realize that the human workforce assets must be protected from harm in order to ensure their maximum value and potential output (Investopedia, 2016). A healthy worker is a productive and happy worker. Worker skills, knowledge, and experience accumulate over time resulting in increased value to their employer.

Companies see the value of healthy workers from the standpoint of productivity or efficiency. In addition to reduced worker compensation costs, there are lower insurance premiums, better employee morale, improved public relations in the community, and fewer potential regulatory penalties for noncompliance to occupational health and safety regulations. Controls put in place in "high-risk" areas or processes to protect workers typically also protect, or reduce the likelihood of damage, to production equipment and process materials. Companies can assess the benefits of reduced economic costs of accidents in terms of production time lost, penalties for missing product delivery timelines, damage to machinery and raw materials, and product liability (Alli, 2008).

Numerous studies have been completed over the years that demonstrate improved corporate profits or efficiencies for companies and organizations that have good safety programs. This can only be demonstrated when the data collected for comparing programs with injury and illness rates are accurate. This is an area where effective reporting and recording mechanisms improve productivity and efficiencies for companies using this data.

Direct costs of injuries and illnesses include monetary payments made directly related to the injury or illness. These include employer payments to insurance funds, wages paid to employees during injury/illness-induced absences, and medical expenses. Indirect expenses include a much broader and less well-defined list and include such items as damage to equipment and materials, production downtime caused by the accident, reduced output or quality due to the absence of the injured worker, costs of replacement staffing, decreased company morale, and administrative overhead time in responding to injuries and illnesses (ILO, 2012b).

8.3 VALUE TO NATIONS

When the U.S. Occupational Safety and Health Administration (OSHA) was created in 1970, one of the first actions for OSHA as directed by Congress was to perform an analysis of how many injuries, illnesses, and fatalities occur in the United States each year (OSHA, 1970). The rationale was to provide a basis for the creation of regulations and standards directed at reducing the risks in the most hazardous U.S. workplaces. It is believed that when accurate and consistent national data on workplace injury, illness, and fatalities are available, it becomes possible to clarify which safety or health programs and regulations are most needed or most effective. The first OSHA statistics were published in 1975, the total number of injuries or illnesses per 100 full-time workers was approximately 9.5. Thirty-eight years later, after the implementation of hundreds of risk-reducing rules and regulations, the development of risk-specific training programs, and special emphasis or regional enforcement programs that utilize the risk data to focus enforcement, injury and illness rates were reduced to around 3.3 in 2013. Similarly, fatalities have reduced from around 6,500 per year in the early 1990s to 4,585 fatalities in the United States in 2013 (Drudi, 2015).

The largest losses from injuries and illnesses are the losses in wages and benefits by employees. In the United States in 2007, these amounted to $139 billion. They were more than double of the medical costs and triple of the lost home production costs (ILO, 2013; Leigh, 2011). Together, these three losses represent about 1.8% of U.S. GDP. It is expected that losses to other nations, particularly economically developing countries (EDCs), are also significant; however, this is an area where research and reporting are lacking.

In addition to the value to nations in terms of worker longevity and morbidity, the value can be measured in monetary value. A significant amount of data is available that shows the reduction in the number of workplace injuries and fatalities improves the financial and economic figures for countries that implement health and safety regulations which are associated with the reduction in the number of injuries, illnesses, and fatalities. Societal costs related to occupational injuries and illnesses include wage replacements for injured workers, increased medical insurance costs, and additional costs associated with social support structures related to work-related injuries and illnesses (ILO, 2012b).

Data on injuries, illness, and fatalities in EDCs are much less available. In one study on Mexico, researchers looked at workplace injuries that were treated in medical centers operated by the Mexican Institute of Social Security. The study showed an injury rate of 2.9 per 100 workers with an average incident cost for medical expenses of approximately $2,000 (Carlos-Rivera, 2009). The results of this study could be extrapolated to the remaining Mexican workforce, but it must be noted that 70% of Mexicans are not covered by the Mexican Institute of Social Security, and they would likely not report injuries or receive treatment at the same rates and costs.

The basic problems about collecting injury, illness, and fatality data in developed countries seem to be exacerbated in EDCs where there are poor communication infrastructure, low education, and poor governance. Due to poor associations between diseases and potential occupational exposures made by healthcare systems, policy makers, and public health officials, even more go unidentified in developing countries. Not only are informal workplace injuries likely to be unreported, but also much higher percentages of workers are in the informal workforce in low economic countries. And perhaps most importantly, lower economically performing countries have weaker and less comprehensive reporting systems and methodologies and enforcement, so fewer injuries are reported in even the formal sectors (ILO, 2013).

8.4 GENERAL SURVEILLANCE OF WORKER HEALTH

Worker health surveillance is a strategy that uses medical evaluations and screening to determine whether workers are capable of safely performing assigned tasks, and not susceptible to potential hazards in specific work conditions (Alli, 2008). A secondary activity of health surveillance is the identification of workplace exposures that are having detrimental effects on worker health, and the development of subsequent exposure control plans and worker health treatment plans (Halperin, 1996).

Pre-employment medical screenings provide a baseline of new employee health and identify preexisting medical conditions. This can be valuable information for an organization moving forward, and specific examinations can be performed to quantify and document the current status of worker health when work begins. For example, a health-care worker can be tested for tuberculosis or hepatitis prior to assignment. This would allow for treatment of the worker and elimination of the potential to spread the disease in the hospital where they are working to other workers or patients. Another example would be the use of pre-assignment audiograms to document a worker's hearing levels when they begin work, to have a baseline to compare against later measurements.

ILO Occupational Health Services Recommendation No. 171 of 1985 says that health surveillance for workers should include health assessment prior to assignment to hazardous tasks, health evaluation at periodic intervals related to hazardous work, health assessment upon the resumption of work after a prolonged absence, and health assessment at the termination of assignments involving hazardous work. The results of health assessments should never be used to refuse employment or dismiss workers who are found to have certain conditions, but rather used to identify illnesses and provide the basis for future treatments and protections for the worker and others. Nor are health assessments a replacement for comprehensive worksite hazard evaluation and control programs (ILO, 1985).

The determination of particular workplace hazards and effective surveillance and control involve active communication and participation of three groups: occupational hygienists, occupational health doctors and nurses, and workers. Occupational hygienists must accurately evaluate, quantify, and report the workplace risks from various exposures to chemical or physical agents. It is also the hygienist's job to report the risks to management and take action to control worker exposures to the hazards so as to minimize the risks. The occupational hygienist must communicate the results of risk evaluations to occupational medicine physicians for consideration in their surveillance activities for the workers. The occupational medicine physician should be familiar with potential health symptoms and outcomes from worker exposures to the hazard. Occupational health physicians and nurses must be able to tie symptoms together with exposures and effectively treat the injury or illness. The workers must be aware of the potential health consequences to the workplace exposures and be able to recognize hazardous exposure conditions, and possible symptoms of overexposure, and where and when to report potentially work-related symptoms. Working together with the occupational hygienists, occupational medicine physicians and nurses, and the workers can develop strategies to minimize future exposures and protect the workers further, when necessary.

For certain workplaces, medical surveillance can be used to identify and quantify worker exposures to particular hazardous agents. Biological Exposure Indices (BEIs) have been developed for numerous chemicals. These chemicals or their metabolites may show up in an exposed worker's blood, urine, hair, breath, and other body parts or excretions that can be used to quantify the worker's exposure at the jobsite. Lead concentrations in urine are one example. Thyroid scanning for gamma rays is commonly used in nuclear power plants to determine whether workers had been exposed to radioactive iodine. Again, working as a team, the knowledge gained from BEI results can be used to identify where workplace exposures may be occurring, and how to better control them. And workers can use the information to better understand that increased vigilance on their part, perhaps in better use of PPE, may be appropriate.

At the national level, policies and mechanisms for recording occupational disease should be created in all branches of economic activity, to provide consistent and reliable information through established channels. Responsibilities for collecting and analyzing data by competent authorities need to be assigned in laws and regulations. Data sources may include death certificates, hospital discharge records, and other disease registries. Accurate data on occupational exposures to a broad variety of hazardous agents including biological, chemical, physical, and psychosocial are also needed. Further, mechanisms need to be coordinated by which various authorities and institutions can access data to conduct research and coordinate such services as social support and health care. A sample organizational structure for a national health surveillance network system is provided in Figure 8.2 (ILO, 2013).

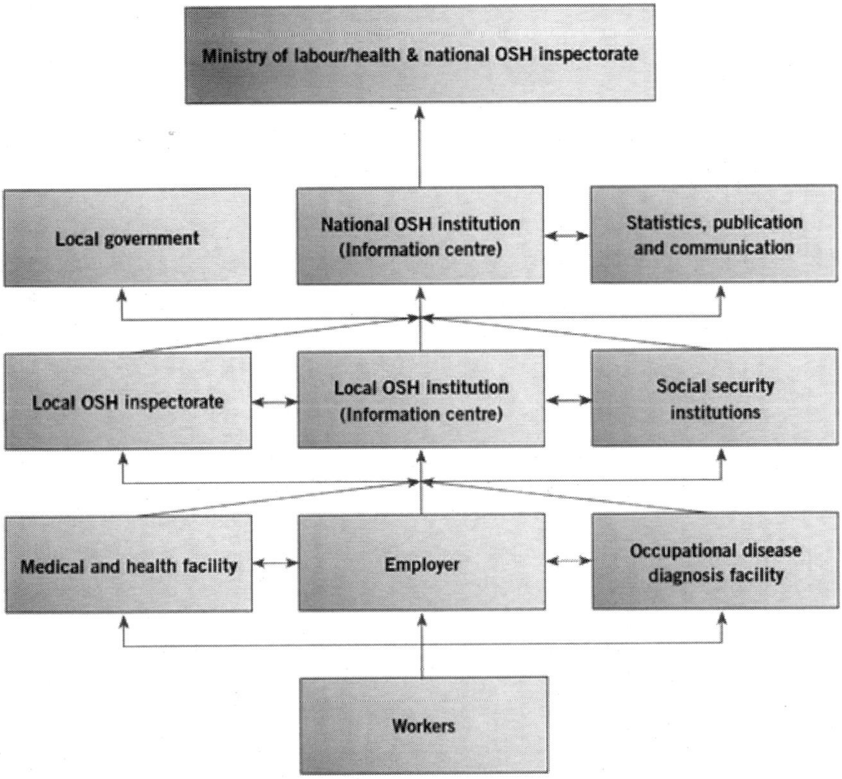

FIGURE 8.2 A model mechanism for recording and notification of occupational diseases. (Courtesy of ILO, 2013. With permission.)

8.5 WHAT IS WORK AND WORK RELATED?

One of the main problems in attaining consistency in measures of risk when comparing different work processes is the determination of when the individual is "working," and if they were exposed to the hazard during work. Perhaps the simplest observation is a worker painting an object on an assembly line. They are clearly at work in the factory at the post. And they can be seen applying the paint, so they are being exposed to the vapors to some extent. And they are also exposed to some musculoskeletal risk since they are standing/or sitting and moving their arms to dip the brush and paint the object. All that remains is a measure of duration on the job doing these tasks and perhaps a concentration of the paint vapors they are exposed to for an assessment of the risk from the chemicals.

The abovementioned example paints a simple picture of a workplace and exposures. But many workplaces are not so well defined. A huge portion of work globally today is in the informal sector, meaning it is not in a traditional enterprise setting. Not only are these types of small businesses less likely to identify injuries as work related, but they are also less likely to record and report them. Some examples might be a childcare center being run at someone's home. If a child bites one of the assistants, and the bite gets infected, would this be likely to be reported? Or if a person is working at home as a call center support staff answering the phone or making telemarketing calls all day, would they be likely to report wrist or neck pain from typing and holding the phone all day? And if they got up to take a restroom break and tripped on the way there and broke their arm, would it be work related, as it might be in a factory?

A direct way to determine work relatedness is to ask whether the person was doing work when they were exposed, and did the exposure have a direct impact on the initiation or progression of the injury or disease? If the answer is "yes" to both questions, then it would be considered work related and should be recorded as such. In general, when a worker is on company property, they are considered to be working. So an employee who slips and breaks their wrist in the parking lot on the way into the office would usually be counted as work related.

In most cases, injuries that occur while commuting on public roads, or other transportation, to and from work are not considered work related. However, injuries that occur as part of a job and traveling would be considered work related. These might include traffic accidents with truck drivers, physical assaults of taxicab drivers, or a bicycle courier who has an accident while delivering a package. Different countries are likely to have different definitions of when work begins which are important to consider. The definitions of "work" and "workplace" become even more vague when activities in the informal sector are included (Nelson, 2005).

8.6 FATALITY REPORTING

The simplest and most accurate workplace statistic is the number of fatalities from accidents. This is of course because it is the easiest to quantify and document in the case of workplace accidents that happen in a discrete time period. There are fewer gray lines, as a fatality is easy to measure and document. Deaths of workers, even in informal settings, are typically recorded at the hospital. And if not in a hospital, even in EDCs, a medical practitioner, or coroner, is commonly responsible for acknowledging a death and determining and recording a cause.

Deaths are much more difficult to identify however, when workplace exposures cause diseases with long latency periods. Many workplace exposures that cause disease, and eventually death, are never identified as workplace related. As a result, many workplace exposures and hazards are greatly underestimated (Hämäläinen, 2006; Takala, 2014). One example would be worker exposures to bloodborne pathogens in health care. A worker may be exposed to hepatitis B or hepatitis C and not become ill or symptomatic, even die from the disease, until many years after exposure, when they may no longer work in health care. In addition, the individual may have been exposed through other, non-workplace pathways, potentially resulting in a misdiagnosis that the condition is not work related. As a result, many healthcare workers who die of liver cancer many years after a workplace exposure to hepatitis B are not identified as work related, burying the true significance of these workplace exposures. Another example includes exposure to workplace chemicals that are carcinogenic with long latency periods. It can be difficult to trace a workers' cancer death to a workplace exposure, when exposures were not measured or recorded while the work was being performed, especially when the work was performed many years earlier (IARC, 2017).

In recent years, the use of Disability Adjusted Years of Life (DALYs) has expanded to allow comparison between injuries/illness and exposure, in limited terms. By weighting the estimated number of years living with a disability, a relative severity can be estimated. The DALY represents the gap between a normal standard life expectancy in perfect health and the morbidity caused by exposures in the workplace (Driscoll, 2005; Nelson, 2005; Ezzati, 2004). DALYs represent the present value of the future years of healthy life lost due to injury or illness, plus the future years of life lost due to premature death (Murray, 1994, 1997).

In one systematic evaluation of the global burden of disease for the years 1990–2016, risk factors were identified for a variety of workplace exposures and risk factors. CRA was used to quantify and compare risks of a variety of exposures to carcinogens, infectious agents, and physical hazards, and relate these exposures to DALYs for a broad range of injury, illness, and fatality outcomes (Gakidou, 2017). Using epidemiological models, population attributable risks can be identified for a specific type of disease and exposure scenario for a given country. Global burdens of disease, including those attributable to occupational exposures, can then be determined and reported (Bikbov, 2014).

8.7 INJURY AND ILLNESS REPORTING

As stated earlier, accurate accident and injury reporting can be a valuable tool for organizations to identify and correct problem areas and processes that are associated with accidents and injuries. Correction of deficiencies can improve profits and worker morale. Nations should have an interest in generating the same sort of data as a means to improve the health of workers, reduce injuries and illnesses that contribute to hindering overall national economic advancement, and become a burden on social and health-care systems. Associations between reporting injuries and illnesses and improved economic outcomes have been routinely demonstrated in many developed nations (UK HSE, 2017). Less current information is available to show benefits in EDCs, but it should be expected that outcomes and benefits would be comparable (ILO, 2012b).

With better national data and recording systems, it becomes possible to identify particular industries or regions with increased injury and illness rates. It would indicate where more advanced exposure assessment or possible exposure reduction controls would be useful. More detailed data could be used to identify particularly hazardous tasks or agents within an industry or region. Injury and illness reporting data may indicate where additional regulations may be needed, or where enforcement of existing regulations may need to be expanded. The data may also support the development and implementation of advanced or special training for particular industries or hazardous agents or activities.

In EDCs, injury and illness data may come from either worker compensation databases or government agencies. Worker compensation systems can be accurate sources of information, as they are tightly bound to the workers' injuries, and time away from work. These can be useful, accurate, and quantifiable measures over time. Government reporting can also be accurate, when required reporting methods and measures are explicit and enforced regularly and consistently.

Problems with injury and illness reporting stem from a variety of inherent weaknesses in data collection systems. These may include vagueness or difference in reporting timeframe requirements or injury and illness terminology. Variability in the collection of data between worker compensation systems can make it difficult to make broad comparisons of working conditions and injury/illness rates. The population attributable risks approach using multifactorial analysis of data to estimate the number of incident cases attributable to occupational exposures can be useful, but typically the lack of baseline data on the incidence of the disease in the working population or the general population can make it difficult to draw conclusions.

Other weaknesses with existing reporting systems include the lack of accurate and quantifiable worker exposure data to associate with the health outcomes. Relatively accurate data are generally available in some industries, such as exposure to noise and the resulting hearing loss. But an exposure–outcome relation can be much more difficult in other industries where possible exposures can come from other community settings, such as tuberculosis or hepatitis B in health-care workers. Or where the disease latency period is extremely long, such as cancer in radiation workers. In industries or jobs where there are long latencies and community exposures, it is especially difficult to identify the disease as work related, such as in occupational exposure to ultraviolet radiation in construction workers and occurrence of a malignant melanoma.

Many nations' accident and injury reporting schemes use their own recording systems that are not easily comparable to other countries (Takala, 2014). Many worker illness reporting directories do not include respiratory diseases, infectious agents, heart disease, or cancer. Inconsistencies in data collection make accurate comparisons between nations difficult. The inaccuracies in the estimates of injury and illness cause the global burden of occupational disease to be underestimated (Driscoll, 2005).

Reporting systems may be exceptionally deficient in several other areas. Some national injury and illness recording systems exclude large segments of the population. In the United States for example, two million workers in government jobs and the military are summarily excluded from reporting (US OPM, 2018). Also in the United States, most agricultural workers, including migrant workers, are not included in injury and illness federal reporting. In one U.S. study, it was determined that as many as two-thirds of all occupational injuries may go unreported (Leigh, 2004; Boden, 2008). This is similarly true in numerous other countries. In many less developed countries, more than 50% of the working population is in the informal sector and these workers are often excluded from official labor force injury and illness estimates (Giuffrida, 2002). According to a study by Probst (2008) the rate of eligible injuries that were not reported to OSHA was 10.9 injuries per 100 employees. In addition, in companies with poor safety climate up to 81 percent of eligible injuries went unreported. In the United States, businesses with fewer than 11 employees are not required to report worker health statistics. EU member states do not do much better, and studies have shown that only between 40% and 50% of non-fatal occupational accidents are ever reported (Hämäläinen, 2009). Underreporting leads to gross inaccuracies in actual national and industry morbidity rates.

In the European Union, an accident is defined as a discrete event that takes place during work and leads to physical harm. This includes cases of acute poisoning and willful acts by other persons such as violence, as well as accidents occurring during work but not on the company's premises. EU definitions of accident exclude deliberate self-inflicted injuries, accidents on the way to and from work (commuting accidents). This includes cases of road traffic accidents in the course of work (EU, 2001 DG employment).

There may be numerous reasons for the lack of reporting injuries and illnesses. In a study by Azaroff (2002), a variety of key factors were identified to include the following: workers not reporting due to fear of reprisals by supervisors or peers, workers not wanting to miss work and loss of pay, lack of insurance for medical care, gaps in coverage or time delays in worker compensation, economic incentives for employers not to report and not to make insurance claims, and the inability of both workers and health-care practitioners to relate illnesses to workplace exposures. Workplace causes for illnesses are particularly difficult to identify, and some studies have estimated that for every fatal accident at work, there are approximately ten deaths caused by work-related diseases (Leigh, 2000; Steenland, 2003).

8.8 SELF-EMPLOYED REPORTING

In light of the large numbers of workers who are either self-employed or work in the informal sector, special attention should be paid to the development of education and reporting mechanisms to collect data on injuries, illnesses, and fatalities in this broad and diverse group. National policies and programs should be developed to provide training to this large workforce segment regarding the definitions of occupational illnesses and diseases, and why it is in everyone's interest that they be reported from all workplaces. Resources should be committed to ensuring the self-employed know what to report, when, and to whom. Methods to report should be made as simple as possible to ensure that the maximum number of notifications is made and that there are no negative consequences to the individuals or organization making a report.

The provision of socially provided health services or worker compensation for the injured or ill self-employed could be one incentive for these workers to report workplace injuries or illnesses. Part of this service would presumably include an assessment of worker exposure, even if only performed post-injury. Various interview and case analysis methods could be used to reconstruct exposures and doses. As more information is collected and disseminated, workplace risks can be better identified and presumably better controlled to prevent future worker exposures and resulting injuries.

8.9 ETHICAL ISSUES

Regardless of the details of the structure of the reporting system, there are certain ethical standards of practice associated with the collection of information and the rights of the workers. Laws or wording may vary from country to country, but the basic rights should include at a minimum:

- The right to confidentiality of personal and medical information
- The right to full explanation of the purposes and results of the surveillance
- The right to refuse invasive medical procedures
- The results of surveillance cannot be used to discriminate against the worker
- The right of the worker or their representative to observe and analyze workplace exposure data related to the worker's assignments (Alli, 2008)

8.10 INTERNATIONAL LABOR ORGANIZATION

The ILO Code of Practice, Recording and Notification of Occupational Accidents and Diseases provides fundamental guidance for the development of reporting programs (ILO, 1995).

Basic ILO Codes of Practice most related to OSH include the following:

- Occupational Cancer Recommendation (No. 147)
- Occupational Safety and Health Convention (No. 155)
- Occupational Health Services Convention (No. 161)
- Recommendation (No. 171)
- Asbestos Convention (No. 162)
- Chemicals Recommendation (No. 177)

The ILO Codes of Practice provide internationally accepted definitions and requirements for reporting, and also provide countries with ready-to-use tools for developing programs and collecting data. The ILO suggests that nations use the ILO definitions within their national legally binding legislation on occupational injury and illness reporting to ensure a maximum level of accuracy and consistency within nations and to provide a broad platform of data for international comparisons. Just as international companies may use injury and illness data to benchmark against competitors, nations could use national data and statistics to evaluate themselves against other nations in terms of worker protections, and economic advancement, as it relates to occupational injuries, illnesses, and fatalities.

As an example of the variation in injury, illness, and fatality statistics between countries, Table 8.1 demonstrates the fatal occupational injury rates per 100,000 workers for the manufacturing section for several different nations.

In just this short excerpt from data published by the ILO for various countries, we can see wide and unexpected variation in the values. In a modern developed country like Switzerland, we might have expected that the fatality rates would be lower than for a less advanced country like Bulgaria, yet the fatality rates in Switzerland are double those of Bulgaria. It is difficult to ascertain how and why the numbers are so different.

Employer-based record-keeping systems should be designed and required at the national level to ensure that injuries, illnesses, and fatalities are reported to a competent authority. The rules and definitions should be explicit to provide as much guidance as possible in order to improve consistency and usefulness of the data at

TABLE 8.1
Injury, Illness, and Fatality Statistics per 100,000 Workers per Year for a Sample of Nations

	Years		
Nation	**2013**	**2014**	**2015**
Australia	1.3	1.5	1.3
Canada	2.5	0.3	0.4
Switzerland	12.3	9.9	6.9
Turkey	7.7	6.2	2.3
United States	2.1	2.3	—

Source: Adapted from ILO, 2018.

the national level. At a minimum, employer-based recording and notification systems should include the following (ILO, 2013):

- Standard forms for recording and notification
- Detailed requirements for what must be reported
- Procedures for making notifications
- Requirements for the validation and verification of recorded and reported cases
- Identification of responsible individuals
- The use of technology in the system
- Training and education regarding recording and notification

The ILO has played a key role in the harmonizing of international definitions and designations of occupational diseases as a means to identify and control them in the workplace since 1925 (Kim, 2013). The ILO lists are added to and updated on a regular basis, when needed. Recent additions to occupational illness designations include expanded definitions of occupational cancer and the addition of musculoskeletal diseases. Inconsistencies in national reporting definitions of disease remain, however, and make an effective international comparison of global injury and illness rates difficult.

8.11 VARIOUS NATIONAL REPORTING SYSTEMS

Reporting methods vary greatly between countries. In the ILO, occupational injury refers to a death or any personal injury or disease that results from an occupational accident. Occupational accidents are events that arise out of or in the course of work that results in fatality or nonfatal occupational injury. The criteria for nonfatal injuries are those where the employee missed 3 or more days of work (74) (ILO, 1996). This methodology makes a comparison to the U.S. Bureau of Labor Statistics methods used in the United States difficult since they record injuries and illnesses by the number of days lost, and do not consider the sudden nature of the illness *per se*.

In 2013, Great Britain updated and expanded requirements for health and safety reporting. These regulations provided detailed definitions of terms to be used in injury and illness reporting. This document requires any workplace to report injuries that occur to workers who are incapacitated and out of work for three consecutive days (including the day of the accident) within 15 days of the accident. Fatalities that occur within one year of a workplace accident must be reported, including those that result from a biological exposure. An exception to reporting of a fatality is made when the deceased is self-employed. And the regulation includes a list of work-related diseases that might result from physical (e.g., carpal tunnel syndrome) and chemical exposures (e.g., cancer) that are to be reported (UK, 2013). The requirements also apply to certain workplaces and activities that occur outside of Great Britain. The UK reporting document requires that injuries and illnesses in workers working offshore must also be included in reports.

In Finland, accidents to farmers and self-employed workers are not included in country statistics. In the United States, employers with fewer than 11 employees, small farms, and federal government workers are not included in national injury and illness statistics. Also in the United States, state and local government workers in 27 states are not included in national injury and illness reporting (Ruser, 2008). Although fatality rates are usually easier to calculate, in many countries accurate estimates of the population do not exist, so it is difficult to calculate national fatality rates (UN, 2017).

A comprehensive study of reporting in New Zealand demonstrated that OSH reporting was often incomplete, inconsistent, unverifiable, and inaccurate (Brown, 2005). It was determined that additional government oversight and direction for the collection of data was needed in order for substantial improvements for injury illness reporting to be made. Voluntary corporate reporting tended to be insufficient and incomplete, and reduced the usefulness of the data at the national social level.

In France, all employees are covered by compulsory occupational medicine. Occupational physicians are familiar with workers and their risks of exposure to hazards in their workplaces. Occupational physicians have an obligation to report work-related diseases, but only 33% of illnesses are reported (Valenty, 2012). And data that are collected by physicians are not useful for epidemiological purposes due to inconsistencies in data collection.

One author estimated that in Malaysia, only 7% of actual occupational accidents were reported. And in sub-Saharan Africa, where there were an estimated 54,000 fatal accidents per year, 0% of the estimated number was actually reported. In the Middle East, only 0.9% of occupational accidents are reported (Hämäläinen, 2006).

In a 2-year South African study of work-related fatalities, it was found that only 15% of the deaths had been reported (Schierhout, 1997). Reporting in rural areas and on farms was particularly poor.

8.12 CONCLUSIONS/RECOMMENDATIONS

On a global scale, CRA estimates for the global burden of occupational work-related deaths are probably significantly underestimated due to unavailability and inconsistency of data. Additionally, work-related diseases are probably also underestimated, stemming from problems recognizing or quantifying occupational exposures to causative agents (Driscoll, 2005).

Due to wide variations and discrepancies in the measures and methods to collect and analyze data used by different countries, comparisons are difficult to make. Extrapolation of morbidity between countries is also difficult due to gross differences in such factors as community health, social support structures, healthcare systems, and environmental exposures.

Research on injury and illness reporting is lacking in developing countries. Accurate measures need to be standardized to ensure consistency and the ability to extrapolate study results to other industries, working conditions, and countries. Linking financial and economic benefits to accurate injury, illness, and fatality reporting systems can become a major incentive for any developing countries to track these events and make fundamental improvements to occupational safety and health at the national level. The ILO has identified several detailed areas where reporting should be improved in all nations and developing countries in particular. Data to be collected necessary to raise OSH development policy include the following:

- Credible estimates of the incidence rates of occupational injuries and, if possible, illnesses
- Disaggregated according to major industry and occupational categories
- Plausible estimates of the costs to employers of occupational injuries and illnesses, taking into account relevant absenteeism and presenteeism multipliers
- Credible estimates of the direct costs to households of occupational injuries and illnesses, including medical expenses and lost income
- Plausible estimates of the indirect costs to households, including lost household production
- Lost market work from unpaid caregivers and detrimental effects on education
- Credible estimates of the role of unanticipated occupational morbidity and, if possible
- Mortality shocks in the propagation of poverty and
- Credible estimates of the costs socialized by government programs not funded by contributions tied to injury or illness events (ILO, 2013)

There is a need for additional survey tools globally to collect data on injuries and illnesses that occur in the informal sectors. Methods need to be developed and piloted to identify weaknesses and make necessary improvements. Then, larger studies using these data collection tools can be used to collect data and make reports. As the methods become more standardized, they could be used to make associations with other industries and informal workers in other countries. The creation of an informal sector reporting tool, for even developed nations, could eventually be modified and used in EDCs.

Since such a large number of injuries and illnesses go unreported by employees, special emphasis and training by employers and the government is needed to encourage workers to report. Young workers in particular often downplay their injuries and tend to blame themselves and thus not report injuries (Tucker, 2014). They need special encouragement and a better understanding of the reasons why reporting is valuable.

With significant problems in data consistency, the emergence of new diseases and occupational risk factors complicates useful data collection even further. As reporting systems are expanded to collect information on new agents such as nanoparticles or human immunodeficiency virus, the databases need to change (Wiatrowski, 2005). Special care must be taken to ensure that the systems not only remain inclusive of original measures for the sake of consistency but also expand to include new diseases for the sake of comprehensiveness and accuracy. Even medical treatments change over time and need to be included in reporting schemes. For example, a new bandage design might require a modification to what is considered first aid. There is also the potential to just "improve" existing systems to be more thorough, such as the inclusion of such characteristics as "race, gender, and ethnicity."

Accurate reporting of injuries, illnesses, and fatalities is closely linked to the improvement in occupational safety and health at corporate levels (Suan, 2017). Communication of issues and risks helps improve the bottom line and even helps to make workers more aware and responsive to OSH initiatives. The inclusion of OSH reporting in corporate social responsibility reporting schemes would be a valuable tool for employers, employees, and community stakeholders (Brown, 2005). More research and development should be done in this area.

REFERENCES

Alli, B. (2008). *Fundamental Principles of Occupational Health and Safety* (2nd ed.). Geneva: International Labor Organization. ISBN 978-92-2-120454-1.

Azaroff, L., Levenstein, C., Wegman, D., Occupational injury and illness surveillance: Conceptual filters explain underreporting, *American Journal of Public Health* (2002) Vol. 92, No. 9, pp. 1421–1429.

Bikbov, B., Perico, N., Remuzzi, G., Mortality landscape in the global burden of diseases, injuries, and risk factors study, *European Journal of Internal Medicine* (2014) Vol. 25, pp. 1–5.

Boden, L. I., Al O., Capture–recapture estimates of nonfatal workplace injuries and illnesses, *Annals of Epidemiology* (2008) Vol. 18, pp. 500–506.

Brown, J., Butcher, F., Reporting on occupational health and safety in annual reports: A look at disclosure practices in New Zealand, *New Zealand Journal of Employment Relations* (2005) Vol. 30, No. 1, pp. 1–22.

Carlos-Rivera, F., Aguilar-Madrid, G., Gómez-Montenegro, P., Juárez-Pérez, C., Sánchez-Román, F., Durcudoy Montandon, J., Borja-Aburto, V., Estimation of health-care costs for work-related injuries in the Mexican institute of social security, *American Journal of Industrial Medicine* (2009) Vol. 52, pp. 195–201.

Driscoll, T., Takala, J., Steenland, K., Corvalan, C., Fingerhut, M., Review of estimates of the global burden of injury and illness due to occupational exposures, *American Journal of Industrial Medicine* (2005) Vol. 48, pp. 491–502.

Drudi, D., The quest for meaningful and accurate occupational health safety statistics, *Monthly Labor Review* (2015) Vol. 138, pp. 1–19.

Embleton, K., Jones, D., Engel, B., Comparative risk assessment primer, *Environmental Software* (1996) Vol. 11, No. 4, pp. 203–207.

Erickson, L., Williams-Evans, S. A., Attitudes of emergency nurses regarding patient assaults, *Journal of Emergency Nursing* (2000) Vol. 26, pp. 210–215.

EU, European Union, European Communities (EC)—DG Employment and Social Affairs. (2001). *European Statistics on Accidents at Work (ESAW) Methodology*. Luxembourg: Office for Official Publications of the European Communities.

Ezzati, M., Lopez, A., Rodgers, A., Murray, C. (2004). *Comparative Quantification of Health Risks: Global and Regional Burden of Disease Attributable to Selected Major Risk Factors*. Geneva: World Health Organization. ISBN 92 4 158031 3.

Gakidou, E., Global, regional, and national comparative risk assessment of 84 behavioural, environmental and occupational, and metabolic risks or clusters of risks, 1990–2016: A systematic analysis for the Global Burden of Disease Study 2016, *The Lancet* (2017) Vol. 390, pp. 1345–1422.

Giuffrida, A., Iunes, R., Savedoff, W., Occupational risks in Latin America and the Caribbean: Economic and health dimensions, *Health Policy Plan* (2002) Vol. 17, pp. 235–246.

Halperin, W., The role of surveillance in the hierarchy of prevention, *American Journal of Industrial Medicine* (1996) Vol. 29, No. 4, pp. 321–323.

Hämäläinen, P., Takala, J., Saarela, K., Global estimates of occupational accidents, *Safety Science* (2006) Vol. 44, pp. 137–156.

Hämäläinen, P., Saarela, K., Takala, J., Global trend according to estimated number of occupational accidents and fatal work-related diseases at region and country level, *Journal of Safety Research* (2009) Vol. 40, pp. 125–139.

IARC. (2017). *Press Release No. 251, Latest Data Show a Global Increase of 13% in Childhood Cancer Incidence Over Two Decades*. Geneva: World Health Organization.

ILO, R171 Occupational Health Services Recommendation, 1985 (No. 171) (1985) www.ilo.org/dyn/normlex/en/f?p=NORMLEXPUB:55:0::NO::P55_TYPE,P55_LANG,P55_DOCUMENT,P55_NODE:REC,en,R171,/Document accessed July 27, 2018.

ILO, Fatal Occupational Injuries Per 100,000 Workers by Economic Activity (%) (1995) www.ilo.org/ilostat/faces/oracle/webcenter/portalapp/pagehierarchy/Page27.jspx?subject=OSH&indicator=INJ_FATL_ECO_RT&datasetCode=A&collectionCode=YI&_afrLoop=180225946367458&_afrWindowMode=0&_afrWindowId=i4g92nc2u_1#!%40%40%3Findicator%3DINJ_FATL_ECO_RT%26_afrWindowId%3Di4g92nc2u_1%26subject%3DOSH%26_afrLoop%3D180225946367458%26datasetCode%3DA%26collectionCode%3DYI%26_afrWindowMode%3D0%26_adf.ctrl-state%3Di4g92nc2u_83 accessed March 21, 2018.

ILO. (1996). *Recording and Notification of Occupational Accidents and Diseases*. Geneva: International Labor Organization, ILO. ISBN 92-2-109451-0.

ILO. (2002) The Economics of Health, Safety and Well Being Assessing the economic value of developing an healthy work environment. http://www.ilo.org/wcmsp5/groups/public/---ed_protect/---protrav/---safework/documents/publication/wcms_110381.pdf accessed December 26, 2018.

ILO. (2012a). *Improvement of National Reporting, Data Collection and Analysis of Occupational Accidents and Diseases*. Geneva: International Labor Office. ISBN 978-92-2-126817-8 www.ilo.org/wcmsp5/groups/public/---ed_protect/---protrav/---safework/documents/publication/wcms_207414.pdf accessed March 22, 2018.

ILO. (2012b). *Estimating the Economic Costs of Occupational Injuries and Illnesses in Developing Countries: Essential Information for Decision-Makers*. Geneva: International Labor Office. ISBN 978-92-2-127016-4 www.ilo.org/safework/info/publications/WCMS_207690/lang--en/index.htm accessed March 23, 2018.

ILO. (2013). *National System for Recording and Notification of Occupational Diseases—Practical Guide*. Geneva: International Labor Organization. ISBN 978-92-2-127057-7 www.ilo.org/wcmsp5/groups/public/---ed_protect/---protrav/---safework/documents/publication/wcms_210950.pdf accessed March 22, 2018.

Investopedia, Top 25 Developed and Developing Countries (2016) www.investopedia.com/updates/top-developing-countries/ accessed July 28, 2018.

Kim, E., Kang, S., Historical review of the list of occupational diseases recommended by the international labor organization, *Annals of Occupational and Environmental Medicine* (2013) Vol. 25, No. 14, pp. 1–10.

Leigh, J., Economic burden of occupational injury and illness in the United States, *The Milbank Quarterly* (2011) Vol. 89, No. 4, pp. 728–772.

Leigh, J., Markowitz, S., Fahs, M., Landrigan, P. (2000). *Costs of Occupational Injuries and Illnesses*. Ann Arbor, MI: University of Michigan Press.

Leigh, J., Marcin, J., Miller, T., An estimate of the US governments undercount of nonfatal occupational injuries, *The Journal of Occupational and Environmental Medicine* (2004) Vol. 46, No. 1, pp. 10–18.

Moll Van Charante, A., Mulder, P., Reporting of Industrial Accidents in the Netherlands, *American Journal of Epidemiology* (1998) Vol. 148, pp. 182–190.

Murray, C. J., Quantifying the burden of disease: The technical basis for disability-adjusted life years, *Bulletin of the World Health Organization* (1994) Vol. 72, No. 3, pp. 429–445.

Murray, C., Acharya, A., Understanding DALYs (disability-adjusted life years), *Journal of Health Economics* (1997) Vol. 16, No. 6, pp. 703–730.

Nelson, D., Concha-Barrientos, M., Driscoll, T., Steenland, N., Fingerhut, M., Prüss-Üstün, A., Corvalan, C., Leigh, J., The global burden of selected occupational disease and injury risks: Methodology and Summary, *American Journal of Industrial Medicine* (2005) Vol. 48, pp. 400–418.

Pransky, G., Snyder, T., Dembe, A., Himmelstein, J., Under-reporting of work-related disorders in the workplace: A case study and review of the literature, *Ergonomics* (2010) Vol. 42, No. 1, pp. 171–182, doi:10.1080/001401399185874.

Probst, T. M., Brubaker, T. L., Barsotti, A. Organizational injury rate underreporting: The moderating effect of organizational safety climate, *Journal of Applied Psychology* (2008) Vol. 93, pp. 1147–1154.

Rosenman, K. D., Kalush, A., Reilly, M. J., Gardiner, J. C., Reeves, M., Luo, Z., How much work-related injury and illness is missed by the current national surveillance system? *Journal of Occupational and Environmental Medicine* (2006) Vol. 48, pp. 357–365.

Ruser, J., Examining evidence on whether BLS undercounts workplace injuries and illnesses, *Monthly Labor Review* (2008) Vol. 131, pp. 20–32.

Schierhout, G., Midgley, A., Myers, J., Occupational fatality under-reporting in rural areas of the Western Cape Province, South Africa, *Safety Science* (1997) Vol. 25, No. 1–3, pp. 113–122.

Shannon, H. S., Lowe, G. S., How many injured workers do not file claims for Workers' Compensation Benefits? *American Journal of Industrial Medicine* (2002) Vol. 42, pp. 467–473.

Steenland, K., Burnett, C., Lalich, N., Ward, E, Hurrell, J., Dying for work: The magnitude of Us mortality from selected causes of death associated with occupation, *American Journal of Industrial Medicine* (2003) Vol. 43, No. 5, pp. 461–482.

Suan, A., A mini review on efficacy of safety management systems in construction, *International Journal of Engineering Science and Computing* (2017) http://ijesc.org/upload/4f44f8a2ec9ac3f6c469bfd04ba8d758.A%20Mini%20Review%20on%20Efficacy%20of%20Safety%20Management%20Systems%20in%20Construction.pdf accessed July 27, 2018.

Takala, J., Hämäläinen, P., Saarela, K., Yun, L., Manickam, K., Jin, T., Heng, P., Tjong, C., Kheng, L., Lim, S., Lin, G., Global estimates of the burden of injury and illness at work in 2012, *Journal of Occupational and Environmental Hygiene* (2014) Vol. 11, No. 5, pp. 326–337.

Tucker, S., Diekrager, D., Turner, N., Kelloway, E., Work-related injury underreporting among young workers: Prevalence, gender differences, and explanations for underreporting, *Journal of Safety Research* (2014) Vol. 50, pp. 67–73.

UK, Statutory Instrument 2013 No. 1471 Health and Safety—The reporting of injuries, diseases and dangerous occurrences, into force October 1, 2013 www.ilo.org/wcmsp5/groups/public/---ed_protect/---protrav/---ilo_aids/documents/legaldocument/wcms_388391.pdf accessed March 22, 2018.

UK HSE, Costs to Britain of Workplace Fatalities and Self-Reported Injuries and Ill Health, 2015/2016, Health and Safety Executive (2017) www.hse.gov.uk/statistics/pdf/cost-to-britain.pdf accessed July 27, 2018.

UN, World Population Prospectus—Methodology of the United Nations Population Estimates and Projections (2017) https://esa.un.org/unpd/wpp/Publications/Files/WPP2017_Methodology.pdf accessed July 27, 2018.

US Office of Personnel Management (OPM), Data Analysis and Documentation Federal Full-Time Employment (2018) www.opm.gov/policy-data-oversight/data-analysis-documentation/federal-employment-reports/reports-publications/full-time-permanent-age-distributions/ accessed March 21, 2018.

US OSHA, Occupational Safety and Health Administration, Public Law 91-596, 84 Stat. 1590 §24 (1970) www.osha.gov/pls/oshaweb/owadisp.show_document?p_id=2743&p_table=OSHACT accessed December 9, 2018.

Valenty, M., Homere, J., Mevel, M., Dourlat, T., Garras, L., Brom, M., Imbernon, E., Surveillance programme of work-related diseases (WRD) in France, *Safety and Health at Work* (2012) Vol. 3, pp. 67–70.

Wiatrowski, W., Occupational safety and health statistics: New data for a new century, *Monthly Labor Review* (October, 2005) Vol. 128, No. 10, pp. 3–10.

9 Introduction to Global Occupational Safety and Health Education and Training

Thomas P. Fuller
Illinois State University

CONTENTS

9.1 INTRODUCTION

Millions of people are injured and killed each year while they are working. Many of these injuries and deaths are due to inadequate training or education in occupational safety and health (OSH). As technology becomes more sophisticated, associated hazards become more complex. Educating and training workers, employers, and even governments on hazards and the effective means to control and minimize them is becoming increasingly challenging. As hazardous industries move to and grow in economically developing countries (EDCs), the need for training and education is ever more urgent. Although developed countries have existing education and training systems, improving international harmonization and inter-coordination might lead to consistent curricula to be used globally. Some international training organizations have arisen to provide educational materials and courses in developing nations. Growth of these programs, and the possible use of online courses, may be a way to provide sufficient numbers of educated and qualified OSH workers, managers, and professionals moving into the future.

OSH professionals use education and experience to anticipate, recognize, evaluate, and control workplace risks. Their goal is to provide workplaces free from hazards that can cause injury, disease, or death. They work to prevent exposure to conditions and agents identified and treated by occupational health nurses and occupational medicine physicians. OSH professionals bring expertise in engineering, ergonomics, toxicology, chemistry, and statistics to help prevent diseases and injuries before they occur.

In the past, OSH professionals received training in one of the core sciences and then morphed into their safety roles through experience or additional coursework. In recent years, however, with the expansion of knowledge and sophistication in safety science and industrial hygiene, entire baccalaureate undergraduate programs have become warranted. In addition, advanced graduate degrees have become commonplace. Graduates of doctoral programs in the safety sciences and industrial hygiene have the ability to analyze hazards and develop highly advanced and innovative controls, even as technology advances at a rapid pace on many other fronts.

Unfortunately, the ever-increasing demands for OSH professionals are significantly outpacing the ability of existing educational infrastructure to supply competent and qualified graduates. Workers in developing countries with expanding economies are still subject to ever-increasing OSH risks. The least developed countries are falling further and further behind (Ahasan, 2001). In addition, the governments and organizations in economically challenged countries are often comparatively unaware of the risks of human exposure to toxic industrial agents and industrial operations. In some developing countries, only 10% of workforce activities are covered by any form of regulation or governmental protections for worker safety (LaDou, 2003).

The majority of the world's workers do not meet the minimum safety standards outlined by the International Labor Organization (ILO) or the World Health Organization (WHO) (Goldstein, 2001). Much of the problem is the lack of professionals educated or trained on the basic fundamentals and principles of OSH (Lucchini, 2014). Many of these shortcomings are associated with the lack of qualified and trained professionals either in the government or in the industry (Parikh, 2003). Even in economically advanced countries, there remain a shortage of qualified and competent OSH professionals, particularly in small to midsized enterprises, and for specialty topics.

The costs in terms of human injury, suffering, and death are staggering. Approximately 2.3 million people die from work-related accidents or diseases each year, or about 6,300 workers each day (ILO, 2017). And millions of workers are injured on the job each day around the world. The global economic burden of workplace accidents and disease-related illnesses is estimated at 4% of global gross domestic product each year (ILO, 2017). If it is assumed that many of these injuries, illnesses, and fatalities could be prevented through effective controls, then more trained professionals are needed to analyze workplace hazards and provide protection.

There is a need to determine where the OSH educational shortfalls are most prominent, and how a greater number of well-trained OSH professionals can be added to the occupational health team supply. This chapter attempts to analyze and explain the existing capabilities for OSH education in individual nations and regions. Then, in the next portion of the chapter, currently available global OSH training programs will be described. Finally, existing systems and shortcomings will be analyzed in an attempt to identify some possible solutions to the shortages, and to identify some possible future directions for human resources capacity building in OSH.

9.2 WORKER TRAINING

In recent years, businesses and organizations have expanded operations and supply chains broadly into other countries and regions of the world. Free markets have become less centralized and consumer access to information has brought increased demand for products and materials from distant parts of the globe. Suppliers need to supply the demands of the consumers and expand their product lines in order to remain competitive (Gerhardsson, 1998).

As global markets expand, employers also feel a need to hire and retain the best and most educated and experienced workers. Globally accepted standards of practice for OSH and protection of the environment have been developed that include a minimum criterion for OSH professionals (ISO, 2018; EU, 2002; BSI, 2007). Human resource managers need to be able to find good workers in any area of the world. Employers need to ensure that their workers are continuously trained and updated on the latest methods and techniques just to stay competitive. Workers in global organizations need to have comparable skills across multinational satellite locations. Employers working in or moving to a developing country need to either bring a competent experienced workforce with them when moving to a new geographic region, or hire and train a workforce locally in the new region. Often, a combination of transported and regional workers is used to fulfill all employee quotas for a given project. New workers hired regionally must be trained by the new employer, and as a result, they become more valuable as workers, and demand for them will increase over time.

Multinational organizations and corporations need to identify their human resource needs and then determine how they will fulfill them in the international locations. The levels and types of workers may be stratified locally, regionally, and internationally. Some OSH capabilities may be required at a local level at all sites. Other OSH capabilities and activities may be provided from a central location with periodic worker travel to regional or international sites. Various internal or external sources of training need to be identified to structure and supplement the global workforce.

Studies have shown that highly educated and skilled professionals are often not motivated to seek out all the latest resources available to them, but rather tend to rely on their initial training and past work experiences to make decisions (Schaafsma, 2004). One study of occupational health physicians demonstrated that doctors did not make significant efforts to seek out the latest scientific information regarding treatments and diagnoses of occupational injuries and diseases (Hugenholtz, 2007). Although there are likely exceptions in every case, the tendency to rely on training and experience, and not seek new ways of doing work, is probably generally true for other sorts of professions, including OSH. Educational or certification credentials that require ongoing professional development and training is one method to ensure professionals remain current in their fields.

Even skilled and experienced OSH professionals with access to high-level sources of information need to be shown where the material is, and how to access it and put it to use. Sources of timely information can include online peer-reviewed journal articles, access to library systems, and publications available through the government, tripartite groups, labor representatives, and professional organizations. Websites and webinars can also be a valuable source of timely information. Professionals may need training on the latest topics and issues in OSH and direction on where to access the current standards of practice on a variety of topics.

9.3 OSH WORK ABROAD

The need for OSH educated and experienced professionals has continued to expand globally. But not all of these professionals have the experience, inclination, or soft skills necessary to succeed on international assignments. OSH professionals need additional credentials in order to be effective working abroad. Experience in working and traveling internationally provides some insight into the aptitude of a potential international candidate. These workers need to be sympathetic to other cultures and regulatory systems. The ability to speak the language of the foreign assignment is often a prerequisite. Specialized consulting firms have begun to provide cultural education services prior to and during international assignments to prepare employees and their families for extended assignments abroad.

Personal characteristics and cultural backgrounds of expatriates have been shown to be closely tied with an individual's success and willingness to work in exotic cultures and geographic locations. A person's willingness to travel physically and psychologically is imperative to success on the job when working abroad (Sullivan, 2006; Forrier, 2009).

Companies with continuing global operations need to include experience and other success factors when considering hiring or career advancement of new and existing workers. Prior international assignments are important determinants in the future success in foreign assignments (Jokinen, 2010). In general, employees with international experience tend to have higher individual career competencies and are more effective working in organizations. Global leadership skills improve from boundary spanning activities such as rotational assignments and directed training on global management issues (Mendenhall, 2012).

Foreign language skills play a key role in the success of professionals working internationally (Itani, 2015). The ability to speak a foreign language differentiates workers from other staff and makes them more successful in international assignments. Speaking the language of the country where they are working allows employees to better understand local regulations, cultural norms, and organizational policies or programs. Workers who speak the local language build better workplace relationships and assimilate more easily into the local culture (Itani, 2015).

Companies that are successful in their global business strategies will be those that have strong international OSH programs and staff experienced in global operations (Nunez, 2011). Competent and qualified OSH staff will continue to be in increasing demand; companies with a steady supply of global OSH workers will have a competitive advantage over other organizations (Nunez, 2011). As workplace technologies and operations become more complex and hazardous, the availability of a competent OSH staff that can adequately address the risks becomes even more important.

As operations of an organization expand globally, there is an increased need for effective coordination of OSH programs, policies, and practices. Management systems need to include the means for standardization of OSH practices, when necessary, and communication of program requirements. When local norms, regulations, or environmental factors require changes to OSH practices, the programs need to consider these and be altered accordingly. Existing international management standards such as ISO 45001 can be used to help structure such international OSH programs. The same OSH program activities essential in homeland operations must be considered in international operations. This includes risk assessment, hazard control, training, and change management (Nunez, 2011). The collection, analysis, and interpretation of OSH data that are important in national operations will be equally important, yet more difficult to obtain and manage, in international activities.

Case

A large multinational paint manufacturer recently opened a facility in Malaysia. One of the job tasks performed outdoors required workers to wear rubber coveralls, a rubber hood, full-face respirator, and heavy rubber boots and gloves for about 45 minutes in order to protect themselves from the dust coming from a hopper when it was filled with a plasticizer in power form. When this activity was performed in Michigan, US, outdoor temperatures were seldom over 24 degrees Celsius, so heat exhaustion or strain was not typically an issue. In Malaysia however, it was not uncommon for temperatures to exceed 30 degrees Celsius much of the year. In addition, the relative humidity in Malaysia is often 70 or 80 percent in certain seasons.

How would the risks to the workers performing these tasks change between the two countries? What types of controls might the international organization incorporate to better protect the workers in Malaysia for the risks that you listed? What are some cultural or regional factors or concerns that you might consider in your answers?

9.4 EXISTING GLOBAL OSH TRAINING MODELS

Existing training programs cannot provide the number of health and safety professionals needed to address increasingly sophisticated technologies and associated OSH hazards globally. The need for professionals trained and qualified to recognize, evaluate, control, and mitigate hazards continues to increase (Kromhout, 2002). Additional training is needed at both the technical and professional levels, in both developed and developing countries. In an attempt to fill the void, a variety of international training models and organizations have arisen.

Professional training in OSH is to be encouraged through continuous learning by participation in various professional organizations offering training through conferences, technical courses, and webinars. Students and professionals with limited resources are encouraged to attend conferences abroad when possible or if not, attend webinars online to expand knowledge or awareness. Economically disadvantaged potential students should always ask for a discount as many organizations will reduce the course rates when need is demonstrated. Operators of these courses should also be sure to market these materials to economically challenged regions and potential students by stating that discounts or various scholarships may be available.

9.4.1 Occupational Hygiene Training Association

The Occupational Hygiene Training Association (OHTA) promotes occupational hygiene standards of practice on an international platform. OHTA develops a variety of educational training materials, including texts, slide presentations, work exercises, and assessment tools. These materials are made available online at the OHTA website and are available for training providers, students, and the general public for free. The organization is the result of collaborations between several international groups and depends on the support of occupational hygiene professional volunteers for a significant amount of the work that is completed (OHTA, 2018).

OHTA is a registered charity in the United Kingdom. Operating costs are kept to a minimum through the use of volunteers and are funded by small fees on approved training providers and OHTA examination fees. Sponsors and supporters also provide funding and support various OHTA projects, such as new course development and translations of course materials to other languages.

Many OHTA courses are taught around the world in EDCs by volunteer occupational hygienists as a way to build capacity in countries and regions with shortages of OSH professionals. Many of these areas such as Botswana, Swaziland, Vietnam, Mozambique, and Tanzania have little OSH capacity and little educational development in the subject. OHTA materials in collaboration with other support organizations, such as the International Occupational Hygiene Association (IOHA) and Workplace Health Without Borders (WHWB), go a long way in improving the capacity of OSH and ensuring the health and safety of workers in these regions.

OHTA courses are also offered by training consultants for fees, and used by multinational corporations to train staff globally. The materials represent the latest methodologies and standards of practice in occupational hygiene and are an excellent training resource for any organization. Approved training providers are listed on the OHTA website.

OHTA modules can be used as stepping stones to professional certification by organizations within the IOHA National Accreditation Recognition program. Completion of six intermediate OHTA modules earns an International Certificate in Occupational Hygiene (ICertOH). The ICertOH in addition to a bachelor's degree in subject area topics (e.g., chemistry, engineering, and biology) and work experience can qualify a candidate for NAR examination and certification.

Current OHTA courses offered include the following

- W201 Basic Principles in Occupational Hygiene
- W501 Measurement of Hazardous Substances
- W502 Thermal Environment
- W503 Noise—Measurement and Its Effects
- W504 Asbestos and Other Fibers
- W505 Control of Hazardous Substances
- W506 Ergonomics Essentials
- W507 Health Effects of Hazardous Substances

9.4.2 Workplace Health Without Borders

WHWB is a nongovernmental nonprofit organization founded in 2011, with a volunteer workforce that conducts various training courses globally. The goal of WHWB is to engage volunteers with expertise in exposure and risk assessment, hazard control, and other technical areas of occupational hygiene to donate time to support projects in underserved worker populations around the world. The focus of the organization is on research, infrastructure development, and training to improve worker health in low-income and low-opportunity countries and regions (WHWB, 2018). WHWB members volunteer their time and often teach week-long OHTA technical courses in EDCs. The host country organization provides a course venue, markets the course to local OH professionals, and collects a nominal fee for course attendees, typically up to 30–35 students. The airfare and ground food and accommodations for the trainer are paid by the host from the course proceeds, and the host still typically makes a profit that may support the host organization or local professional societies. WHWB has already provided OHTA and other courses in Mozambique, Tanzania, Vietnam, South Africa, Botswana, and Swaziland. A recent OHTA training course conducted by WHWB in Swaziland is shown in Figure 9.1.

FIGURE 9.1 An OHTA course being taught by WHWB in Swaziland. (Photograph courtesy of Thomas P. Fuller.)

9.4.3 International Occupational Hygiene Association

The IOHA is a nongovernmental and not-for-profit global community of occupational hygienists and professionals who are dedicated to the discipline and application of the inherent principles used to protect workers from hazards to reduce injury and illness (IOHA, 2018). Activities include the development of comprehensive health and safety training materials and holding international conferences to support the exchange of information and ideas. With more than 13,000 chartered safety and health practitioners, IOHA has more certified professionals than any other organization (IOHA, 2018).

9.4.4 International Training Center of the ILO

The International Training Center of the ILO (ITCILO), located in Turin, Italy, provides a variety of courses as a means to improve human resource capacity and capabilities in a broad range of areas, including OSH. In March 2017, the International Training Center (ITC) conducted a 10-day course on inspection of OSH onsite for 3,600 euros, including accommodations. Also in 2017, the ITC conducted a distance learning online course on OSH in eight modules covering a variety of topics. The course tuition costs 3,000 euros (ITCILO, 2017).

9.4.5 Institute of Occupational Safety and Health

The Institute of Occupational Safety and Health (IOSH) is an organization of safety and health professionals working together to try and improve worker health internationally. This group provides fundamental courses in safety management at various locations and times throughout the year. IOSH provides guidelines of ethical conduct for practicing safety professionals and a basic credentialing scheme based on attendance and graduation from specific academic institutions in the United Kingdom and other universities approved by the organization (IOSH, 2017).

9.4.6 World Health Organization

The WHO has a variety of projects and publications on occupational health and safety on a broad range of topics. (WHO, 2013) They also conduct seminars and conferences to discuss current issues and future directions. In the 2007 Plan of Action for the WHO (2007), they identified training of workers and employers on occupational health as a priority (WHO, 2007). The WHO website has a variety of training modules in several languages available to the public for free.

9.4.7 European Network Education and Training in Occupational Safety and Health

The European Network Education and Training in Occupational Safety and Health (ENETOSH) is a network of more than 80 partners in 33 different countries. The goal of ENETOSH is to mainstream OSH into all levels of education and training as a means to develop a culture of OSH injury and illness prevention in society. The network is coordinated by the German Social Accident Insurance (DGUV) Institute for Work and Health (IAG). The organization is governed by a steering committee comprising representatives from network members.

In order to affect all social levels, ENETOSH believes that a lifelong learning approach is integral to effective and lasting OSH education and training. ENETOSH programs cover all areas of education from kindergarten, primary school, secondary school, colleges, universities, and continuing vocational training. Courses and projects result from collaborations between professionals with expertise in all aspects of OSH and education. Target groups for courses include educational, insurance, policy makers, governments, nongovernmental, and tripartite organizations globally.

One of the current ENETOSH projects is the systematic collection and dissemination of good practice for OSH education at all levels. The quality management system for the project includes a list of criteria for selecting examples of good practice, a data collection system, a coding system, program statutes, and a code of practice of users of the ENETOSH platform. Selected examples of "good practice" are available on the ENETOSH website where users can perform searches for certain keywords and selection criteria. The website includes examples of how to mainstream OSH training into course content and curricula at all educational levels.

Another ongoing ENETOSH project is the development of a consensus standard of competence for teachers in OSH. This project focuses on the improvement of competencies of teachers, lecturers, and trainers.

ENETOSH is an open network, and all experts dealing with OSH and education or training are welcome to join. Some of the existing network organizations are shown in the following list:

- European Center for the Development of Vocational Training (CEDEFOP)
- European Map of Intergenerational Learning
- European Network for Workplace Health Promotion (ENWHP)
- European Network of Safety and Health Professional Organisations (ENSHPO)
- The International Network of Safety & Health Practitioner Organisations (INSHPO)
- World Health Organization Global Network Collaborating Centres for Occupational Health

9.5 FORMAL OSH EDUCATIONAL SYSTEMS

OSH educational approaches vary greatly between countries. The profession of occupational hygiene is perhaps most similar between the United Kingdom, Australia, and the United States; however, even in these countries, the training of OSH students is quite different. In Australia, occupational hygiene is taught at the master's level at three universities, and the doctorate is offered at one. In Great Britain, occupational hygiene is also taught at the master's level. In the United States, however, several universities offer bachelor's degree in occupational hygiene, in addition to master's and doctoral degrees. In the United States, the American Board of Engineering Technology (ABET) accredits university programs in a broad variety of areas including occupational health, environmental health, safety, and occupational hygiene. This accreditation is a measure of program effectiveness and other categories (ABET, 2018).

In the United States, the first environmental health program was established in 1913 at the Harvard—Massachusetts Institute of Technology School for Health Officers. Many of the initial advances in occupational hygiene stemmed from the research conducted in these universities including development of ventilation and engineering principles in the protection of workers from airborne hazards, and the advanced understanding of the importance of the respiratory tract as the route of absorption for toxic aerosols. For over a century, the Harvard Department of Industrial Hygiene and other associated divisions such as Environmental Health and Engineering Sciences have continued to make advances in research and education in occupational hygiene, safety, industrial medicine, and other related subjects (Harvard, 2018).

The National Institute of Occupational Safety and Health (NIOSH) has created a program of Educational Research Centers (ERCs) with the goal of providing training and educational programs to build capacity in OSH. The ERCs focus on technological advances, globalization, and new and emerging risks to prepare workers and graduates of ERC programs. The training and associated research are conducted at 18 official ERC sites and numerous other affiliated universities and organizations. The existing NIOSH ERCs are listed in Table 9.1.

European educational approaches to OSH vary considerably by country. Academic programs in OSH in Europe began in the 1970s. These programs did not grow significantly; however, until around the 1980s or 1990s when there was an increased awareness of the value of OSH after several significant industrial accidents in Europe, these programs demonstrated the need for a better understanding of risk

TABLE 9.1
NIOSH Educational Research Centers

University of Alabama at Birmingham
University of California, Berkeley
University of California, Los Angeles
University of Cincinnati
University of Colorado Denver
Harvard T.H. Chan School of Public Health
University of Illinois at Chicago
University of Iowa
Johns Hopkins Bloomberg School of Public Health
University of Kentucky
University of Michigan
University of Minnesota
Mount Sinai School of Medicine
University of North Carolina at Chapel Hill
University of South Florida
University of Texas Health Science Center at Houston
University of Utah
University of Washington

assessment and hazard control. In general, European occupational hygiene programs are less structured than those in the United States. Rather than taught as a discreet field of inquiry, universities in Europe tend to teach OSH in a fragmented manner where various other technological, managerial, and scientific fields such as engineering and medicine curricula include coursework in OSH (Hale, 1997). Students in master's programs generally enter with bachelor's degrees in other relevant sciences such as chemistry, engineering, or physics. In a study of postgraduate courses in OSH in 18 European countries, courses in OSH have been increasing consistently over the past decade. Programs include courses in safety science, occupational health, risk assessment, management, prevention, and environmental science. Programs are typically run under engineering, applied science, or management departments. But despite continuing attempts to harmonize educational curricula and outcomes, OSH programs, course structures, and course content continue to vary (Arezes, 2012).

European higher education of OSH includes a variety of methods. In the United Kingdom, OSH is integrated into risk education and engineering. In Portugal, OSH is integrated into engineering graduate and postgraduate programs. In Austria, safety prevention, emergency response, biological hazards, and radioactive materials are covered in laboratory safety courses. And in Spain, postgraduate courses are offered in the management of occupational risks (Reinhold, 2014). In France and the United Kingdom, safety and occupational hygiene courses occur at the master's levels as part of other sciences and are commonly aimed at risk assessment and reliability rather than industrial hygiene (Arezes, 2012). There are also master's and doctoral programs in ergonomics, human factors engineering, and OHS management.

In Germany, two main branches deliver OSH training. One branch is done by accident insurers and is closely affiliated with safety engineering. The second branch is offered to physicians and advanced training in occupational medicine and is provided by specialized colleges and universities. Unfortunately, at this time, harmonized training and standardized curricula for occupational hygiene are lacking in Germany. Most safety engineers arrive from the bachelor degree university level. The occupational hygiene profession and standards or practice are not well defined in Germany at this time.

In Belgium, occupational hygiene is one of the welfare domains as included in the Act on well-being of workers in the performance of their work (04/08/1996) and the Royal Decree on the policy of well-being of workers at work (27/03/1998). The Internal and External services for Prevention and Protection at work requires employers to have prevention officers and experts with various skills, including occupational hygiene. Occupational hygienists collaborate with other prevention officers in a multidisciplinary way.

There are two training courses in which the Belgian Society for Occupational Hygiene (BSOH) participates substantively. One program is a Dutch specialist training course for prevention officer occupational hygiene that is offered in collaboration with KU Leuven, Odisee, and BSOH. The Dutch course runs over two academic years and is mainly given at the KU Leuven Gasthuisberg campus. The second program offered is a French specialist training course for prevention officer occupational hygiene taught in collaboration with Universite Catholique de Louvain (UCL), Centre d'enseignment superior (CPFB), Ecole Polytechnique de Louvain, the European Confederation of Independent Trade Unions, and

BSOH. The French training course runs over two academic years and is taught at Louvain-la-Neuve-UCL.

The purpose of this training is to prepare students for the daily interpretation of a job as occupational hygienist, based on a thorough theoretical knowledge combined with practice-oriented applications. One learns to identify and assess chemical, physical, and biological hazards at the workplace. For the different topics, a similar course structure is followed, consisting of the following:

1. Organization, legislation, and limit values
2. Agents with specific hazard properties
3. Toxicology and routes of exposure
4. Health effects
5. Measurement possibilities: external and internal doses
6. Preventive measures

The training aims that at the end of the course, one has the required knowledge and skills to take necessary preventive measures in specific work situations, leading to better hygiene and health at work. This is done in consultation with a multidisciplinary team of prevention officers and safety experts (engineers, occupational physicians, ergonomists, psychologists, and environmentalists). This training is meant for all interested parties for prevention and well-being of workers in companies and organizations. The program is conducted in accordance with the prescribed requirements, as specified in the Royal Decree on the expertise of the prevention officers of the external services for prevention and protection at work (5/12/2003). Prevention officers that are experts in occupational hygiene are required to have successfully completed the specialization module "Occupational Hygiene."

In order to meet the upcoming demand for OSH professionals, new and more educational programs need to be developed and provided in a more structured and formalized approach (Arezes, 2013). The growth of internationally standardized models for occupational hygiene training and curricula will help guide the future development of education systems in Europe and other regions of the world. International recognition of OSH education credentials and other professional certifications, like that of IOHA, that require specified training in OSH will also help push the future development of the occupational hygiene training models. As the expectations of governments and global employers become more sophisticated and harmonized in what is expected in OSH staff, educational approaches will need to advance to satisfy the demand.

9.6 OCCUPATIONAL SAFETY AND HEALTH EDUCATIONAL CURRICULA

In a study of potential employers, the capabilities shown to be most important in recent OSH graduates included the following:

- Hazard identification and control
- Understanding relationships between exposures and health outcomes
- Interpretation of laws and regulations

- Ability to describe physical, chemical, and biological aspects of hazards
- Collecting, managing, and analyzing data
- Interpreting and applying scientific findings
- Exposure assessment and risk characterization

In addition, employers preferred to have graduates who have learned how to work in teams, speak publicly to conduct training and make presentations, and communicate effectively in writing (Brosseau, 2009). It is also important for new OSH graduates to have decision-making, personal, management, and social skills in order to be successful professionals.

As OSH professionals gain education and experience, they reach a level where certification by a respected credentialing body is a way to demonstrate expertise in their respective profession (Adams, 2004). The most reputable credentials are those that require graduation from an approved academic program, a minimum number of years of practical work experience in OSH, and passing a comprehensive and rigorous certification examination. In an effort to standardize the credentialing bodies operating around the world, the IOHA has created an evaluation rubric to identify the various comparable OSH certifications internationally. The rubric includes detailed information regarding educational level and course content for related accepted university degrees. The following list represents the credentials currently seen as comparable by IOHA:

- American Board of Industrial Hygiene (ABIH)
- Australian Institute of Occupational Hygienists (AIOH)
- British Occupational Hygiene Society (BOHS)
- Canadian Registration Board of Occupational Hygienists (CRBOH)
- Dutch Occupational Hygiene Society—NVVA
- French Occupational Hygienists Society (SOFHYT) (Document in French and English)
- German Society for Occupational Hygiene (DGAH)
- Hong Kong Institute of Occupational and Environmental Hygiene
- Institute of the Certification of the Figures of Prevention (Italy) (ICFP)
- Japan Association For Working Environment Measurement (JAWE)
- Malaysian Industrial Hygiene Association (MIHA)
- Norwegian Occupational Hygiene Association (NYF)
- Swedish Occupational and Environmental Certification Board (SOECB)
- Southern African Institute for Occupational Hygiene (SAIOH)
- Swiss Society of Occupational Hygiene (SSOH)

9.7 EXISTING LIMITATIONS AND FUTURE DIRECTIONS IN GLOBAL OSH EDUCATION

As technology advances, occupational hazards evolve and become more complex. To some extent, this has led to further compartmentalization of OSH practice. Occupational health physicians (doctors who prevent and treat diseases and injuries that occur in the workplace) who play a significant role in many national OSH regulatory systems, and may have in the past been more involved in workplace assessments, now must spend too much time keeping up with expansion and advances in medicine, such as toxicological consequences of exposure to new chemicals or materials such as nanoparticles, to spend time in the field in physical workplaces. As a result, and in addition, industrial hygienists, ergonomists, and general OSH professionals must now increase their capacities for anticipation, evaluation, and control of hazards in numerous new and old industries.

Due to the growth of industry around the world and particularly in EDCs, formal educational institutions cannot keep up with the need for OSH professionals or qualified graduates from OSH programs. In addition, there is a significant amount of global variability in OSH curricula between academic institutions and credentialing.

Initiatives to provide training at the worker level are even more disparate. Although a large amount of training materials are available on a broad range of topics from such reputable and prestigious agencies such as the WHO and ILO, there is no clear direction or pathway for workers, employers, or governments to follow towards human resource development and ultimately credentialing and certification of OSH workers internationally. Perhaps the closest system is that provided by the OHTA where students take an examination after their courses and receive a certificate of recognition for each course, and a series of course certificates can lead to another level of recognition. Educational certificate programs such as these have been shown to be one of the best ways to develop OSH capacity in an existing workforce (Rosen, 2014). Other than university degrees, the OHTA system seems to be one of the only international programs that include examinations and such certificates along with their courses.

Although many of the ILO and WHO documents are available in multiple languages, most other training materials available online are generally only in English, or sometimes Spanish or French. Perhaps a weakness of this study and analysis is that little research can be conducted in China, Russia, and India without working knowledge of those languages. But this is also a weakness in getting information on western advances in OSH to those countries.

Moving forward, it seems that an optimal area for the advancement of modern OSH principles globally will be through the use of the Internet for training (Van Dijk, 2015). As training tools become more advanced and easier to use, it may be possible for students in Africa to not only access OHTA training modules and materials, but actually take courses and earn certificates online. It seems a logical next step for the advancement of worker capabilities and rights.

There is a great need for further development of OSH programs of study and the harmonization of curricula so as to increase the numbers of competent professionals, and ensure that competencies are more comparable and interchangeable between countries. These improvements could be used for models of expansion globally, to improve the numbers and quality of OSH professionals in expanded regions.

A bit of variety in college curricula and credentialing rubrics will be inevitable moving forward. Each country has different cultures, capabilities, and workplace settings. But hopefully communication between tripartite, professional, educational, and labor organizations will improve, and they will work towards a consensus to clarify the OSH professional role in worker health and safety, and educational requirements. This would go a long way in helping workers and organizations in EDCs continue to improve their understanding and capabilities in OSH.

REFERENCES

ABET, Accredited Program Search (2018) http://main.abet.org/aps/Accreditedprogramsearch.aspx accessed June 25, 2018.

Adams, P., Brauer, R., Karas, B., Bresnahan, T., Murphey, H., Professional certification, *Professional Safety* (December 2004) Vol. 49, No. (12), pp. 26–31.

Ahasan, M., Partanen, T., Occupational health and safety in the least developed countries: A simple case of neglect, *Journal of Epidemiology* (2001) Vol. 11, pp. 74–80.

Arezes, P., Swuste, P., Occupational health and safety post-graduation courses in Europe: A general overview, *Safety Science* (2012) Vol. 50, pp. 433–442.

Arezes, P., Swuste, P., The emergence of (post) academic courses in occupational safety and health: the example of Portugal, *Industrial and Commercial Training* (2013) Vol. 45, No. 3, pp. 171–179.

Brosseau, L., Fredrickson, A., Assessing outcomes of industrial hygiene graduate education, *Journal of Occupational and Environmental Hygiene* (2009) Vol. 6, pp. 257–266.

BSI, British Standards Institute. (2007). *Occupational Health and Safety Management Systems – Specification*. London: BSI OHSAS 18001:1999 and OHSAS 18001:2007.

EU, European Union, European Agency for Safety and Health at Work. (2002). *OSH Systems and Programmes: The Use of Occupational Safety and Health Management Systems in the Member States of the European Union*. Luxembourg: Office for Official Publications of the European Communities.

Forrier, A., Sels, L., Stynen, D., Career mobility at the intersection between agent and structure: A conceptual model, *Journal of Occupational and Organizational Psychology* (2009) Vol. 82, No. (4), pp. 739–759.

Gerhardsson, G., How to boost occupational health and safety's visibility worldwide, *AIHA Journal* (January 1998) Vol. 59, No. (1), p. 6.

Goldstein, G., Helmer, R., Fingerhut, M., The WHO global strategy on occupational health and safety, *African Newsletter on Occupational Health and Safety* (2001) Vol. 11, pp. 56–60.

Hale, A., Kroes, J., System in safety, 10 years of the chair in safety science at the Delft University of Technology, *Safety Science* (1997) Vol. 26, No. (1/2), pp. 3–19.

Harvard University, T.H. Chan School of Public Health Department History (2018) www.hsph.harvard.edu/environmental-health/department-overview/department-history/ accessed online June 3, 2016.

Hugenholtz, N., Schreinemakers, J., A-Tjak, M., Van Dijk, F., Knowledge infrastructure needed for occupational health, *Industrial Health* (2007) Vol. 48, pp. 13–18.

ILO, Safety and Health at Work (2017) www.ilo.org/global/topics/safety-and-health-at-work/lang--en/index.htm accessed August 8, 2017.

Institute of Occupational Safety and Health, IOSH About Us (2017) www.iosh.co.uk/About-us.aspx.

International Occupational Hygiene Association, Our Vision and Mission (2018) http://ioha.net/objectives accessed December 10, 2018.

International Training Center of the International Labor Organization, Homepage (2017) www.itcilo.org/en.

ISO, International Organization for Standardization. (2018). *Occupational Health and Safety Management Systems – Requirements with Guidance for Use*. Geneva, Switzerland: ISO 45001:2018.

Itani, S., Jarlstrom, M., Piekkari, R., The meaning of language skills for career mobility in the new career landscape, *Journal of World Business* (2015) Vol. 50, pp. 368–378.

Jokinen, T., Development of career capital through international assignments and it's transferability to new contexts, *Thunderbird International Business Review* (2010) Vol. 52, No. (4), pp. 325–336.

Kromhout, H., An international perspective on occupational health and hygiene, *International Journal of Occupational and Environmental Health* (2002) Vol. 8, pp. 111–112.

LaDou, J., International occupational health, *International Journal of Environmental Health* (2003) Vol. 206, pp. 303–313.

Lucchini, R., London, L., Global occupational health: Current challenges and the need for urgent action, *Annals of Global Health* (2014) Vol. 80, pp. 252–256.

Mendenhall, M., Reiche, B., Bird, A., Osland, J., Defining the "global" in global leadership, *Journal of World Business* (2012) Vol. 47, pp. 493–503.

Nunez, I., Villanueva, M., Professional practice safety capital: The management of organizational knowledge on occupational health and safety, *Journal of Workplace Learning* (2011) Vol. 23, No. 1, pp. 56–71.

OHTA, Occupational Health Training Association, The Occupational Health Training Association (2018) www.ohlearning.com/about-ohta/purpose-and-principles.aspx accessed June 30, 2016.

Parikh, D., Saiyed, H., Education and training in occupational hygiene in India, *Indian Journal of Occupational and Environmental Medicine* (2003) Vol. 7, No. (1), pp. 11–15.

Reinhold, K., Siirak, V., Tint, P., The development of higher education in occupational health and safety in Estonia and selected EU countries, *Procedia-Social and Behavioral Sciences* (2014) Vol. 143, pp. 52–56.

Rosen, M., Koshy, K., Patel, M., Safety and health certification programs: Practical applications beyond training, *Journal of Safety, Health and Environmental Research* (2014) Vol. 10, No. (2), pp. 162–168.

Schaafsma, F., Hulshof, C., Van Dijk, F., Verbeek, J., Information demands of occupational health physicians and their attitude towards evidence-based medicine, *Scandinavian Journal of Work, Environment & Health* (2004) Vol. 30, pp. 327–330.

Sullivan, S., Arthur, M., The evolution of the boundary less career concept: Examining physical and psychological mobility, *Journal of Vocational Behavior* (2006) Vol. 69, No. (1), pp. 19–29.

Van Dijk, F., Bubas, M., Smits, P., Evaluation studies on education in occupational safety and health: Inspiration for developing economies, *Anals of Global Health* (2015) Vol. 81, No. (4), pp. 548–560.

Workplace Health Without Boarders, About WHWB (2018) www.whwb.org/about-us accessed December 10, 2018.

World Health Organization. (May 23, 2007). *Worker's Health: Global Plan of Action, Sixtieth World Health Assembly*, Agenda item 12.13. Geneva: World Health Organization.

World Health Organization. (April 2013). *WHO Global Plan of Action on Workers' Health (2008–2017): Baseline for Implementation*. Geneva: WHO, www.who.int.

10 Credentialing Occupational Hygiene

Thomas P. Fuller
Illinois State University

CONTENTS

10.1 INTRODUCTION

Broadly defined, credentials are anything that provides a basis for confidence, belief, or credit in the capabilities of someone or something. They may provide a form of status or entitlement. Credentials may be based upon education, experience, or other past actions that indicate the level of a person's commitment, understanding, and capabilities. They often represent a certain level of achievement in personal qualities or capabilities, and they are sometimes supported by documents or certifications awarded by organizations or institutions. It should be noted that many of the terms used and described in this chapter have significantly different meanings and definitions in various fields and professions. Credentialing has a range of meanings from knowledge-based training-based certificates to credentials based on actual practice and experiences.

Over the past several decades, credentialing and licensing of professions have increased significantly in many countries. In the United States, licensing has increased about 30% since the 1950s (Kleiner, 2013). Credentialing required by both governmental and nongovernmental organizations has increased significantly in the past decades and plays a major role in the supply and costs of services in many occupations (Sweetman, 2015). In a 1992 study, there were more than 800 different occupations licensed in at least one state, and more than 1,100 occupations either licensed, certified, or registered (Brinegar, 1992). In the United States, licensed occupations represent between 18% and 38% of the workforce (Kleiner, 2000, 2010).

In the United Kingdom, the percentage of workforce that requires a government license doubled between 1998 and 2010 (Bryson, 2010). In one study of 29 European and eastern European countries, 19 require the appointment of safety professionals, prepared through specific education, training, or credentialing, to protect workers (Hale, 2008).

In general, licensing tends to be more common in more educated workers, union members, and government employees. Professional credentialing can be offered by self-regulatory colleges, boards, or associations. Or they may be provided as a form of licensure by local, state, or federal governmental regulating bodies. In general, as the licensure requirements become more stringent, the benefits become more monopolistic.

Registration is generally considered the simplest form of credential, with often only requiring the provision of basic background information. This may also be referred to as "conformance licensing." Registration may include some basic educational minimum criteria, statements of agreement with ethical principles or codes of conduct, and possible criminal record assessment. In many professions, it is possible to practice the profession without registration, and in many professions, registration is voluntary. Registration in its simplest form involves having one's name placed on a list. At the other end of the spectrum, registration requires having education through an accredited academic program, having supervised work experiences (in some cases also through and accredited program), and passing a standardized examination [e.g., Registered Nurses (RNs), Registered Pharmacists (RPHs), Registered Dietitians (RDNs), and Radiologic Technologists (RTs).]

Higher-level forms of registration often overlap with definitions of "certification," which typically require more stringent educational qualifications, experience levels, apprenticeships, or examinations. Maintenance of professional certifications may require ongoing professional development in the form of experience and education maintenance and associated periodic review by the accrediting body. Certification tends to be implemented and enforced through the control over the use of title (Hemphill, 2016).

Licensure is typically the most stringent form of professional credentialing. A license may be required to perform the given professional activities. Licensed professions may be broad and for fairly simple activities such as taxi driving, or they may be highly specific and complex such as prescribing pharmaceuticals. Licensure may be used as an efficient way to ensure the quality and accuracy of professional services, and as a way to build government revenues. Licensing is commonly used in economically developing countries as a replacement for regulatory oversight because of weak government oversight and inspection of industries (Ogus, 2005).

10.2 DEFINING PROFESSION

Professions can be described as work or vocation that requires certain specific training and experience and follows a particular code of practice in accordance with agreed upon standards of quality and efficiency.

Occupational hygiene is the anticipation, evaluation, measurement, and control of the work environment with the goal of minimizing safety and health risks to workers (Nash, 1953). Occupational hygienists work as part of a team with occupational medicine physicians, occupational health nurses, toxicologists, and industrial engineers to evaluate potential health effects to workers and minimize the associated risks.

10.3 THE GOALS OF PROFESSIONAL CREDENTIALING

We tend to think that the primary goal of professional credentialing, especially in occupational safety and health (OSH), would be to ensure professionals are competent to assess workplace risks and take actions to protect the workers from harm. This is true, but there are many other reasons for registration and licensing that are worth noting.

One of the most pervasively regulated industries in many cultures is for hairstylists. Although one might be able to find someone who died from a bad hairdo, safety is not a typical concern for licensing this profession. Although hairstylists come in intimate contact with the public, and there is some rationale to ensure certain levels of awareness regarding sanitation, other similar industries (e.g., tattooists and body modification professionals) do not have such requirements for licensing. Partly, hairdresser licensing is to ensure minimum levels of quality as a consumer protection for the public. It is also used to ensure ethical codes of practice for stylists that they comply with common fair business practices and codes of conduct for their clients. This concept is worth noting and remembering because even safety professionals need to practice codes of conduct according to professional ethical standards.

Regulations and laws are often created for reasons of safety, health, and protection of the environment. In the private sector, organizations also find value in creating systems to evaluate conformity to enable comparability and ensure competitiveness. Accepted standards of practice are evaluated by conformity assessment bodies (CABs) to conduct inspection, testing, and certification. CAB assessments have become important to societal stakeholders including the public, regulators, businesses, and consumers. Accreditation bodies evaluate the CAB capabilities to perform assessments based on particular standards and normative guidelines, and award accreditations to competent CABs (ISO/IEC, 2004). A conceptual diagram of conformity assessment and accreditation processes is shown in Figure 10.1.

In ancient times, self-enforced guilds and third-party regulations were created to shift business from personal to impersonal transactions, and set requirements for competence and honesty. Regulations were seen as an economic benefit to society as a way to protect the public without clogging the judicial system with suits and torts after the fact. Regulations remain fundamental to modern economies today, and it might be acknowledged that the most advanced societies have the most advanced and numerous regulatory or credentialing systems. The credentials of professionals such as health-care providers, architects, and restaurant inspectors are just a few examples of what it takes to ensure a safe and healthy society.

In a U.S. study of low- to moderate-income occupations that required licensure, the average fee for a license was $209 and it required passing an examination, 9 months of education, and minimum age levels. In many cases, the strength of the licensure requirements was not particularly related to health and safety but instead was related to the lobbying prowess of practitioners in securing laws to shut out competition (Carpenter, 2015).

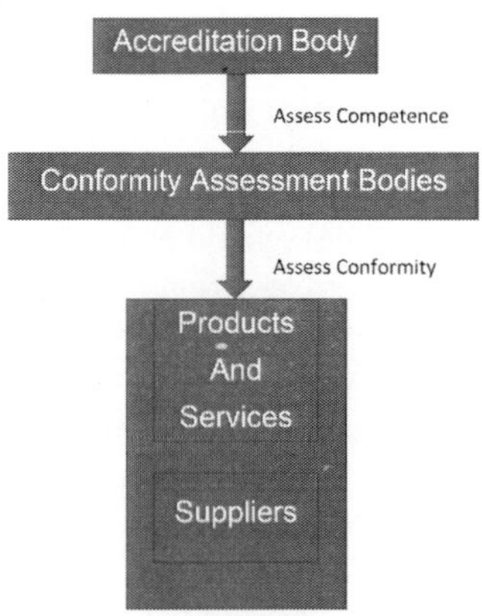

FIGURE 10.1 Conceptual diagram of conformity assessment and accreditation processes. (Adapted from ISO/IEC, 2004.)

CASE

In one study it was shown that emergency medical technicians, who save people's lives, fell behind 66 other professions in terms of licensure requirements. These ambulance workers charged with the responsibility of treating injured patients and keeping them alive on their way to the hospital, had fewer licensure burdens than locksmiths, landscape workers, barbers, and manicurists. By comparison, the average cosmetologist requires 372 days of training, whereas the average EMT only needs 33 days of training.

(Carpenter, 2015).

10.3.1 Job Security and Higher Wages

One of the major reasons that professional organizations and their members appreciate credentialing and licensure is for protection and security of their jobs and positions. Access to the profession by potential entrants can be greatly reduced. Stringent credentialing requirements tend to restrict entrance into a given profession, thus reducing the supply of candidates. Another outcome is the support of higher wages for the profession by about 15% (Kleiner, 2013, 2010; Meehan, 2015; Timmons, 2008; Pagliero, 2013). In a study about de-licensing occupations in the United States, it was noted that attempts were met with stiff resistance by both those professionals who held licenses and the boards who granted the license. Each group found the license valuable in terms of finances and status (BLS, 2015).

Higher wages for licensed workers may be partly due to the monopolistic nature of the profession and barriers to entry that reduce supply. But it can also be argued that the added educational and experiential requirements for license holders ensure higher quality overall for those workers holding a license in a given profession. In a study by Kleiner (2010), it was demonstrated that occupational licensing is positively associated with higher educational levels as shown in Figure 10.2.

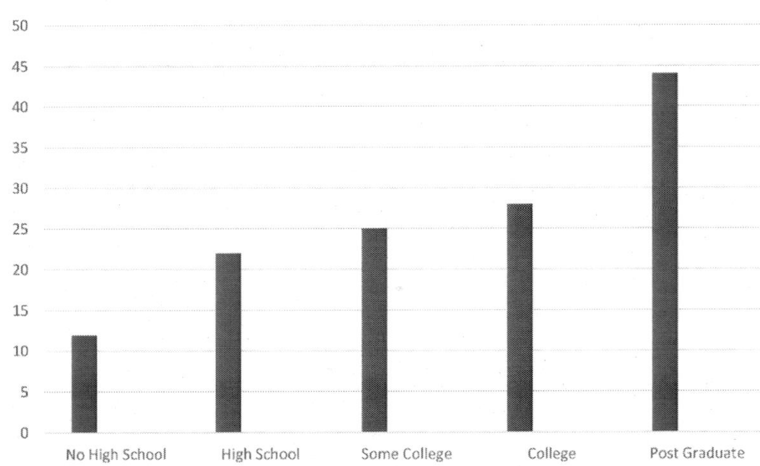

FIGURE 10.2 Percentage of professionals with occupational license by education. (Adapted from Kleiner, 2010.)

Despite professionals with higher educational levels, it could be assumed that the reduced pool of available professionals in this field would also then reciprocally increase costs to clients and employers.

In general, as the demand for professionals with certain credentials increases, the salaries to those professionals will increase. That said, the number of students enrolling into associated educational programs, or people entering the profession, would also be expected to increase. Salaries would be expected to adjust continually.

10.3.2 Clarification of the Profession

Credentialing can be a way to delineate the role and activities of a given profession. It can help clarify what the professionals are expected to know and be able to do. Credentials that are designed for public protection codify the role and activities of the profession. True credentials should be based on what professionals are doing currently through a job-task analysis process. A person gains the work experience and earns the associated credential. It is a way to communicate with the public and provide information on what should be expected for various professionals and what they should be able to do. This can also be useful for government officials. Through credentialing publications and professional meetings, the roles and actions, such as ethical responsibilities, of professionals in the field can be determined and clarified for its members.

10.3.3 Documentation

Credentialing and licensure programs can improve collaboration/identify transferable skills across professions/better connections/clarify responsibilities. In Australia, the development of a national accreditation body for several major fields such as medicine, nursing, engineering, and accounting has greatly increased the ability of workers to bypass regional requirements and migrate to different states to fill worker shortages (Hawthorne, 2011).

Within a profession, credentialing can be a means to ensure worker readiness for the job. Benchmarks of experience and education can be clearly delineated. Both foundational- and occupation-specific skills necessary to complete tasks expertly and safely can be identified within credentialing constructs. At the higher end of credentialing, it can be used to provide direction to a profession and create a vision for the future. Credentialing can be used as a form of leadership and help to mold future leaders.

Credentialing can be a means to differentiate between various levels of experience and competence in professionals. The American Association of Occupational Health Nurses (AAOHN) has identified three different competency levels for professionals using the categories of competent, proficient, and expert. Workers and employers can use the definitions of the competencies for each level to make determinations for assignments and career development and advancement and to indicate areas where more training may be needed (AAOHN, 2007).

As a benefit to society, credentialing can act as a guideline to the public for the minimum qualifications of quality. Licensure or certification of a group can provide valuable information about the qualifications of a professional that can be difficult to ascertain otherwise.

By helping to clarify and create individual professions of practice, credentialing can also provide a means to complete segmented research on the group. By understanding the practices and beliefs of the professionals in the group, it can reinforce standards of practice and ethical standards. It can provide valuable data about social needs as a means to track the working population in a field. It can be used to identify the growth of specialty areas. Information learned from research can be disseminated, furthering the growth of the profession. When a profession is distinct enough and large enough, it will result in both professional and technical journals. Professional journals can be used to communicate with the professional community. Technical journals in the field can be used to spread scientific and policy research results. As a profession becomes more advanced, it can lead to profession-specific educational pathways/programs. Credentialing can set standards for lifelong learning and professional growth. This is done in conjunction with accreditation in advanced professions. It can require and create opportunities for various forms of continuing education for practicing professionals in the field.

Credentialing and certification requirements can also work against migrants when external credentials are not recognized across borders. This often keeps potential migrant workers from entering existing job pools and reduces their overall earning potential (Augustine, 2015).

As licensure requirements can be a barrier to entry, it creates monopolistic conditions for those professionals with the credential. In some cases where licensed activities of one group overlap with another, there has been the potential to lead to turf wars between regulatory bodies and professional organizations (Brockman, 1996). Examples might include ophthalmologists and optometrists, or occupational therapists and physical therapists.

Training certificates may also be considered a form of credentialing, which often requires lower levels of formal education and shorter durations of associated training. Certificates in "short courses" on specific topics in OSH can be a way to bolster the qualifications and experience of a worker to be better prepared in OSH on the worksite. Certificate training programs offered by colleges, universities, and even for-profit consulting firms can be a way to improve hiring potential of graduates and raise wages and pay (Rosen, 2014). Typical college certification programs run 12–24 months. Training certificates have been identified as a practical, realistic, and necessary way to increase the numbers of practicing safety professionals (Alesbury, 2013).

10.4 ASSESSMENT OF CREDENTIALING

The International Standards Organization (ISO) is a specialized system of standardization whereby national bodies participate in the development of international standards of practice through expert technical committees established in various fields of endeavor. Various governmental, nongovernmental, and professional organizations collaborate in areas of mutual interest to derive international standards and guidelines of recommended practice in various areas.

ISO/IEC 17024:201 Conformity assessment—General requirements for bodies operating certification of persons is the ISO standard created to create globally accepted benchmarks for organizations that award personal professional credentials (ISO, 2012). The main objective of the conformity assessments is to ensure objective criteria are used to measure competence and scoring of certification criteria and associated examinations. Conformity assessments such as ISO 17024 help to assure that well-planned and well-structured certification examinations are administered impartially, with minimal risk of conflict of interest.

The ISO certification via assessment can serve as a basis for the recognition of certification bodies, and the certification scheme under which persons become certified. Conformity assessment of the credentialing body under ISO facilitates recognition and acceptance at the national and international levels. Additionally, the international development and use of these conformity assessments and accreditations enhance harmonization of technical fields, including occupational hygiene. It helps in the development of recognition of the profession internationally and aids in the global exchange of professional personnel in the OSH labor force. The ISO standard and associated conformity assessments can serve, and often do serve, as the basis for government recognition of professionals.

10.5 INTERNATIONAL RECOGNITION OF ACCREDITING BODIES

The International Occupational Hygiene Association (IOHA) maintains a standing National Accreditation Recognition (NAR) Committee charged with the responsibility of establishing an international accreditation scheme of occupational hygienists. The purpose for the accreditation scheme is to promote respect for and recognition of occupational hygiene certification as a means to promote worker health and safety globally.

The IOHA philosophy is that there is a need for practicing occupational hygienists to meet and maintain a given minimum level of competency in order to safely practice the profession. Competency is achieved through the attainment of knowledge. Knowledge can be achieved through a combination of working experience and education. Through documentation of work experience and education, occupational hygiene professionals can demonstrate that they have met the minimum level of competent professional practice (IOHA, 2018a).

Demonstration of the mastery of occupational hygiene practice builds upon the minimum competencies and assures that the practicing professional is proficient in a wide variety of occupational hygiene topics. Certification methods can take different forms. It may require written examinations, oral examinations, or other review of experience and education. The NAR Committee reviews national certification schemes and processes and determines whether they meet the IOHA minimum standards. The NAR Committee does not review, however, the quality or content of the individual certification schemes. There are presently 16 countries approved by the IOHA accreditation scheme.

The primary areas of occupational hygiene IOHA competencies are exposure assessment, risk assessment, risk control, and occupational hygiene management. The basic educational requirements include a review of official transcripts of academic study and coursework to the level of bachelor of science (or equivalent) in a relevant science or occupational/industrial hygiene. Professional experience requirements include thorough review of records, references, work histories, and detailed portfolios that indicate the individual has attained minimum levels of competent practice (IOHA, 2018b).

National accreditation processes and bodies currently recognized by IOHA include the following organizations: the American Board of Industrial Hygiene (ABIH), the Australian Institute of Occupational Hygiene (AIOH), the British Occupational Hygiene Society (BOHS), the Canadian Registration Board of Occupational Hygienists (CRBOH), the Dutch Occupational Hygiene Society (NVVA), the French Occupational Hygiene Society (SOFHYT), the German Society of Occupational Hygiene (DGAH), the Hong Kong Institute of Occupational and Environmental Hygiene (HKIOEH), the Institute of the Certification of the Figures of Prevention (ICFP), the Japan Association for Working Environment (JAWE), Malaysian Industrial Hygiene Association (MIHA), the Norwegian Occupational Hygiene Society (NYF), the Swedish Occupational and Environmental Certification Board (SOECB), the South African Institute for Occupational Hygiene (SAIOH), and the Swiss Society of Occupational Hygiene (SSHT).

10.5.1 American Board of Industrial Hygiene

The ABIH was created in 1960 as a means to evaluate the knowledge of practicing industrial hygienists. It is meant to provide a valid, reliable, and rigorous process to ensure competency of practitioners and thus the safety of workers and the public. The certification awarded by the ABIH, known as "Certified Industrial Hygienist (CIH)," is also a means for employers to ensure their employees meet the highest possible standards.

The CIH credentialing program by the ABIH is accredited by the American National Standards Institute (ANSI) under the standard ISO/IEC 17024—Conformity assessment—General requirements for bodies operating certification of persons. The ABIH CIH program is also accredited by the National Commission for Certifying Agencies (NCAA) under the Institute for Credentialing Excellence. These two designations are indications of the high quality of the CIH credential in terms of management, conformity, auditing, fairness, and other important aspects of the credentialing process.

The ABIH requires all CIH recipients to practice a particular code of ethics established by the organization. The code requires CIHs to follow specific practices and responsibilities towards clients, employers, the public, and to the profession itself. The ethical codes cover conduct regarding professional and legal standards, education and experience competency, and conflicts of interest.

To obtain a CIH, applicants must satisfy specific education and experience requirements, and then pass a written examination. The examination has 180 multiple-choice questions on a broad range of industrial hygiene topics. The test duration is 5.0h. The examination pass rate for the years 2003 through 2017 ranged from 36.2% to 57.9% (ABIH, 2018a).

The ABIH CIH program was accredited to the ISO/IEC 17024 standard by conformity assessment completed by the ANSI March 13, 2009. ABIH has also received accreditation by the NCAA. NCCA is the accreditation body of the Institute for Credentialing Excellence. The NCCA accreditation process is used to ensure that professional certification programs meet modern standards of practice (ABIH, 2018b).

10.5.2 Australian Institute of Occupational Hygiene

The AIOH is a group of approximately 1,500 occupational hygiene professionals working to protect worker health and safety in Australia. The organization has a variety of technical committees that work on various projects and topics. The AIOH produces practice position papers and practical publications meant to assist working health and safety professionals to protect the well-being of Australian workers.

One function of the AIOH is the accreditation of university master's educational degree programs in Australia. There are currently three accredited master's degree programs in occupational hygiene in the country.

Another function of AIOH is the certification of occupational hygienists in Australia. The designation of Certified Occupational Hygienist (COH) recognizes that a person is professionally competent to practice occupational hygiene independently. This certification is recognized by the IOHA and benchmarked with other similar national credentials. Members who receive this credential must meet a strict set of educational, experiential, and competency requirements including an oral examination of their occupational hygiene capabilities (AIOH, 2018).

10.5.3 British Occupational Hygiene Society

The BOHS is the only UK-based examining board for qualifications in occupational hygiene. The highest qualification offered is the Diploma of Professional Competence in Occupational Hygiene (DipOH). In order to achieve the DipOH, one must complete 5 years of comprehensive professional experience in occupational hygiene and also hold a BOHS Certificate of Occupational Competence (CertOH), or hold a BOHS accredited postgraduate qualification (BOHS, 2018).

In order to attain a CertOH, a candidate must have 3 years of professional experience and complete six modules of Occupational Hygiene Training Association (OHTA), complete a Personal Learning Portfolio, and pass a 1-h oral examination. If a candidate has a degree from a BOHS-approved university in the United Kingdom, the OHTA certificate requirement is waived. A third option for candidates that have a degree from a non-BOHS-accredited university is to have 3 years of experience, complete a Personal Learning Portfolio, pass a 1-h oral examination, and then pass a 6-h core written examination (BOHS, 2018). The written examination is comprised of short-answer and essay questions. Subjects covered on both written and oral CertOH examinations include hazard recognition, risk assessment, measurement equipment and methods, interpretation of data, and control methods. These examinations also include questions regarding UK-specific occupational hygiene legislation (BOHS, 2016).

10.5.4 Canadian Registration Board of Occupational Hygienists

The CRBOH is a not-for-profit organization that implements a system to establish professional competence in the practice of occupational hygiene. The CRBOH confers two designations: the registered occupational hygienist (ROH) and the registered occupational hygiene technologist, indicating two different levels of professional competency.

The CRBOH promotes a particular code of ethics for professionals carrying its credentials. And the organization strives to improve and expand awareness and understanding of the practice of occupational hygiene. The overall goal is to provide safe working conditions for all Canadians and minimize workplace injury and illness.

Eligibility to become registered is dependent upon a specific set of requirements for formal university and postgraduate education and related combinations of experience. All candidates must sit for a half-day multiple-choice examination followed by a half-day essay format examination. Once passing the written examination, the candidates then sit for a half-day essay format examination (Verma, 1994). After successfully passing the written portions of the examination, candidates take a 1-h oral examination. Special exceptions to the normal examination process are allowed for professionals who hold certifications from other credentialing organizations endorsed by the IOHA NAR Committee and for graduates of particular Canadian postsecondary programs in occupational hygiene (CRBOH, 2017).

10.5.5 Dutch Occupational Hygiene Society

The NVVA is a professional group dedicated to safeguarding the health of workers and their descendants. They promote the science of occupational hygiene and raise awareness of the specialty.

Professionals can achieve a ROH designation based upon educational requirements and professional experience. The certification process in the Netherlands does not require a separate certification examination (NVVA, 2018).

10.5.6 French Occupational Hygiene Society

SOFHYT is the largest organization of occupational hygiene professionals in France and was created in 1991. They conduct regular meetings to present and discuss technical topics in occupational hygiene and to promote the understanding and awareness of the value of occupational hygiene in France. They have been a member of IOHA since 1991 and are also affiliated with the OHTA.

SOFHYT has been offering a professional certification in Occupational Hygiene Professional Certification since 2008. This credential was then approved and accepted to meet IOHA NAR designation in 2010. Candidates must meet a minimum level of education and experience before sitting for a written examination. In 2015, SOFHYT created an intermediate professional designation of "Certificate in Occupational Hygiene" (SOFHYT, 2015).

10.5.7 German Society of Occupational Hygiene

DGAH promotes knowledge and recognition of occupational hygiene and the maintenance of safe and healthy working conditions. The organization of professionals provides education and communication between professionals and organizations practicing occupational hygiene. The DGAH also defines a code of ethical practice for the profession.

Membership in DGAH requires the demonstration of completion of a strict set of educational and experiential competencies. Candidates must also successfully complete a written and oral examination before receiving the designation of occupational hygienists and membership to DGAH (DGAH, 2018). The DGAH professional designation and credentialing process is recognized by the IOHA NAR Committee.

10.5.8 Hong Kong Institute of Occupational and Environmental Hygiene

The HKIOEH administers a Certificate of Registered Professional Hygienist for work with hazardous chemicals, agents, and processes. Candidates must pass both written and oral examinations, possess a degree in a relevant subject, and have 5 years of experience to take the oral examination (HKIOEH, 2016a, b).

10.5.9 Institute of the Certification of the Figures of Prevention

The ICFP is the Italian occupational hygiene credential (ROHs) recognized by the IOHA NAR program. This program credentialing process includes specific requirements for education, experience, professional references, an oral examination, and a written examination (ICFP, 2018).

The Italian certification institution is accredited by the Italian National Unification Body. This is a private nonprofit association recognized by the European Union that develops and publishes voluntary technical standards for industrial and commercial sectors (UNI, 2018). The ICFP is reviewed and certified by the Unente Italiano di Normazione (UNI) to have met and be in compliance with ISO/IEC 17024:2003: General requirements for certifying bodies, published by the ISO.

The Italian credentialing body is not directly associated with the Italian Association of Industrial Hygiene (AIDII) and is an independent organization.

10.5.10 Japan Association for Working Environment

The JAWE is a nonprofit professional organization created in 1979 meant to provide a tripartite collaboration between employers, workers, and occupational hygiene experts to ensure safe and healthful working conditions in Japan. In 2003, JAWE became a member of IOHA.

JAWE offers credentials for a Work Environment Measurement Professional at two different levels. In order to receive the credential, the professional must pass an examination and complete program-associated registration courses. Topics covered by the Type 1. Course and examination include designing, sampling, and analyzing environmental hazards. The Type 2. Professional level includes the evaluation of work designs, in addition to sampling and analysis (JAWE, 2018). It is not clear what the examination entails, whether prior work experience is required, or whether the examination includes other OSH topics such as risk assessment, management, or ergonomics. The JAWE accreditation system is recognized by the IOHA NAR system.

10.5.11 Malaysian Industrial Hygiene Association

The MIHA is a nonprofit professional organization created in 2003. With more than 100 members, it is one of the largest worker health and safety organizations in Southeast Asia. Its goals are to promote professional development of industrial hygienists and provide timely and relevant ongoing education and training in occupational hygiene.

The MIHA offers a professional designation credential called the Certified Professional Industrial Hygienist (CPIH). The CPIH is governed by a board headed by the current MIHA President and follows a per-defined set of criterion. In order to achieve the CPIH designation, professionals must be full members of MIHA, be currently in the practice of industrial hygiene, meet academic and professional experience requirements, and pass an oral examination. The MIHA CPIH is recognized by the IOHA, and the MIHA is an affiliate member of IOHA (MIHA, 2017).

10.5.12 Norwegian Occupational Hygiene Society

The NYF is a professional association of occupational hygienists with the goal of protecting worker health and safety in Norway. They disseminate information about the benefits and practice of occupational hygiene and develop a set of ethical standards for practicing professionals in the field.

Within the NYF, a board called the Foundation Norwegian Occupation Hygiene Certification conducts a professional certification and credentialing scheme. Certification is based upon educational requirements for a minimum of 3-year technical-scientific studies at an approved university or academy, a minimum of 7-year work experience in occupational hygiene, and completion of both a written and an oral examination (NYF, 2018). This certification scheme has been acknowledged by the IOHA NAR.

10.5.13 Swedish Occupational and Environmental Certification Board

The SOECB is an independent certification agency that was established in 2001. The board provides a means to guarantee the qualifications of occupational and environmental hygienists according to criteria adopted by the Swedish Association of Occupational and Environmental Hygienists (SAOECB, 2003).

10.5.14 South African Institute for Occupational Hygiene

The SAIOH was created in 1983 to promote awareness and recognition of occupational hygiene as a professional discipline. In 1993, the Institute for Occupational Hygienists of South Africa (IOHSA) was created as a means to provide professional credentials to individuals who met appropriate standards for knowledge and practical experience in the area of occupational hygiene. In 2000, these two organizations merged to form the South African Institute for Occupational Hygiene (SAIOH).

Membership in SAIOH is provided at a variety of levels depending on education, experience, and examination. General membership is allowed for those interested in occupational hygiene, but not practicing. Then, levels progress from occupational hygiene assistant, occupational hygiene technologist to the occupational hygienist designation. The highest level of occupational hygienist requires a bachelor's degree and 5 years of experience, or a master's or doctoral degree, and 4 years of experience. In addition, the successful completion of a 3-h written examination followed by a 1-h oral examination is required to receive the occupational hygienist designation (SAIOH, 2018).

10.5.15 Swiss Society of Occupational Hygiene

The SSHT represents professionals who practice the discipline of anticipation, determination, verification, and communication of hazardous working conditions. They are dedicated to the wellness and safety of workers and the general community (SSHT, 2018a, b).

The Committee de Certification des Hygiénistes du Travail (CCHT) is a group within SSHT responsible for the certification of hygienists in Switzerland. This committee is charged with administering written certification examinations and maintaining a list of hygienists that hold the SSHT certification (SSHT, 2018a, b).

10.6 OTHER REGULATED PROFESSIONS AND OCCUPATIONS RELATED TO OCCUPATIONAL HYGIENE

In this chapter about global credentialing of occupational hygiene, it is worth noting other closely related professions that interact or overlap with occupational hygiene credentialing.

10.6.1 Occupational Safety

10.6.1.1 European Network of Safety and Health Professional Organizations

European harmonization of safety educations and the profession began as early as 1970 initiated by the International Social Security Association (ISSA) Safety Training Section. This harmonization process has since been assumed by the European Network of Safety and Health Professional Organizations (ENSHPO).

The ENSHPO was established in 2001 in order to bring together health and safety professional organizations from across Europe. The main objectives of ENSHPO are to

- Ensure participation from all of the professional organizations across Europe and represent the views, opinions, and concerns of this group.
- Operate as a dialogue partner with relevant national and international authorities.
- Cooperate with other organizations, institutions, and federations within Europe and beyond.
- Act as a forum where practitioners can exchange information, experiences, and good practice on a wide variety of pertinent topics.
- Develop a European-wide recognition of OSH practitioner qualifications and training.

Current ENSHPO members include the following (ENSHPO, 2016b):

- AEPSAL (Asociacion de Especialistas en Prevencion y Salud Laboral)—Spain
- AIAS (Associazionne professionale Italiana Ambiente e Sicurezza)—Italy
- Arbejdsmiljoraadgiverne (Danish Association of Occupational Health and Safety Consultants)—Denmark
- ARSSM (Romanian Association for Occupational Health and Safety)—Romania
- CIVOP (Occupational Safety and Health and Fire Prevention Chamber of the Czech Republic)—Czech Republic
- CySHA (The Cyprus Safety and Health Association)—Cyprus
- HZZZSR (Croatian Institute for Health Protection and Safety at Work)
- IOSH (Institution of Occupational Safety and Health)—the United Kingdom and Ireland
- IMRSS (Institute of Risk Management and Occupational Health & Safety)—Romania
- MOSHPA (Malta Occupational Safety and Health Practitioners Association)—Malta
- Suissepro (SGAS, SGIG, SwissErgo, SGARM, SGAH, GRMST)—Switzerland
- Tyoturvallisuuskeskus Centre for Occupational Safety—Finland
- VDSI Verband für Sicherheit, Gesundheit und Umweltschutz bei der Arbeit—Germany

Part of the ENSHPO focus has been on the safety and health educational programs in the European Union. Based on UK National Occupational Standards for health and safety, detailed competencies have been delineated for OSH professionals. These have been used as learning outcomes for professional courses under the European Qualifications Framework. These qualification schemes are now being compared and harmonized with health and safety codes of practice in North America, Asia, and Pacific countries (Hale, 2012).

ENSHPO has created an OSH certification standard as a means to ensure a minimum level of competence for practicing professionals in Europe. The goal is to provide employers with a mutually recognized system of competencies and qualification for occupational health professionals (ENSHPO, 2016a).

Certification in the ENSHPO scheme does not require passing an examination. There are two professional levels: the European Safety and Health Manager and the European Safety and Health Technician. Registration fees range from 250 to 450 euros. Certification is based primarily on meeting certain levels of experience and education. It also accepts designations from other recognized certification schemes from other countries that have applied to their program.

10.6.1.2 Board of Certified Safety Professionals

The Board of Certified Safety Professionals (BCSP) offers a variety of occupational safety registrations and certifications based upon experience, education, and examination. The highest level designation is the Certified Safety Professional, which requires 4 years of safety experience, graduation from 4 years of academic institution with a degree in safety engineering or closely related field, and passing a comprehensive set of examinations on occupational safety and management. Covering slightly different rubrics than the CIH, this credential is also highly recognized globally (BCSP, 2018).

10.6.1.3 Occupational Health/Nursing/Medicine

Occupational health nursing is a profession arguably one of the closest to the practice of occupational hygiene. Occupational health nursing focuses on the promotion and restoration of health, prevention of illness and injury, case management, worker compensation programs, and protection from occupational hazards (AAOHN, 2007). The necessary competencies for these nursing professionals are set by the AAOHN. This specialty practice is built upon the foundations of nursing sciences, medical science, public health, safety, toxicology, ergonomics, and industrial hygiene.

The American Board of Occupational Health Nurses (ABOHN) was created in 1972 as a means to set standards for occupational health nurses. The board awards three different credentials:

- Certified Occupational Health Nurse (COHN)
- Certified Occupational Health Nurse—Specialist (COHN-S)
- Case Management (CM)

Certification is based upon a review of candidate professional experience and education in addition to passing a written examination. There are currently approximated 4,000 active certificants (ABOHN, 2018).

10.6.1.4 Radiation Protection

The American Board of Health Physics (ABHP) grants professional certification in the field of radiation protection and is accredited by the Council of Engineering and Scientific Specialty Boards. The Certified Health Physicist provides protection to workers from hazardous radiation sources. Certification examinations administered by the ABHP evaluate candidate's abilities in the areas of radiation measurement, selection of detection instruments, analytical techniques for radiation sampling, mathematical modeling of radiation exposure and control, analysis of data, and preparation of reports. Other general topics include the development of standard operating procedures for radiation fieldwork, emergency response, record keeping, and applicable regulations (ABHP, 2018).

10.6.1.5 Ergonomics

The Board of Certified Professional Ergonomists (BCPE) was incorporated in 1990 and is a nonprofit organization and is endorsed by the International Ergonomics Association (IEA). Federated societies of professional ergonomics, representing more than 50 countries, comprise the council governing body of IEA. BCPE provides professional certification for practitioners of human factors/ergonomics/user experience.

In Europe, the Centre for the Registration of European Ergonomists (CREE) harmonizes the certification practices of the ergonomics societies within numerous nations. The national certification methods are each approved by the CREE council, and the certificates are recognized internationally. There are currently 438 registered

FIGURE 10.3 Countries currently members of CREE. (www.eurerg.eu/ homepage, accessed July 4, 2018).

ergonomists in more than 30 CREE member countries (CREE, 2018). Countries that are currently members of CREE are shown in Figure 10.3. The structure and operations of CREE comply with the International Standard ISO/IEC 17024:2012(en).

Other countries with ergonomics credentialing bodies include the Canadian College for the Certification of Professional Ergonomists (CCCPE) (CCCPE, 2018) that has a certification process that leads to the Certified Canadian Professional Ergonomist (ACE, 2018). In Japan, the Japanese Ergonomics Society (JES) that was founded in 1964 and has more than 2,000 members also offers a certification credentialing process. JES is a member of the IEA (JES, 2018).

10.6.1.6 Laser Safety

The Board of Laser Safety offers the Certified Laser Safety Officer designation to professionals who have demonstrated skills, education, and experience in the practice of occupational laser safety. Related laser safety certification examinations include the topics of laser/optics fundamentals, radiation bioeffects, non-beam hazards, control measures, regulations and standards, hazard evaluation, laser measurements, and laser safety program administration (BLS, 2018).

10.7 CONCLUSIONS/RECOMMENDATIONS

Licensing and credentialing are important tools to support the development and clarification of a profession and the workers that practice the field. In the past few years, major advances have occurred in the field of occupational hygiene through the development of international recognition schemes. In order to take full advantage of these internationally recognized certifications, more work needs to be done to standardize the rubrics for each certification scheme, and ensure the administrative means are in place to verify that each certification is truly equal.

REFERENCES

AAOHN, American Association of Occupational Health Nurses, Competencies in occupational and environmental health nursing, *AAOHN Journal: Official Journal of the American Association of Occupational Health Nurses* (November 2007) Vol. 55, No. 11, pp. 442–447.

ABHP, American Board of Health Physics, Examination Topics (2018) www.hps1.org/aahp/boardweb/pgsec4.html accessed July 3, 2018.

ABIH, Become Certified (2018a) www.abih.org/become-certified/prepare-exam/exam-pass-rate accessed January 28, 2019.

ABIH, Accreditation (2018b) www.abih.org/about-abih/accreditation accessed March 31, 2018.

ABOHN, American Board of Occupational Health Nurses, About Us (2018) www.abohn.org/about-abohn/about-us accessed January 29, 2018.

ACE, Association of Canadian Ergonomists, Professional Certification (2018) https://ace-ergocanada.ca/about/certification.html accessed July 4, 2018.

AIOH, What is Certification – Australian Institute of Occupational Hygiene Webpage (2018) www.aioh.org.au/membership-information/certification accessed March 29, 2018.

Alesbury, R., Bailey, S., Addressing the needs for international training, qualifications, and career development in occupational hygiene, *Annals of Occupational Hygiene* (2013) Vol. 58, No. (2), 140–151.

Augustine, J., Immigrant professionals and alternative routes to licensing: Policy implications for regulators and Government, *Canadian Public Policy/Analyse de Politiques* (2015) Vol. 41(Supplement 1), pp. S14–S27.

BCSP, Board of Certified Safety Professionals (2018) www.bcsp.org accessed March 2018.

BLS, Bureau of Labor and Statistics, The de-licensing of occupations in the United States, Monthly Labor Review (May 2015) www.bls.gov/opub/mlr/2015/article/the-de-licensing-of-occupations-in-the-united-states.htm accessed December 10, 2018.

BLS, CLSO Examination Information (2018) www.lasersafety.org/certification-requirements/certification-exam/clso-areas-of-practice/ accessed July 4, 2018.

BOHS, PQ Qualification Guide for CertOH, v1 (December 12, 2016) Document Reference: PQC-POL001 www.bohs.org/wp-content/uploads/2018/03/PQC-POL001-Qualification-Guide-for-CertOH-and-iCertOH-v2.pdf accessed online March 31, 2018.

BOHS, Diploma of Professional Competence in Occupational Hygiene Qualification Overview (2018) www.bohs.org/membership/benefits-of-membership/professional-development/diploma-in-occupational-hygiene/ accessed on March 30, 2018.

Brinegar, P., Schmitt, K. (1992). State occupational and professional licensure. In *The Book of the States, 1992–1993*. Lexington, KY: Council of State Governments, pp. 567–580 http://knowledgecenter.csg.org/kc/content/book-states-archive-1935-2009 accessed December 10, 2018.

Brockman, J., Dismantling or fortifying professional monopolies? On regulating professions and occupations, *Manitoba Law Journal* (1996) Vol. 24, pp. 301–310.

Bryson, A., Kleiner, M., The regulation of occupations, *British Journal of Industrial Relations* (2010) Vol. 48, No. (4), pp. 670–675.

Carpenter, D., Knopper, L., Erickson, A., Ross, J., Regulation work: Measuring the scope and burden of occupational licensure among low and moderate income occupations in the United States, *Economic Affairs* (2015) Vol. 35, No. 1, pp. 3–20.

CCCPE, Homepage (2018) www.cccpe.ca/ accessed July 4, 2018.

CRBOH, Canadian Registration Board of Occupational Hygienists, ROH Examination Details (2017) www.crboh.ca accessed September 24, 2017.

CREE, About CREE (2018) www.eurerg.eu/about-cree/ accessed July 4, 2018) The structure and operations of CREE complies with the International Standard ISO/IEC 17024:2012(en).

DGAH (2018) German Occupational Hygiene Society, http://ioha.net/files/2015/11/German-OHS-IOHA-NAR-ApplicationDGAH-MOU-June-2014.pdf accessed January 16, 2016.

ENSHPO, European Network of Safety & Health Professional Organisations, Certification Standard for European Safety & Health Managers (EurOSHM) (2016a) www.euroshm.org/full.php accessed June 13, 2016.

ENSHPO, About ENSHPO (2016b) www.enshpo.eu/userfiles/2015_09_22%20%20About%20ENSHPO%20.pdf accessed July 1, 2016.

Hale, A., Harvey, H. (September 2012). Certification of safety professionals: Emerging trends of internationalization, *Proceedings of the 6th International Conference of Working on Safety Network "Towards Safety Through Advanced Solutions"*, Sopot, Poland, pp. 11–14.

Hale, A.R., Harvey, H., Mennozi, M., Miguel, S., Bianchi, G. (2008). Certification of safety professionals at the European level, *Proceedings of the 4th International Working on Safety Conference*, Heraklion, Crete.

Hawthorne, L. (2011). Australia-In recognition of qualifications and competencies of migrants. In A. Schuster, M. Vincenza Desider, G. Urso (Eds.), *International Organization for Migration* (pp. 201–228). Brussels: IOM. www. labourmigration.eu/research/report/20-recognition-of- qualifications-and-competences-of-migrants.

Hemphill, T., Carpenter, D. (Fall 2016). *Occupations: A Hierarchy of Regulatory Options*, Regulation. Washington, DC: Cato Institute, pp. 20–24.

HKIOEH, Registered Professional Hygienist Registration and Examination Board (2016a) www.hkioeh.org.hk/Documents/RPH2017/Registration%20Board%20of%20Professional%20Hygienist%20(2016%20version)%20(v3.5%20drafted%20on%2010.12.2016)%20(clean).pdf accessed April, 2018.

HKIOEH, HKIOEH Registration of Professional Occupational Hygienist Examination Information (2016b) www.hkioeh.org.hk/Documents/RPH2017/RPH%20Exam%20Info%20(2016%20version)%20draft%20on%2010.12.2016%20(clean).pdf accessed April, 2018.

ICFP, Institute of the Certification of the Figures of Prevention, Registration Board of Occupational Hygienists Summary of the Process of Accreditation (2018) http://ioha.net/files/2015/11/Institute-of-the-Certification-of-the-Figures-of-Prevention-Italy-ICFP.pdf accessed March, 2018.

IOHA, National Accreditation Recognition Committee (2018a) http://ioha.net/national-accreditation-recognition-nar-committee/ accessed December 10, 2018.

IOHA, Occupational/Industrial Hygiene Knowledge and Competency Requirements (2018b) http://ioha.net/files/2016/10/Knowledge-Competence-Final.pdf accessed December 10, 2018.

ISO, ISO/IEC 17024:201 (2012) *Conformity Assessment—General Requirements for Bodies Operating Certification of Persons*. Geneva: International Standards Organization, www.iso.org/obp/ui/#iso:std:iso-iec:17024:ed-2:v1:en accessed March 31, 2018.

ISO/IEC, ISO/IEC 17011 (2004) *Conformity Assessment—General Requirements for Accreditation Bodies Accrediting Conformity Assessment Bodies*. Geneva: ISO/IEC 17011:2004(E) International Standards Organizaation.

JAWE, The Way to Becoming a Work Environment Measurement Professional (2018) www.jawe.or.jp/kosyu/sagyokankyomichinori.htm#shiken accessed March 2018.

JES, Homepage (2018) www.ergonomics.jp/e_index.html accessed July 4, 2018.

Kleiner, M., Occupational licensing, *Journal of Economic Perspectives* (2000) Vol. 14, No. 4, pp. 189–202.

Kleiner, M., Krueger, A., The prevalence and effects of occupational licensing, *British Journal of Industrial Relations* (2010) Vol. 48, No. 4, 676–687.

Kleiner, M. M., Krueger, A. B., Analyzing the extent and influence of occupational licensing on the labor market, *Journal of Labor Economics* (2013) Vol. 31, No. 2, pp. S173–S202.

Malaysian Industrial Hygiene Association (2017) http://miha2u.org accessed September 23, 2017.

Meehan, B., The impact of licensing requirements on industrial organization and labor: Evidence from the U.S. private security market, *International Review of Law and Economics* (2015) Vol. 42, pp. 113–121.

Nash, P., Sherwood, R., An occupational hygiene team, *British Journal of Industrial Medicine* (May 10, 1953), pp. 201–206.

NYF, Becoming Certified – Criterion for Certification (2018) www.nyf.no/sertifiseringer/sertifiseringsordningen accessed March 29, 2018.

NVVA, Dutch Scheme – Summary of the Dutch Occupational Hygiene Certification Scheme (2018) http://ioha.net/files/2015/11/Dutch-Occupational-Hygiene-Society-NVVA.pdf accessed January 16, 2018.

Ogus, A., Zhang, Q., Licensing regimes east and west, *International Review of Law and Economics* (March, 2005) Vol. 25, pp. 124–142.

Pagliero, M., The impact of potential labor supply on licensing exam difficulty, *Labour Economics* (2013) Vol. 25, pp. 141–152.

Rosen, M., Koshy, K., Patel, M., Safety and health certificate programs: Practical application beyond training, *Journal of Safety, Health & Environmental Research* (2014) Vol. 10, No. 2, pp. 162–168.

SAIOH, South African Institute of Occupational Hygiene, South African Scheme – Summary of the Certification Process (2018) http://ioha.net/files/2015/11/Southern-African-Institute-for-Occupational-Hygiene-SAIOH.pdf accessed January 16, 2018.

SAOECB, Swedish Occupational and Environmental Certification, Document SOECB C-Bylaws, Version 3, 2009-01-20, adopted by the Board of SAOECB April 28, 2003
www.symf.nu/wp-content/uploads/2017/12/SOECB-Byelaws_20090130.doc accessed December 10, 2018.

SOFHYT, Certification professionnelle en hygiene du travail (2015) www.sofhyt.fr/certification-professionnelle-hygiene-du-travail accessed September 24, 2017.

SSHT, Swiss Society of Occupational Hygiene, SSHT Homepage (2018a) www.sgah.ch/fr/ accessed January 16, 2018.

SSHT, SSHT Status (2018b) www.sgah.ch/attachments/article/104/CCHT%20Statuts%20310509.pdf accessed January 16, 2018.

Sweetman, A., McDonald, J., Hawthorne, L., Occupational regulation and foreign qualification recognition: An overview, *Canadian Public Policy/Analyse de politiques* (August 2015), Vol. 41, pp. S1–S13.

Timmons, E., Thornton, R., The effects of licensing on the wages of radiologic technologists, *Journal of Labor Research* (2008) Vol. 29, No. (4), pp. 333–346.

UNI, Unente Italiano di Normazione, Who we are (2018) www.uni.com/index.php?option=com_content&view=article&id=141&Itemid=2422 accessed March, 2018.

Verma, D., Sass-Kortsak, A., Gaylor, D., Evaluation of professional competency in occupational hygiene in Canada – The CRBOH program, *American Industrial Hygiene Journal* (April 1994) Vol. 55, No. 4, p. 362.

11 Regional and National Occupational Safety and Health Profiles

Thomas P. Fuller
Illinois State University

Frederique Parrot
Sanofi

CONTENTS

11.1 INTRODUCTION

11.2 INTERNATIONAL LABOR ORGANIZATION LEGOSH

The International Labor Organization Global Database on Occupational Safety and Healthlegislation (LEGOSH) provides an overview of the major national regulations for health and safety at work for member countries around the globe. It provides the ability to monitor trends and make comparisons between countries. The information made available online assists governments, researchers, companies, worker representatives, and other professional organizations to analyze policies and practices regarding worker health and safety.

The LEGOSH database structure is based on a comprehensive set of themes that follow the key provisions of International Labor Organization (ILO) standards including the Occupational Safety and Health Convention, 1981 (No. 155), and its accompanying Recommendation (No. 164), and the Promotional Framework for Occupational Safety and Health Convention, 2006 (No. 187). Each country review includes information on the following (ILO, 2016a):

1. The basic national regulatory framework
2. Regulation scope and exclusions
3. Institutions with administrative and enforcement authority
4. Employer duties and responsibilities
5. Management principles and practices
6. Worker responsibilities and rights
7. Specific hazards and risks

8. Recording and reporting of work-related accidents and diseases
9. Inspection and enforcement programs

The transparency and comprehensiveness of this database makes it an invaluable tool for developing countries to use as models in program and regulatory development. The resource provides an excellent foundation for organizations and national governments to create laws and develop agencies and programs. Legislation is often an important step, if not the first step, in creating a regulatory framework that provides safe working conditions for all workers.

11.3 MISCELLANEOUS NATIONAL ORGANIZATIONS, LAWS, AND ENFORCEMENT

11.3.1 France

11.3.1.1 Demography of France

In July 2017, France has just more than 67 million inhabitants with the largest population in Europe, of which 3.18% is based overseas. In 2014, life expectancy at birth is high, with difference between male (78.8 years) and female (85.2 years) (Indexmundi, 2018).

11.3.1.2 Work Structure and Statistics in France

As many countries in Europe, France has been strongly impacted by the industrialization in the 20th century. Urban areas grew and agricultural activities were replaced by heavy industries of manufacturing. The tertiary and commercial activities still continue to concentrate people in urban areas. The French economy has been seriously affected by the global financial and economic crisis in 2007 (EU, 2015).

Today, the repartition of total employment is 2.7% for agriculture, 20.1% for industries, and 75.9% for services. Twenty-three percent of the workforce are government employees in public administration. With Organization for Economic Cooperation and Development (OECD) annual average hours in 2015 of 1,766 h, French workers are on the lower range with 1,482 h annually (OECD, 2017). Trade Union density is 11%, with 9% in private companies and 19 in public institutions (DARES, 2016).

The French work code details the fundamental role of unions to negotiate at the state level, by sectors of activities and in companies. This role is currently under discussion to be developed or reduced. At this time, 98% of French employees are covered by collective bargaining representation. This compares to 6.7% in the United States for private sector workers (USBLS, 2016). France has the second largest economy in Europe, behind Germany.

11.3.1.3 Health and Safety—Performance in France

The standardized incidence rate of accidents at work is high in France, close to two times the average European Union (EU) rate. The French rate was 3,172.9, compared to the EU average rate of 1662.9 in 2013 (EC, 2018). France represents 20% of EU occupational injuries, with only 13% of the total population. France is also the EU the leader for the rate of musculoskeletal disorders in occupational diseases (460 cases for 100,000 workers).

According to the French social security database for more than 18 million workers, covering occupational diseases and injuries, excluding home/job transport injuries, there were 926 deaths, 624,500 injuries, and 50,960 diseases reported in 2015 (France, 2015). There were 2.7 days lost per year and per worker, based on 18.5 million workers. And occupational diseases represent 8% of event but 41% of permanent disability and death, and 22% of lost working days.

11.3.1.4 Health and Safety History in France

History shows that economically stressed periods and wars are moments in France when there are large changes in occupational health and safety norms.

- 1898: first act for compensing occupational injury (France, 1898)
- 1919: addition of occupational disease to the compensing act (France, 1919)
- 1936: first regulation fully dedicated to hygiene and safety in industries
- 1936: after an intensive mobilization, an accord is signed that creates 2 weeks of holidays, 40 h per week work, "Collective Agreements" Act and Act on "Conciliation and arbitration procedures in collective labor disputes"
- 1941: reorganization of inspection of Labor and Manpower, which significantly strengthens the workforce of the Labor Inspectorate and creates the body of labor medical inspectors
- 1942: a social service and a medical service for companies with more than 250 employees Loi du 28 juillet 1942 relative à l'organisation des services médicaux et sociaux du travail
- 1945: creation of social security
- 1946: status for employees' representatives
- 1946: formal creation of the medical surveillance for all workers in private companies. Previously, occupational physicians was placed in mines (beginning XIX), industries for army (1941), and other industries case by case (France, 1946)
- 1947: creation of the health and safety committees
- 1956: third week of vacation
- 1965: organization for professional training during working life
- 1971: fourth week of vacation
- 1971: status for safety delegates in mines
- 1976: law on the development of occupational safety (France, 1991a)
- 1981, 1982: 5-week holidays, 39 h of work per week, protection of occupational injured worker, retirement benefit at 60 years old, hygiene safety, and work condition committee
- 1982: application of regulation on health and safety to public workers Décret n°82-453 du 28 mai 1982 relatif à l'hygiène et à la sécurité du travail ainsi qu'à la prévention médicale dans la fonction publique
- 1991: application of the European directive 89/391/EE requiring the fundamentals approach: assess all risks for health and safety covering physical and mental effects on workers (France, 1991b)
- 2001: permission of night shift for women in industries (France, 2001)

- 2008: work code change from section hygiene and safety to health and safety (France, 2008)
- 2009: creation of the decree to support work inspectors by technical competences; it includes all work risk assessments (France, 2009)
- Since 2010: few acts dedicated to arduous work. The occupational medicine organization is progressively modified

11.3.1.5 Current Regulatory Framework

All European directives are related to health and safety in application of the Directive 89/391/EEC—OSH "Framework Directive." EU countries are required to develop national regulations and standards that meet the minimum criterion of the EU directives. (EU, 1989; EU, 2015)

The Labour Ministry (Ministère du Travail, de l'Emploi, de la Formation professionnelle et du Dialogue social) is responsible for developing and enforcing regulations for occupational safety and health. The Labour Ministry supervises the actions of the agency for the improvement of working conditions in providing guidance, conducting research, and conducting investigations following serious accidents. The main legislation on occupational safety and health is Part IV of the Labor Code entitled "Health and Safety at Work." The code covers topics including work duration, daily rest period, respect of fundamental freedoms, bullying, sexual harassment, discrimination, and worker rights. The French Labour Ministry also produces guidelines on various topics relating to how to meet given requirements. The Labour Code also incorporates the treaty requirements of the EU legislation such as for risk assessment and hazard control.

Under the French system, an "accident" is defined as any workplace occurrence that results in an injury or illness of a worker. Under this definition, there is a clear acceptance of a lag in time between exposure and ill health outcomes, sometime in terms of years. In addition, even depression, post-traumatic stress, or work-related suicide can be considered health effects or "accidents" that result from workplace bullying.

Occupational health and safety research in France is conducted under the National Institute for Health and Medical Research (INSERM) and the National Institute for Research and Safety (INRS). The state Social Security Organization funds support for worker health and safety research through the Caisse d'Assurance Retraite et de la Sante au Travail (CARSAT) regionally throughout France and Caisse Assurance Maladie d'Ile de France (CRAMIF) in the Paris region, in a manner similar to worker compensation in the United States. In addition, however, these organizations also develop exposure assessment methodologies and perform laboratory analyses as a service to companies. They also develop and provide occupational safety training materials to the public.

Within companies, management has the authority and resources to determine equipment and procedures to ensure a safe workplace to meet the requirements assigned by the occupational physician. A safety department or professional may be called upon to provide assistance.

In France, occupational safety and health programs are primarily managed by occupational health physicians, rather than occupational health specialists. For example, federal directives require that occupational physicians perform medical examinations on any and all employees that are likely to be exposed to carcinogens. These physicals must be repeated every 6 months, and the records must be kept 40 years postexposure (Aubrun, 1999). If the employee moves to another company, the medical records must be forwarded to the new occupational physician. Appropriate safety training is provided to employees at the advice of the occupational physician.

Within companies, management has the authority and resources to determine equipment and procedures to ensure a safe workplace to meet the requirements assigned by the occupational physician. A safety department or professional may be called upon to provide assistance.

The third leg of the French occupational safety system includes mandated Health and Safety Working Conditions Committees (CHSCTs) that meet at the direction of the occupational health physician and include representatives from employees, management, occupational safety, and occupational medicine.

The French Agency for food, environmental and occupational health and safety (ANSES) Agence nationale de sécurité sanitaire de l'alimentation, de l'environnement et du travail has 1,383 full-time workers with an annual budget of 142 million euros. Its objective is to contribute to ensuring human health and safety in the environment, work, and food. Among other responsibilities, this group makes recommendations for occupational exposure limits (OELs) to the Ministry of Labour. However, in 2015, only 1% of the research conducted by ANSES (2015) involved occupational health.

The French National Agency for the Improvement of Working Conditions (ANACT) is charged with improving the quality of working conditions and promoting organizational efficiency and communication. The network encourages companies to consider occupational safety and health in business terms. It fosters the coordination of activities and projects between all stakeholders (management and employees). Its aim is to help businesses to conceive, promote, and implement public incentive policies, tools, and innovative methods (ILO, 2016b).

Beyond requiring employers to provide safe working places free from injury and illness causing exposures, the French system has a couple of different approaches to occupational health and safety. One of the major characteristics of the French system is the requirement for occupational health and social service professionals to be an integral, if not the integral, aspect of the health and safety program. Persons with the above credentials are assigned the responsibility for creating and implementing the occupational safety program. Professionals with specific training in occupational safety or hygiene are ancillary to the basic structure. In addition, the French system requires the formation of a site safety committee for workplaces with more than 50 employees, also led by the occupational health professionals. This committee is charged with negotiations of pay, work duration, quality of work life, career management, and workplace safety.

Ministry of Labour has the authority to perform inspections of workplaces that fall under its jurisdiction. Based upon the results of findings, they can issue citations and notices for violations against regulations. Based upon the findings, the labor prosecutor may impose financial penalties, revoke or suspend operating licenses, or require cessation of dangerous work. In France, violations of occupational safety and health regulations can be prosecuted under the criminal code. This includes up to 5 years of imprisonment. Workplace bullying is punishable under criminal codes and is punishable by up to 2 years of imprisonment and a fine of up to 30,000 euros.

11.3.1.6 Legislative Updates

Recent changes in French workplace safety legislation include methods to be used to evaluate risks from exposure to artificial optical radiation. The law went into effect on April 1, 2016, and requires employers to measure or calculate worker exposures for a variety of new radiation sources including artificial sources of ultraviolet and infrared radiation (France, 2016a). In addition, a new directive will change the OEL for styrene effective on January 1, 2019, to 23.3 ppm or 100 mg/m^3 (France, 2016b).

11.3.1.7 Some Specific aspects of OSH in France

In France, the social relationship involved in the work impacts strongly the management of occupational risks. Also, the occupational physicians have a strong position in the prevention of risks at work. The following sections briefly describe some of the details about OSH in France:

- *Occupational medicine*: Initiated during the industrialization period, the role of occupational physicians in OSH has been in place for 50 years. The position is covered by a "protected employee" status, and 20% of French employees have access to an occupational medicine physician as a company's employee. For 80% of employees, the physician is outsourced. The occupational physician has a duty to advise both the employees and the employers. Progressively, their role changes due to the reduction in the number of doctors and the growth in the complexity of OSH science. The goal in France is to reduce the individual medical surveillance and to develop the role of nurse and other actors of the occupational health department.
- *Health and safety managers*: The role was fixed by the framework directive in 1989, article 7 "the employer shall designate one or more workers to carry out activities related to the protection and prevention of occupational risks for the undertaking and/ or establishment… the workers designated must have the necessary capabilities and the necessary means…" (EU, 1989). The text had to bring into force in European countries by the end of 1992. It was initially completed in France in 2002 and then revised in 2015. Some companies designate an internal health and safety manager (France, 2011). For many small to midsized companies and a large number of French companies, the main source of OSH support is now implemented in the outsourced occupational health services. In these cases, all OSH programs and activities conducted in workplaces are directed by the authority of the occupational physician.
- *Arduous work*: The concept was introduced during negotiation (union, employer, state) on prolongation of working period before retirement. Since 2010, a list of constraints has established and described by intensity and by duration. Each employer must include arduous work in job hazard assessments. In cases that exceed arduous limits, the employee wins a maximum

of four yearly points. The first 20 points are available for training in order for employee to find a new job less exposed. Also, the rest of points can offer the maximum of 2 years of early retirement or 2 years at a part-time job. Both benefits require the employee to be paid at their initial rate. The program is covered by an added tax of all companies. Arduous work includes jobs that expose the worker to excessive noise, night work, shift work, extreme pressure, hazardous chemicals, repetitive actions, vibrations, extreme temperatures, manual handling of loads, and arduous postures.

- *Occupational disease*: Related to strong medical surveillance of workers, the recognition of occupational disease is highly developed. For the majority of employees, 98 different tables of occupational diseases are available. They contain the description of diseases, delay for medical action, and conditions that may be involved in work sources. Also, a parallel process is available for diseases that are not described in the list. Stress and psychosocial causes are becoming more recognized as occupational disease in France.

11.3.1.8 Conferences and Meetings

A variety of conferences are held annually in France on various health and safety topics. Many of these represent the several federal agencies with responsibilities for health and safety. For example, on June 29, 2016, several organizations met to discuss health insurance arrangements and worker compensation approaches. The following groups met on June 29, 2016:

- Caisse Primaire d'Assurance Maladie (CPAM)
- CRAMIF
- Union s de recouvrement des cotisations de sécurité sociale et d'allocations familiales (URSSAF)
- Service d'Appui au Maintien dans l'Emploi des Travailleurs Handicapés (SAMETH)
- Caisse Nationale d'Assurance Vieillesse des Travailleurs salaries (CNAVTS)

The French Society of Occupational Hygiene (SOFHYT) is the primary professional organization in France and has approximately 100 members. The group conducts a 1-day professional conference each year at a different city in France. Presentations include a broad range of OSH topics.

11.3.1.9 Plans for the Future

In a recent document published by the Ministère du Travail, de l'Emploi, and de la Formation professionnelle et du Dialogue social, they announced the plans for occupational health and safety for 2016–2020. The goals of the program are to be accomplished in two phases: (1) coordinating the approaches and activities of the five major governmental organizations with roles in workplace safety and (2) coordinating the roles and activities of nine nongovernmental agencies and organizations with roles in occupational health and safety (France, 2016c). The goal of the program is to the focus on a shift from a treatment of disability or illness-oriented system to a more preventative approach to worker health and safety. The overall goal is to shift the thought processes of many of these government agencies that currently emphasize programs to treat injured or ill workers, and switch to a more preventative OSH approach that will protect workers before they become injured or ill. A large part of this process involves improved communication, or spread of the understanding of the benefits of a preventative OSH approach. In a country where the profession of occupational health revolves around physicians and health care, this is a groundbreaking initiative. And getting these numerous federal organizations with existing missions and goals, labor organizations, employer groups, and the professional community to work towards a common goal of a preventative approach to occupational health in France will be a monumental task.

11.3.2 Spain

11.3.2.1 Demography of Spain

Spain is located in the southwest corner of Europe between the Atlantic Ocean and Mediterranean Sea. It has a population of 40 million and a per capital income of $14,500 per year.

11.3.2.2 Work structure

By the year 2002, Spain has modernized and working populations have shifted from agriculture (4.4%) to services (57.9%), industry (30.2%), and construction (7.5%) (Sese, 2002). Unfortunately during the years 2008–2013, the economy shrank and the number of small to midsized enterprises (SMEs) fell by 56%. By 2014, the unemployment rate in Spain was twice the average of the EU at 23% (EU Eurostat, 2017; Trading Economics, 2018).

With 18% of Spain's population living below the poverty line, many formal jobs have shifted to the informal economy (WB, 2015). Unfortunately, starting a business in Spain is extremely difficult due to layers of bureaucracy at several overlapping levels of government. The time need and expense of starting an SME are considerably higher than in other EU countries and vary greatly across Spain itself. Fortunately for workers, Spain's Social Insurance Institution guarantees health care and access to social services for all Spaniards. Benefits cover unemployment, illness, pensions, and family protection. The program is funded by employers, employees, and other federal government allocations.

11.3.2.3 Regulatory Framework

The first documented regulation regarding worker safety was in 1900 and was called the Law of Work Accidents in Spain. It required employers to indemnify their employees who were injured at work and created the fund to support the law in the Mutual of Work Accidents. This was followed in 1932 by the Law of Work Accidents that mandated insurance also covered occupational diseases, accidents, and death of workers. In 1963, the Spanish Social Security system was established and also covered accident insurance for workers. Early on, the Mutual of Work began to establish workplace risk prevention efforts. They subsequently worked closely with the Department of Labor and Social Affairs to establish safe practice guidelines and plans of action for worker health and safety.

11.3.3 Latin America, South America, Central American, and the Caribbean

Different terms used by organizations complicate discussion of Latin American OSH issues. Some sources describe South America as being comprised of 12 countries (Argentina, Bolivia, Brazil, Chile, Colombia, Ecuador, Guyana, Paraguay, Peru, Suriname, Uruguay, and Venezuela) Additionally, there are three United Kingdom Overseas Territories (Falkland Islands, South Georges, and South Sandwich Islands) and a region of France named French Guiana in the region. The combined population is approximately 428 million (World Population Review, 2018; Worldometers, 2018a, b). Brazil comprises just over half of the population on the continent.

Other reputable sources often include Mexico, the Caribbean, and Central America in a broader term of Latin America to describe the region (Worldatlas, 2018; Britanica, 2018).

11.3.3.1 Demography of Latin America

The variety of industry and economic advancement mirrors the grand diversity of terrain and climate across the continent. Countries rich in mineral resources such as aluminum, gold, silver, copper, and petroleum are heavily developed in the mining industries. Agriculture is another well-developed industry based on sugarcane, bananas, cattle, sheep, and coffee. In recent years, forestry has become a large industry in large regions of South America.

All South American countries have gross income inequalities across the social spectrum. In most Latin American countries, the richest 10% of the population receive between 40% and 47% of the income. The poorest 20% receive only 2%–4% (De Ferranti, 2004). Even the countries with the least income inequality in South America are worse than any other country listed by the OECD except for South Africa. The income differences for Brazil and Columbia were 55% in 2016 (OECD, 2016).

11.3.3.2 Work Structure

Much of the South American workforce remains in the informal sector. Workers in the informal sector comprised 48% of the workforce and represents approximately 130 million workers. Up to 10% of the workforce has no social protections (Lora, 1998; ILO, 2014). Seventy-two percent of these informal workers fall below the poverty line, and 63% have only a primary school education. Although most informal work is in the agricultural sector, considerable informal work is a significant portion of other industries as well: construction (69%) trade, restaurants, hotels (56%) (ILO, 2014). Attempts to reverse the trend of informality by creating new jobs in the formal sector are under way, but push against other economic factors such as low growth, unemployment, job insecurity, and poverty.

Partly due to the large percentages of workers in the informal sector, and partly due to the lack of regulations, the occupational injury, illness, and fatality rates for South America are extremely difficult to quantify accurately. Many countries do not require companies to report injuries, illnesses, and fatality statistics. In those that do, reporting is inconsistent and not well inspected or enforced (Giuffrida, 2001). In a 2012 report, Hong demonstrated that workplace risks of injuries and fatalities are higher than other regions, and 30 times higher than those in the United States (Hong, 2012). In a 2001 study, it was determined that manufacturing, utilities, and construction activities contributed to most fatal (43%) and nonfatal (50%) occupational accidents (Giuffrida, 2001). In a 1999 workshop of occupational health, the top three most important occupational health problems for South America were fatal occupational injuries, pesticide poisoning, and low back pain (Choi, 2001). Other widely reported diseases included hearing loss and skin and respiratory diseases (Giuffrida, 2001).

In an analysis of the South Americans' occupational hygiene situation, Giuffrida (2001) found the poor workplace health and safety conditions and related injury and illness statistics to be a result of three major factors. These were the general lack of hazard awareness by workers and management, incomplete data on causes of injuries, illnesses, and fatalities that tend to underestimate the problems, and the inability to develop and enforce workplace health and safety regulations due to the lack of infrastructure and educational resources and professional expertise (Giuffrida, 2001). In addition, the lack of understanding regarding relationships between unsafe and unhealthy working conditions and the impacts on society and the economy were not well understood.

11.3.3.3 Mexico

11.3.3.3.1 Demography

The Mexican population is 124 million as of July 2017. The median age of inhabitants is 28.3 years. The average life expectancy is 76.1 years (Index Mundi, 2018).

11.3.3.3.2 Work Structure

Although only 8% of Mexico's economy comes from agriculture, this sector's employees constitute 23% of its labor force. Conversely, manufacturing employees constitute about 11% of the workforce yet produces 23% of the GDP. Chief products include motor vehicles, iron, steel, chemicals, electronics, and synthetic fibers (Scholastic, 2018). In 2010, services contributed to 62.5% of Mexico's GDP led by tourism and banking (Economy Watch, 2010). Unfortunately, tourism is recently falling off due to increasing crime in Mexico.

11.3.3.3.3 OSH Performance and Statistics

In 2015, Mexico had a fatality rate of 7.9 per 100,000 workers. In 2016, this number rose to 8.2 (ILO, 2018a). Occupational diseases registered most frequently in Mexico include hearing loss, chemical bronchitis and pneumoconiosis, dermatitis, and musculoskeletal injuries due to cumulative trauma (Sanchez-Roman, 2006). The highest rates of occupational accidents are found in the food and beverage preparation and services industries. The metallic products industries have the highest proportion of permanent disability in workers, and ground transportation industries have the highest occupational fatality rates in Mexico.

In an important study of aging, workforce populations in Mexico determined that poor physical work conditions during their working careers reduced socioeconomic levels and mobility later in life. Workers at the lower end of the economic spectrum that spent their careers in jobs that required excessive workloads, standing, carrying heavy objects, working in hazardous environments, and sitting in moving vehicles for extended periods tended to be at a lower socioeconomic level, need more medical services, and be less mobile later in life (Beltran-Sanchez, 2017). These results were of particular concern to Mexico where government subsidized publicly funded free health insurance and coverage is provided to many formal sector workers.

11.3.3.3.4 Regulatory Framework

Standards and principles to be followed by employers for the field of occupational health are found in Article 123 of the Constitution of Mexico. Article 123 of the Political Constitution of the United States of Mexico establishes worker and employee rights and obligations and also separates workers into two groups: Group A is composed of private-sector workers, and Group B includes government workers. The Federal Regulation on Occupational Safety and Hygiene and the Working Environment was adopted in 1997 and includes most related laws on OSH. The main goals of the regulation are to prevent accidents and ensure safe and healthy working conditions.

In 2012, the Federal Labor Law was reformed to include provisions on OSH. In addition, several other Mexican regulations include sections that pertain to OSH activities. The duties of the General Director of Safety and Health at Work are described in the Regulations of the Secretariat of Labor and Social Welfare of 2008, and the General Regulations on the Inspection and Application of Sanctions concerning Labor Legislation Violations of 1998 contains other OSH provisions (ILO, 2013). The Federal Regulations for Occupational Safety and Hygiene is an extensive list of basic safety principles and performance standards that apply in industrial settings. In addition, the Secretariat of Labor issues directives on workplace safety that include a combination of performance-based and explicit standards.

Currently, Mexican OSH is governed and overseen by three major national agencies. The Secretariat of Labor and Social Welfare develops enforceable safety and health standards, conducts inspection, requires the use of joint OSH committees, collects accident and injury data, endorses research directions, and disseminates relevant OSH information. Specific workplace rules are called Official Mexican Norms (NOMs) and do not need to be approved by the legislature, but provide Federal agencies authority to inspect and enforce the rules. The NOMs include a combination of safety, health, and structural standards. The Mexican Institute of Social Security conducts the main worker compensation program in Mexico. And the National Advisory Commission on Occupational Safety and Health conducts studies and develops prevention and control measures to minimize risk and disseminates information on OSH topics.

Mexican federal labor law establishes three levels of joint committees that operate for OSH. Workplace committees are required for various sizes and types of operations and must be composed of equal numbers of employer and employee representatives that monitor compliance with regulations and support government inspections. These committees investigate accidents, prepare reports, and propose corrective actions (Offshore Group, 2017). As a member of the ILO, Mexico has ratified most occupational hygiene-associated agreements and recommendations.

11.3.3.3.5 Specific Aspects of Mexican OSH

Education and training on occupational hygiene, occupational safety, occupational health, and occupational medicine is lacking in Mexico and does not meet the needs of human resources (Sanchez-Roman, 2006). Sources of funding for OSH research are also scarce in Mexico, and there is no national agenda regarding research in occupational hygiene due to the general lack of interest of support for OSH in general.

OSH professionals and technicians tend to be employed in business, government, institutions of higher education, and as independent consultants (Sanchez-Roman, 2006). Businesses that understand the value of OSH services in decreased absenteeism, reduced costs of insurance, and other general business values are most likely employers of OSH staff.

As one of the largest trading partners to the United States, Mexico is one of the best example of some of the main issues confronting global health and safety issues today. Taking advantage of the inexpensive labor in Mexico, and the North Atlantic Free Trade Agreement (NAFTA) of 1994, thousands of U.S. production facilities, referred to as maquiladoras, were built along the borders of Mexico and the United States, and throughout Mexico in other regions. As of 2000, these companies employed more than 950,000 workers (Brown, 2000).

The total population of the border communities is estimated at between 11 and 20 million. The average daily earnings were under $4.00 per day (CorpWatch, 1999). Products of the manufacturing facilities include textiles, chemicals, electronics, machinery, shoes, and food products.

Although occupational safety and health regulations and protections for workers are nearly comparable to world standards on paper, very little is done to enforce the standards in the maquiladoras factories. Companies are encouraged to invest in the region, and there is a conception on the part of the Mexican regulatory agencies that stricter enforcement of health and safety standards would have a negative effect on the growth of the area. Companies operating in this intense production area are not just U.S. based. Other countries such as Korea, Japan, and Spain operate facilities in the region. Companies with these facilities follow a variety of safety practices that may reflect standards from their home countries or Mexican regulations; however, most of the time they do not (Brown, 2000).

11.3.3.3.6 Plans for the Future

Work-related accidents are severely underreported in Mexico due to poor administrative procedures and lack of worker representation. In addition, disability and illness statistics are also downplayed due to early expulsion of unhealthy workers before the completion of their normal work life. Underreporting of cancer, cardiovascular, and leukemia cases go unreported as occupationally related (Sanchez-Roman, 2006).

11.3.3.4 Brazil

Activities concerned with occupational health in Brazil are connected to three federal ministries. The Ministry of Labour enacts legislation on safety, hygiene, and occupational health. They Ministry of Labour also conducts inspections of workplaces through its regional units and also conducts research on health and safety. The Brazilian Ministry of Health provides medical care for work-injured workers or those diagnosed with occupational diseases. The Ministry of Health conducts certain worker surveillance programs for a specific list of occupational diseases. The Ministry of Social Security provides rehabilitation and compensation for injured and occupationally diseased workers (Bedrikow, 1997).

Unfortunately, formal education and training on occupational health and safety is severely lacking in Latin America. Formal research programs are also not extensive or well developed in all regions. The only federal government institution that conducts research and supports occupational safety and health education of the Portuguese-speaking countries is the Fudacentro organization under the ministries of health of Brazil. This organization supports a master's program in OSH and conducts other training, education program development, and dissemination of OSH knowledge (Fundacentro, 2018). In addition, the University of Sao Paulo conducts master's and doctoral programs in occupational hygiene.

In a study of occupational health research trends in Brazil, 1,025 thesis or dissertation reports were completed between 1970 and 2004, 866 of which were completed between 1990 and 2004. Most studies involved topics such as public health, musculoskeletal disorders, and health workers. Considering the known risks, relatively few research articles studied construction safety or occupational cancer (Santana, 2006).

11.3.3.5 Venezuela

Venezuela is the second largest country in South America in terms of land mass and population. It has the largest oil reserves in the world and is a leading oil exporter. Other natural resources in Venezuela include minerals, water, diamonds, gold, and aluminum. Yet with the increasing political turmoil in the past decade and the declining economy, nearly half the population lives in extremely poor conditions (El Universal, 2015).

Also in this timeframe, Venezuelan work became increasingly informal and precarious. The informal employment rate in 2010 was 50% and has probably risen since this study (Perazzi, 2010). As a result of their precarious working conditions, these workers often work in hazardous conditions and are excluded from social support systems including access to health care (Kim, 2010). Moreover, in 2005, 5.4% of Venezuelan children of 5–14 years old worked (USDOL, 2008).

In 2008, the working population was distributed between commerce (38%), manufacturing (18%), hotels and restaurants (9%), transportation (8%), social services (8%), real estate (6%), construction (5%), health and social services (4%), education (3%), and electricity, gas, and water (1%) (Instituto Nacional, 2010). In 2006, Venezuela was reported to have an occupational fatality rate of 15.6 per 100,000 people. This is relatively high compared with the United Kingdom (0.8) and Sweden (1.4) (ILO, 2006). This number is also expected to be grossly underestimated because of the significant amount of workers in the informal sectors, which do not generally report injuries or illnesses.

Political turmoil and economic downturns have kept Venezuela from fully implementing existing OSH regulations. Injury and illness rates for workers are unusually high despite gaps in the data for a large percentage of the workforce (Caraballo-Arias, 2015).

11.3.3.6 Peru

Peru is located on the central Pacific coast of South America. The population of approximately 30 million lives in a surface area of 1.2 million square kilometers.

The Peruvian economy depends largely on the extraction of resources in mining and forestry. Partly due to increased demand for raw materials, Peru's economy has been growing at about 3% per year, above the average of other South American countries. In addition, poverty declined about 50% over the period from 2001 to 2011, partly due to economic growth and partly as a result of other social measures (Cruz, 2015). The top areas of employment are construction, transportation, communications, agriculture, and hospitality. Each of these occupational sectors is characterized by high levels of informal work (Cruz, 2015).

The first workplace health and safety regulations were established in Peru in 1908. Since that time, more than 80 regulations have been promulgated and 70 international conventions have been agreed to, including 30 ILO recommendations (Cruz, 2015). The National System of Health and Safety at Work consists of the National Council for Safety and Health at Work and the Regional Council for Safety and Health at Work. In 2001, a General Labor Inspectorate was created in the government, and the first OSH regulations from this group went into effect in 2005. In 2011, these laws were boosted with Law 29783 on Safety and Health at work that called for a culture of prevention and compliance with federal laws through inspection and worker participation. Since the adoption of the 29873 Law, more companies are expanding their OSH management programs to address the new rules and potential for worksite inspections (Cruz, 2015).

OSH education in Peru consists of courses in postgraduate programs. There are master's programs in occupational and environmental medicine, occupational and environmental health, ergonomics, occupational nursing, and occupational hazard prevention (Cruz, 2015).

11.3.4 India

11.3.4.1 Demography

In 1984, a plant operated by a U.S. firm, Union Carbide, released 41 tonnes of deadly methylisocyanate into the air at night and killed more than 3,000 people in Bhopal, India. Another 300,000 people suffered lifelong injury and illness from their exposures (Gupta, 2002).

At this plant, financial problems led to operational cutbacks that affected the safe operations of the plant. In addition, due to the lack of housing codes or failure to enforce them, a large population density grew very close to the plant property line so there was no distance between the release and the sleeping people. There were no emergency response procedures or systems in place to notify the local population of an eminent release, so people were on their own to figure out what to do and how to escape.

This disaster led to significant changes in the regulations for plant operations, emergency response, and environmental protection in India, the United States, and around the world. It is perhaps the best representation of how technology transfer from developed countries can go terribly wrong when implemented in those with limited resources.

Today, India is one of the world's fastest growing economies, and it is also the world's largest democracy. In 2017, India had 1.3 million people (Statista, 2018a, b). Sixty-eight percent of India's population is of working age (Indian Express, 2017). Ninety percent of India's workforce is in the informal economy, mostly in agriculture and services, and 60% of this is self-employed (Pingle, 2012).

11.3.4.2 Health and Safety Performance

Major occupational risks in India are accidents, pneumoconiosis, silicosis, musculoskeletal injuries, chronic obstructive lung diseases, pesticide poisoning, byssinosis, asbestosis, noise-induced hearing loss, and workplace stress (Pingle, 2012). According to a report by the Indian statistics organization Directorate General, Factory Advice Service, and Labour Institutes (DGFASLI), there were 29,837 injuries and 1,433 fatalities in factories in India in 2011 (India DGFASLI, 2018). It would appear that these injuries and fatalities would be somewhat underreported since the entire working population for the period was 11,634,070 total workers.

11.3.4.3 Regulatory Framework

In 2008, the India Ministry of Labour and Employment updated national regulations and created a National Policy on Safety, Health and Environment at Workplace (Indian IMLE, 2008). The policies lay down directive principles meant to guide basic regulatory programs for worker safety and health. The policies are based on the belief that sound safety and health incentives and regulations for employers will benefit national economics, individuals, and society by reducing the incidence of work-related injuries, illnesses, and fatalities. It is also believed that the rules will also benefit the environment. The goals of the statutes will be to design a control system of compliance, inspection, and enforcement, in addition to research and data collection that will promote and improve safe and healthy work environments. The objectives of the policy and associated programs will also include the promotion of health and safety principles to workers and the public, and expansion of safety and health education and training for professionals.

Policies and programs for the informal sectors including agriculture and services have yet to be developed in India. It is hoped that legislative advances in the formal sectors of mining, docks, and construction with trickle into the broader public consciousness begin to impact workers in other sectors.

One of the greatest challenges ahead for India is the shortage of occupational safety and health professionals. There are currently around 1,000 occupational health physicians and industrial hygienists, when 10,000 would be warranted in the organized sectors alone (Pingle, 2012). There is a need for both short-term training courses to address specific topics, and full university programs in OSH and occupational medicine. In the long range, national accreditation standards for OSH professionals need to be developed.

11.3.5 Africa

11.3.5.1 Demography

The African continent comprises 54 independent nations. Twenty-six of the countries are classified as middle-income countries by the World Bank. Seven of these countries are considered least developed nations, two are "fragile," and nine are regarded as "highly indebted poor countries" (ILO, 2016c). These independent nations with different government systems, economies, languages, and cultures also provide very different outlooks of the rights of workers. In many African countries, governments are still grappling with poverty, pandemics, climate change, religious intolerance, and political instability. The rights of workers and their health and well-being have not risen to the forefront of the regulatory landscape.

With a population of 633 million, Africa represents a huge number of workers exposed to a multitude of workplace hazards in every imaginable industry (ILO, 2016c). And according to the ILO, African workplaces claim over 59,000 lives each year. Studies have shown that in Africa, work-related injuries, illnesses, and fatalities cause twice the level of disability-adjusted life years (DALY) than in North America and Europe (Ezatti, 2002). But it is also assumed that these numbers are largely underrepresented in that many injuries, illnesses, and fatalities that result from workplace exposures go unreported.

Many countries in Africa continue to be challenged by high rates of infectious disease, including human immunodeficiency virus, tuberculosis, Ebola, and others. These infectious agents continue to drag down labor productivity and the capital stock of human resources. Infectious disease has also been shown to play a role in food availability and overall longevity of the populations (Adedeji, 2016). When considering risk and hazard assessment in Africa, it is important to consider the role of infectious disease in the work environment in many African workplaces.

11.3.5.2 Work Structure

The African workforce is in a particularly precarious position due to the lack of continuous and stable work and climate instability for which such a large part of the workforce is dependent. Where large segments of the workforce are migrants, and the remaining have few rights or regulations to protect them, employers generally see health and safety as an expense going against production and profit (Marie, 2006). Workers are willing to take hazardous work and are reluctant to voice concerns about safety for fear of losing their job (Hilgert, 2013). The large number of African workers working in informal sectors are particularly vulnerable due to the lack of regulatory oversight and the instability of the work (Benach, 2007). Worker vulnerabilities in Africa are not only a result of the power differentials between the employer and worker, but many other factors including race, class, ethnicity, and gender all contribute to the work landscape. The levels of vulnerability for African laborers are far greater than those experiences in other countries, particularly western nations. Historical and cultural conditions across the continent make it difficult to ameliorate the worker vulnerabilities strictly through the promulgation of regulations, even if they were enacted (London, 2014).

Despite the significant impact on worker health, relatively little peer-reviewed research is published regarding African workplaces. High hazard industries of agriculture, mining, and informal labor make up over 65% of the jobs in Africa. Agriculture, which is one of the largest employers in most African countries, is a particularly hazardous sector to workers, being seasonal, low paid, and isolated work. And unlike mining and industry, which have received regulation in some nations, agriculture remains largely informal and unregulated (London, 2014). There is a particular need for research that demonstrates the economic value to a nation when occupational health and safety is regulated and worker health and longevity are improved; thus, there is a need for comprehensive workplace regulation. Evidence shows that Africa has also been a recipient of the transfer of dangerous industries and toxic chemicals shifting away from developed and western/northern countries (Swuste, 2002). Figure 11.1 shows construction workers in Morocco working on insufficient scaffolding.

Africa is a continent of migration. Population growth is estimated to add 380 million workers to the job market by 2030, and 220 million of those jobs are expected to be in rural development and agriculture (FAO, 2018a, b). Seventy-five percent of migration occurs within Africa by people looking for work. Directions shift periodically between rural and urban areas and are greatly affected by crops, weather, and even climate change. As migrants, these workers often do not have the same worker rights, understanding of language, receive the same safety training, have access to medical care, or are likely to report unsafe conditions or injuries to their employer,

FIGURE 11.1 Construction workers in Morocco. (Photographer: Thomas P. Fuller.)

compared to nonimmigrant workers (Shelley, 2007). Migrants have been shown to work longer hours and in more hazardous jobs than nonmigrants (Eurofound, 2007; Basok, 2002). All factors put this work group at greater risk of injury or fatality in their jobs (Liebman, 2013).

Africa is also a continent of child labor. According to the United Nations, Africa contributes 47% of child labor in the world. There are between 59 and 72 million child laborers in Africa, and 83% of those are under the age of 12 (FAO, 2018a, b, ILO, 2018b).

11.3.5.3 Health and Safety Performance

It has been estimated that occupational injuries and illnesses in Africa cause twice as many years (DALYs) per thousand people as in Europe or North America (Ezatti, 2002). In a review by Moyo (2015), several shortcomings were identified in the status of a group of South African countries (South Africa, Zimbabwe, Zambia, and Botswana). Moyo noted significant inadequacies in terms of resource capital for OSH initiatives, and a shortage of training in occupational health services. Unfortunately, much of the article focused on occupational medicine and the treatment of workers after they are injured or ill from workplace exposures. The article was oriented towards the outdated, physician-oriented, reactive approach to OSH, rather than the preventative approach provided by the practice of occupational hygiene. This orientation demonstrates the remaining lack of awareness in South Africa regarding what occupational hygiene is, and how it can benefit workers. There is a great need for additional OSH training, research, and collaboration in the field of OSH in South Africa (Joubert, 2002).

Although OSH research in Africa is sparse, there have been a handful of studies or reviews. In one study of OSH practice in Kenya, it concluded that in government-operated mortuaries, there was an overall failure to implement universal safety precautions for hazardous exposures to infectious agents (Okoth-Okelloh, 2015). The study found a lack of safety culture that included inadequate administrative programs, engineering controls, and personal protective equipment (PPE) to protect workers. In a study of construction workers in Ethiopia, it was concluded that a significant loss of productivity and increased worker injuries occurred due to shortcomings in OSH training and the use of PPE (Tadesse, 2016).

11.3.5.4 Regulatory Framework

South Africa has the most advanced systems of regulatory development and enforcement. In addition, the Witwatersrand University in South Africa offers education in occupational hygiene through the doctoral level. Yet with these advances, there remains a gap between occupational health professionals and large numbers of workers that remain in unhealthy and unsafe working conditions (Loewenson, 2004).

In South Africa, occupational safety and health falls under the Department of Labor through the Occupational Safety and Health Act and Amendments (South Africa, 1993). The Department writes regulations and employs workplace inspectors that police and enforce occupational hygiene legislation.

Nigeria was the first country to host a seminar on Occupational Health for Developing Countries in 1968. In 2005, there was an OHS meeting in Benin.

TABLE 11.1
ILO Conventions at Work

Convention	Number of Ratifications
Forced labor convention no. 29—1930	53
Abolition of forced labor convention no. 105—1957	53
Discrimination convention no. 111—1958	53
Right to organize and collective bargaining convention no.98—1949	52
Worst forms of child labor convention no. 182	50
Minimum age convention no. 138—1973	49
Occupational safety and health convention no. 155—1981	13
Occupational cancer convention no. 139—1974	2

Source: ILO (2018c, d, e).

Also in 2005, the International Occupational Health Association conducted their sixth triennial conference in Pilanesberg, South Africa.

Many countries see a value in added worker protections in terms of public health and social benefits, in addition to national economic prosperity. However, development of strong national regulations is labor intensive and dissemination of information and enforcement remains difficult to implement with limited knowledgeable staff and limited resources in most African nations.

Most African nations have ratified many of the ILO conventions for work. Out of the 54 nations, the numbers in Table 11.1 indicate which conventions have been ratified.

With greatly expanding international investment in Africa, many companies are forcing implementation of higher safety standards for workers through corporate social responsibility programs. Investors want to be sure that worksites follow international safety standards and comply with international treaties. Many developed countries require global companies to follow the same OSH regulations for workers abroad. And the definitions can often include the nationals hired to perform work on international projects. Consequently, more African states and companies are realizing the need for OSH programs and educated professionals in the area, as a means to remain competitive.

Unfortunately, many specifically African companies have not yet come to realize the economic and financial advantages to having good OSH programs in place. In addition, many international companies coming from countries, such as China, do not require safety standards for international workers and do not adhere to ILO treaties on worker safety.

11.3.6 Japan

Japan has a long history of laws regarding worker safety, beginning in 1916 with the national Factory Law meant to protect workers. This was followed by the Labour

Standard Law introduced in 1947 that included more broad regulations on occupational health and safety. Other laws that followed included the following:

- Pneumoconiosis Law—1960
- Ordinance on Prevention of Organic Solvent Poisoning—1960
- Ordinance on Safety and Health at Work under High Pressure—1961
- Ordinance on the Prevention of Lead Poisoning—1967
- Ordinance on Prevention of Hazards due to Specified Chemical Substances—1971
- Industrial Safety and Health Act—2006

In 1972, all of the above ordinances were made to conform with a new overarching regulation with the enactment of the Industrial Safety and Health Law (Sakurai, 2012).

Since the implementation of the Industrial Safety and Health Law in 1972, occupational injury and illness rates have continued to decline, while the working population has increased. In 2010, the main occupational illnesses in Japan were low back pain (61.2%), heat disorders (10.1%), pneumoconiosis (6.4%), chemical injuries (2.9%), and biological agents (1.6%) (Sakurai, 2012). Overwork and long working hours has long been a recognized occupational hazard in Japan (Hori, 2012). The related mental stress due to overwork has been associated with cerebrovascular disease, ischemic heart disease, and suicide.

In Japan, the Occupational Medicine Physician assumes a prominent role in occupational and environmental health. In 1978, the University of Occupational and Environmental Health was established for the purpose of training occupational physicians, nurses, and hygienists.

Research on OSH in Japan is conducted by the Japanese National Institute of Occupational Safety and Health (JNIOSH), which was established in 1942. JNIOSH publishes the oldest OSH journal in Japan called Industrial Health. Other related OSH organizations in Japan include the Japan Occupational Hygiene Association and the Japan Society of Occupational Health.

Concerns for the future of OSH in Japan largely involve the slow change from large industrial manufacturing environments to smaller companies. The incidence of injury of workplaces with less than 50 employees is about two times higher than larger enterprises (Sakurai, 2012). It is anticipated that the breakdown of large conglomerates, the rapid decrease in population, the increase in nonstandard forms of employment, and the increasing intensification of work will all be contributing factors to additional worker risks in Japan moving forward.

11.3.7 China

11.3.7.1 Demography

The current population of China is 1.4 billion and comprises 18% of the people on the globe. The median age is 37.3 years, and 59.3% of the citizens live in urban centers (Worldometers, 2018a, b; Statista, 2018a, b).

11.3.7.2 Work Structure

According to the ILO, China had approximately 779 million workers. Roughly 34 percent of these work in the informal sector (ILO, 2012). The numbers of workers in various sectors are shown in Table 11.2.

11.3.7.3 Health and Safety Performance

China's 2001 workplace fatality rate has been reported as 11.1 per 100,000 workers (Zhang, 2010). Pneumoconiosis continues to be a significant occupational health problem in China, with as many as 12 workers exposed to silica containing dusts, and 15,000 new cases of pneumoconiosis diagnosed annually. Most of the exposures occurred in the mining industries, followed by construction, manufacturing, metals, and machinery industries (Liang, 2004). Workplace fatalities in coal mining have been reported as high as 173.88 deaths per 100,000 workers. In construction, workplace fatalities are listed at 31 per 100,000 workers (ILO, 2012).

Up to 20% of occupational illness in China comes from exposure to chemicals and other toxic agents. Acute poisoning (4.2%) and chronic poisoning (11.26%) were the most common causes of occupational illness in China (Ji, 2008). Carbon monoxide, hydrogen sulfide, and ammonia made up about 30% of acute poisonings, of which 50% were fatal. The main industries where these accidents occurred were chemical, manufacturing, water treatment, mining, and construction. Other common occupational diseases in China involved musculoskeletal and psychological hazards. Job stress is a common psychological work-related disease in China with 20% of the working population suffering from mental health problems (Fan, 2004).

In 2000, it was reported that the routine occupational health service coverage provided for township enterprises was only 1.4%–36% (Zhi, 2000). In a 2012 study of occupational diseases in China by the ILO (2012), the top five occupational health problems were as follows:

- Excessive numbers of occupational diseases are reported.
- Pneumoconiosis comprises 87.42% of all occupational diseases.

TABLE 11.2
Number of Employees in Different Sectors in China (in million)

Sector	2008
Total	121.9
Manufacturing	34.3
Construction	10.7
Transportation	6.3
Mining	5.4
Agriculture, forestry, fisheries	4.1

Source: Adapted from ILO (2012).

- There is broad variation in the types of diseases occurring and the numbers of industries where they occur.
- Many of the occupational diseases and poisonings are difficult to cure by the time they are diagnosed.
- There are a large number of occupational poisonings that occur, and they often affect large numbers of workers at once.

11.3.7.4 Regulatory Framework

The Chinese constitution provides the foundation for occupational health and safety legislation in China. Subsequent laws are enacted by the National People's Congress and Standing Committees. Administrative regulations are enacted by the State Council. Local regulations are promulgated by the People's Congress or their Standing Committees at the provincial, regional, and municipal levels of government. The Standards Administration of China is responsible for creating national standards of practice throughout the country, and particular standards for various industries are set by the State Council (ILO, 2012).

The Law of the People's Republic of China on Work Safety became effective in 2002. It includes provisions for work safety, duties of employers, accident response, conducting investigations, and other legal responsibilities. Other laws related to safety include the law on safety in mines which came into effect in 1993, and the law on the prevention of occupational diseases promulgated in 2002.

Regulations for safety management of construction projects came into effect in 2004. These laws include safety responsibilities for owners, contractors, and management of construction projects. Regulations for the safe administration and handling of hazardous chemicals became effective in 2002 and describe requirements for storage, use, and transportation. In 2011, hazardous chemical regulations were revised to include requirements for licensing, additional administrative controls, reduce requirements on shipping poisonous chemicals and for transportation of hazardous chemicals on inland waters, improve methods of chemical identification, and increase punishments for violations of laws.

Tripartite coordination between China Enterprises Confederation (employers), the All-China Federation of Trade Unions (workers), and the Ministry of Human Resources and Social Security (government) is required under the Chinese Trade Union Law of the People's Republic of China and the Labour Law of the People's Republic of China. The purpose is to provide participation and democratic management to discuss and resolve major OSH problems and issues. Research is meant to inform regulatory development and other improvements. The research also includes public safety, industrial safety, occupational hazards, and chemical safety.

Some relevant websites for Chinese regulatory organizations are provided in the following list:

Ministry of Human Resources and Social Security—www.mohrss.gov.cnds
Ministry of Health—www.moh.gov.cn
State Administration of Work Safety—www.chinasafety.gov.cn
China Academy of Safety Science and Technology—www.chinasafety.ac.cn

Chinese Center for Disease Control and Prevention—www.chinacdc.cn/en
Secretariat of the National Safety Standardization Technical Committee—www.chinasafety.ac.cn

11.3.7.5 Research

The China Academy of Safety Science and Technology (CASST) is a national research institute that conducts research on accident prevention, safety, and emergency response. This includes safety science and technology (CASST, 2018).

In the past, China offered prime locations for epidemiological evaluations of worker exposures due to the relatively long-term employment and consistent exposure conditions. In recent years however, occupational migration and work in the informal sectors have increased dramatically, making these evaluations more difficult and less likely.

11.3.7.6 Legislative Updates

China advances in work safety continue to lag, and there remain a high number of workplace accidents occurring nationwide. Foundations of safety organization remain weak, and several industries continue to operate at high levels of risk. Support of regulatory systems remains weak, and many agencies cannot fully implement existing programs; as a result, loopholes allow unsafe systems and practices to continue. In addition, outdated safety technology, aging equipment, and low capacity in safety management continue to endanger both workers and the general public (ILO, 2012).

In 2006, a national network of disease reporting was initiated. Occupational disease is reported to the Institute of Occupational Health and Poisoning Control in the Chinese Center for Disease Control and Prevention. The system includes (1) standard reporting indicators, (2) better means to track data, (3) real-time data to be used to evaluate prevalence trends, and (4) specific reporting requirements (Chen, 2008).

11.3.7.7 Plans for the Future

Occupational Safety and Health Management Systems (OSHMSs) have grown in popularity and usefulness in China in recent years. By August 2011, 75 OSHMS certification bodies had been accredited in China, and hundreds of technical consulting services now provide program development to a wide variety of industries including electric power, construction, petrochemical, and manufacturing (ILO, 2012). By the end of 2011, there were more than 30,000 enterprises accredited with certification.

Despite the creations of new regulations and government organizations with responsibilities for worker health and safety in China, there remain significant worker exposures to hazardous agents and working conditions, and associated risks of illness and injury. Considering the population and size of the country, research on OSH topics is lacking in many growing and historic areas. In addition, the shortage of OSH educated and experienced professionals curtails full implementation of proposed OSH goals that have been promoted. OSH capacity must be developed by expanding related research and educational opportunities.

REFERENCES

Adedeji, A., Akinlo, A., Infectious diseases, longevity and labour productivity in Sub-Saharan Africa, *African Development Review* (2016) Vol. 28, No. 1, pp. 127–139.

ANSES, Agence nationale de sécurité sanitaire de l'alimentation, de l'environnement et du travail (2015) Annual Report 2015, www.anses.fr/fr/system/files/RA2015.pdf/ accessed June 8, 2016.

Aubrun, J., Binet, S., Bozec, C., Brochard, S., Fontaine, B., Guenel, P., Luce, D., Martinet, Y., Moulin, J., Mur, J., Pietruszynaski, M., Vallayer, C., Occupational cancer in France: Epidemiology, toxicology, prevention, and compensation, *Environmental Health Perspectives* (May 1999) Vol. 107, No. Supplement 2, pp. 245–252.

Basok, T. (2002). *Tortillas and Tomatoes. Mexican Transmigrant Harvesters in Canada*. Montréal: McGill-Queen's University Press.

Bedrikow, S., Algranti, E., Buschinelli, J., Morrone, L., Occupational health in Brazil, *International Archives of Occupational and Environmental Health* (1997) Vol. 70, No. (4), pp. 215–221.

Beltran-Sanchez, H., Pebley, A., Goldman, N., Links between primary occupation and functional limitations among older adults in Mexico, *SSS Population Health* (2017) Vol. 3, pp. 382–392.

Benach, J., Muntaner, C., Santana, V. (2007). *Employment Conditions and Health Inequalities. Final Report to the WHO Commission on Social Determinants of health (CSDH). In Employment Conditions Knowledge Network (EMCONET)*. Geneva: World Health Organisation.

Britanica, List of countries in Latin America (2018) www.britannica.com/topic/list-of-countries-in-Latin-America-2061416/ accessed August 6, 2018.

Brown, G. (November 2000). *Double Standards: U.S. Manufacturers Exploit Lax Occupational Safety and Health Enforcement in Mexico's Maquiladoras*. Washington, DC: Multinational Monitor, pp. 24–28.

Caraballo-Arias, Y., Occupational safety and health in Venezuela, *Annals of Global Health* (July–August 2015) Vol. 84, No. 4, pp. 512–21.

China Academy of Safety Science and Technology (CASST), Homepage (2018) www.china-safety.ac.cn accessed July 22, 2018.

Chen, S., Wang, D., Zhang, M., Report of occupational diseases in internet era (in Chinese), *Chinese Journal of Industrial Hygiene Occupational Diseases* (2008) Vol. 11, pp. 673–675.

Choi, B. C. K., Tennassee, L. M., Eiikemans, G. J. M., Developing regional workplace health and hazard surveillance in the Americas, *Panam Salud Publica* (2001) Vol. 10, No. (6), pp. 376–381.

CorpWatch, Maquiladoras at a glance, CorpWatch website (1999) http://s3.amazonaws.com/corpwatch.org/downloads/maqfacts.pdf accessed June 13, 2016.

Cruz, I., Huerta-Mercado, R., Occupational safety and health in Peru, *Annals of Global Health* (2015) Vol. 81, No. 4, pp. 568–575, ISSN 2214-9996.

DARES La syndicalisation en France—Des salaries deux fois plus syndiqué dans la function publique, Direction de l'Animation de la Recherche, des Etudes et des Statistiques, (2016) http://dares.travail-emploi.gouv.fr/IMG/pdf/2016-025.pdf accessed July 21, 2018.

De Ferranti, D., Perry, G. E., Ferreira, F. H. G., Walton, M. (2004). *Inequality in Latin America*. Washington, DC: World Bank. EC, ECHI Data Tool, European Commission European Union http://ec.europa.eu/health/dyna/echi/datatool/index.cfm?indlist=31 accessed July 21, 2018.

Economy Watch, Mexico Industry Sectors, (March 24, 2010) www.economywatch.com/world_economy/mexico/industry-sector-industries.html accessed May 22, 2018.

El Universal, Al finalizar 2015 la pobreza en Venezuela será de 55%. Caracas: EU (2015) www.eluniversal.com/nacional-y-politica/150615/al-finalizar-2015-la-pobreza-envenezuela-sera-de-55; 2015 accessed July 20, 2015.

EU, Council Directive 89/391/EEC (1989) http://eur-lex.europa.eu/legal-content/EN/TXT/PDF/?uri=CELEX:01989L0391-20081211&from=EN accessed July 21, 2018.

EU, Eurofound research, data and analyses—Living and Working in France (2015) www.eurofound.europa.eu/countries/France accessed July 20, 2018.

EU Eurostat (2017) http://appsso.eurostat.ec.europa.eu/nui/show.do?dataset=lfsi_long_e01&lang=en accessed July 31, 2018. As of July 2018, the unemployment rate had rebounded to 15.3 percent.

Eurofound (May 2007). *Employment and Working Conditions of Migrant Workers*. Dublin: EWCO, p. 37.

EC, European Commission, ECHI Data Tool, Health Status Indicators, Workplace Injuries (2018) http://ec.europa.eu/health/dyna/echi/datatool/index.cfm?indlist=31 accessed December 11, 2018.

Ezatti, M., Lopez, A., Rodgers, A., Vander Hoorn, S., Murray, C., Selected major risk factors and global and regional burden of disease, *Lancet* (2002) Vol. 360, pp. 1347–1360.

Fan, X., Epidemiological study of job stress, *Occupational Health and Emergency Rescue* (2004) Vol. 1, p. 53.

FAO, Food and Agriculture Organization of the United Nations, Rural Africa in Motion (2018a) www.fao.org/news/story/en/item/1054009/icode/ accessed March 2018.

FAO, Food and Agriculture Organization of the United Nations, The international partnership for cooperation on child labor in agriculture on the new ILO global estimates of child labor (2018b) www.fao.org/rural-employment/resources/detail/en/c/1038934/ accessed March 10, 2018.

France, Loi du 9 avril, 1898 sur les responsibilities des accidents don't les ouvriers sont victims dans leur travail (Bulletin de l'Inspection du travail, No. 2, 1898) loi du 9-04-1898 sur les accidents du travail (1898) http://travail-emploi.gouv.fr/IMG/pdf/Loi_du_9_avril_1898.pdf accessed July 21, 2018.

France, JORF du 27 octobre 1919 page 11973 Loi du 25 octobre 1919 Etend aux maladies D'Origine Professionnelle la Loi du 9-04-1898 sur les Accidents du Travail (1919) www.legifrance.gouv.fr/affichTexte.do;jsessionid=3247667DA8168B7B-F86035791ADD54DF.tpdila16v_1?cidTexte=JORFTEXT000000869167&dateTexte= accessed July 21, 2018.

France, Service Medicaux du travail, Journal Officiel de la Republique Francaise, JORF du 12 octobre 1946 page 8638 Loi n°46-2195 du 11 octobre 1946 organisation des services médicaux (1946) https://gallica.bnf.fr/ark:/12148/bpt6k65338354/f12.image.texteImage accessed July 21, 2018.

France, Loi n ° 91-1414 du 31 décembre 1991 modifiant le code du travail et le code de la santé publique en vue de favoriser la prévention des risques professionnels et portant transposition de directives européennes relatives à la santé et à la sécurité du travail (1991a) www.legifrance.gouv.fr/affichTexte.do?cidTexte=JORFTEXT000000173965&categorieLien=id accessed July 21, 2018.)

France, Directive 89/391/EEC-OSH "Framework Directive" (1991b) https://osha.europa. eu/fr/legislation/directives/the-osh-framework-directive/1 accessed July 21, 2018.

France, Law n ° 2001-397 of 9 May 2001 on professional equality between women and men (2001) www.legifrance.gouv.fr/affichTexte.do?cidTexte=JORFTEXT000000756495 accessed July 21, 2018.

France, Décret n° 2008-244 du 7 mars 2008 relatif au code du travail (2008) www.legifrance.gouv.fr/affichTexte.do?cidTexte=JORFTEXT000018254394 accessed July 21, 2018.

France, Décret n° 2009-1377 du 10 novembre 2009 relatif à l'organisation et aux missions des directions régionales des entreprises, de la concurrence, de la consommation, du travail et de l'emploi (2009) www.legifrance.gouv.fr/affichTexte.do?cidTexte=JORFTEXT000021259245 accessed July 21, 2018.

France, Law n° 2002-73 du 17 janvier 2002 revised by loi n° 2011-867 du 20 juillet 2011, Santé, solidarité, sécurité sociale (2011) www.legifrance.gouv.fr/affichTexte.do?cidTexte=JORFTEXT000000408905 accessed July 21, 2018.

France, l'Assurance Maladie, Risque accident du travail: Statistiques sur la sinistralité de l'année 2015, suivant la nomenclature d'activités française (NAF) (2015) www.risquesprofessionnels.ameli.fr/fileadmin/user_upload/document_PDF_a_telecharger/etudes_statistiques/AT_2015/AT%202015-%20par%20NAF%20(n-2016-100).pdf accessed July 21, 2018.

France, Code de Travaille, Legifrance, JORF No.0066 du 1er mars 2016 texte No. 30, Arrêté du 1er mars 2016 relatif aux modalités de l'évaluation des risques résultant de l'exposition aux rayonnements optiques artificiels en milieu de travail (2016a) www.ilo.org/dyn/natlex/docs/ELECTRONIC/101468/122280/F-1470284102/FRA-101468.pdf, accessed July 22, 2016.

France, Code de Travaille, Legifrance, JORF No.0071 du 24 mars 2016 texte No. 24, Décret n° 2016-344 du 23 mars 2016 fixant une valeur limite d'exposition professionnelle contraignante pour le styrène (2016b) www.ilo.org/dyn/natlex/docs/ELECTRONIC/101636/122555/F-647956008/FRA-101636. pdf accessed July 22, 2016.

France, Ministère du Travail (2016c) Plan du santé au travail 2016-2020, http://travail-emploi.gouv.fr/IMG/pdf/exe_pst_2016-2020_ok_v7_web.pdf accessed July 23, 2016.

Fundacentro, Postgraduate studies (2018) www.fundacentro.gov.br/pos-graduacao/inicio accessed March 12, 2018.

Giuffrida, A., Iunes, R., Savedoff, W. (May 2001). *Economic and Health Effects of Occupational Hazards in Latin America and the Caribbean*. Washington, DC: Inter-American Development Bank, Pub. No. SOC-121.

Gupta, J., The Bhopal gas tragedy: Could it have happened in a developed country? *Journal of Loss Prevention in the Process Industries* (2002) Vol. 15, pp. 1–4.

Hilgert, J. (2013). *Hazard or Hardship. Crafting Global Norms on the Right to Refuse Unsafe Work*. New York: Cornell University Press.

Hong, O., Exploring occupational health nursing in South America through Brazilian experience, *Workplace Health and Safety* (2012) Vol. 60, No. (3), pp. 115–121.

Hori, M. (2012). The trend and issues of occupational safety and health in Japan. *Procedia Engineering, International Symposium on Safety Science and Engineering in China, 2012 (ISSSE-2110)* Vol. 43, p. 610614.

International Labor Organization (ILO), Organización Internacional del Trabajo. Seguridad y salud en el trabajo Perfiles nacionales. Perfiles nacionales: OIT (2006) www.ilo.org/legacy/english/protection/safework/worldday/products06/report06_profiles_sp.pdf; 2006 accessed March 13, 2018.

International Labor Organization (ILO), International Labor Organization, National Profile Report on Occupational Safety and Health in China (March 2012) (2012) www.ilo.org/wcmsp5/groups/public/---ed_protect/---protrav/---safework/documents/policy/wcms_186991.pdf accessed July 23, 2018.

International Labor Organization (ILO), Mexico Description of national OSH regulatory framework LRGOSH (2013) www.ilo.org/dyn/legosh/en/f?p=14100:1100:0::NO:1100:P1100_ISO_CODE3,P1100_SUBCODE_CODE,P1100_YEAR:MEX,2013:NO accessed August 6, 2018.

International Labor Organization (ILO), 2013 Labor Overview Latin America and the Caribbean (2014) ISSN: 2305-025X (web pdf version) www.ilo.org/wcmsp5/groups/public/---americas/---ro-lima/documents/publication/wcms_242634.pdf accessed March 12, 2018.

International Labor Organization (ILO), International Labor Organization, LEGOSH Global database on occupational safety and health legislation (2016a) www.ilo.org/safework/info/publications/WCMS_217849/lang--en/index.htm accessed June 29, 2016.

International Labor Organization (ILO), LEGOSH Regulatory Framework France 2015 (2016b) www.ilo.org/dyn/legosh/en/f?p=14100:1100:0::NO:1100:P1100_ISO_CODE3,P1100_YEAR:FRA,2015:NO accessed July 23, 2016.

International Labor Organization (ILO), Decent work in middle-income countries in Africa—A framework for ILO engagement, ISBN 978-92-2-130978-9 (web pdf) (2016c) www.ilo.org/wcmsp5/groups/public/---africa/---ro-addis_ababa/documents/publication/wcms_533498.pdf accessed March 10, 2018.

International Labor Organization (ILO), Fatal occupational injuries per 100,000 workers by sex and migrant status (2018a) (%)www.ilo.org/ilostat/faces/oracle/webcenter/portalapp/pagehierarchy/Page27.jspx?subject=OSH&indicator=INJ_FATL_SEX_MIG_RT&datasetCode=A&collectionCode=YI&_afrLoop=1260730701916965&_afrWindowMode=0&_afrWindowId=185a2lc3h7_1#!%40%40%3Findicator%3DINJ_FATL_SEX_MIG_RT%26_afrWindowId%3D185a2lc3h7_1%26subject%3DOSH%26_afrLoop%3D1260730701916965%26datasetCode%3DA%26collectionCode%3DYI%26_afrWindowMode%3D0%26_adf.ctrl-state%3D185a2lc3h7_45 accessed July 22, 2018.

International Labor Organization (ILO), Child labour in Africa (2018b) www.ilo.org/addisababa/areas-of-work/child-labour/lang--en/index.htm accessed March 10, 20180.

International Labor Organization (ILO), Labor Standards in Africa (2018c) www.ilo.org/addisababa/areas-of-work/labour-standards/lang--en/index.htm accessed March 10, 2018.

International Labor Organization (ILO), Ratifications of Occupational Cancer Convention No. 139—1974 (2018d) www.ilo.org/dyn/normlex/en/f?p=1000:11300:0::NO:11300:P11300_INSTRUMENT_ID:312284 accessed March 10, 2018.

International Labor Organization (ILO), Ratification of Occupational Safety and Health Convention No. 155-1981 (2018e) www.ilo.org/dyn/normlex/en/f?p=1000:11300:0::NO:11300:P11300_INSTRUMENT_ID:312300 accessed March 10, 2018.

indexmundi, France Demographics Profile (2018) www.indexmundi.com/france/demographics_profile.html accessed August 6, 2018.

Index Mundi, Mexico Demographic Profile (2018) www.indexmundi.com/mexico/demographics_profile.html accessed July 22, 2018.

India DGFASLI, Directorate General, Factory Advice Service and Labour Institutes (DGFASLI) (2018) DGFASLI Chapter 8 Statistics on status of compliance with statutory provisions based on data collected by DGFASLI, www.dgfasli.nic.in/info1.htm accessed March 5, 2018.

Indian Ministry of Labor and Employment (India IMLE), National Policy on Safety, Health and Environment at Workplace (2008) www.dgfasli.nic.in/info1.htm accessed March 5, 2018.

Indian Express, India to dominate working-age population growth in Asia Pacific by 2050: UN (2017) http://indianexpress.com/article/india/india-news-india/india-working-population-united-nations-report-2772203/ accessed March 5, 2018.

Instituto Nacional de Estadísticas (INE), IV Censo Económico 2007–2008 (2010) Caracas: INE. www.ine.gov.ve/documentos/Economia/IVCensoEconomico/pdf/InformeIVCE.pdf accessed June 16, 2015.

Ji, F., Xia, Z. L. China's occupational poisoning hazards and highrisk toxic workplace management (in Chinese), *Ind Med* (2008) Vol. 6, pp. 404–406.

Joubert, D., Occupational health challenges and success in developing countries: A South Africa perspective, *International Journal of Occupational and Environmental Health* (2002) Vol. 8, No. 2, p.119124.

Kim, I., Muntaner, C., Chung, H., Benach, J., EMCONET network. Case studies on employment-related health inequalities in countries representing different types of labor markets, *International Journal of Health Services* (2010) Vol. 40, No. (2), p. 255e67.

Liang, Y. X., Xiang, Q.Y. Occupational health services in PR China, *Toxicology* (2004) Vol. 198, pp. 45–54.

Liebman, A. K., Wiggins, M. F., Fraser, C., Levin, J., Sidebottom, J., Arcury, T. A. Occupational health policy and immigrant workers in the agriculture, forestry, and fishing sector. *American Journal of Industrial Medicine* (2013) Vol. 56, pp. 975–984.

Loewenson, R., Global challenges to equity in safety and health at work: Struggles for fair work in Southern Africa, *Perspectives on Global Development and Technology* (2004) Vol. 3, No. 1–2, pp. 163–170.

London, L., Tangwa, G., Maatchaba-Hove, R., Mkhize, R., Nyika, A., Westerholm, P., Ethics in occupational health deliberations of an international workgroup addressing challenges in an African context, *BMC Medical Ethics* (2014) Vol. 15, No. 48, pp. 1–11.

Lora, E. (1998). *The Employment Problem in Latin America: Perceptions and Stylized Facts*. Office of the Chief Economist Working Paper 371. Washington, DC: Inter-American Development Bank.

Marie, J., OSH and globalisation, challenges for today. *La Medicina del lavoro* (2006) Vol. 97, pp. 125–131.

Moyo, D., Zungu, M., Kgalamono, S., Mwila, C., Review of occupational health and safety organization in expanding economies: The case of Southern Africa, *Annals of Global Health* (2015) Vol. 81, No. 4, pp. 495–502.

OECD, Society at a Glance 2016 (2016) ISBN 978-92-64-26148-8 (PDF), www.oecd-ilibrary.org/docserver/download/8116131e.pdf?expires=1520834046&id=id&accname=guest&checksum=583FD71C646BB20973FA43198AE3ED07 accessed March 12, 2018.

OECD, Heures travaillées, Organization for Economic Cooperation and Development (2017) https://data.oecd.org/fr/emp/heures-travaillees.htm accessed July 21, 2018.

Offshore Group, Mexico's Industrial Health and Safety Regulations (November 16, 2017), https://insights.offshoregroup.com/mexicos-industrial-health-and-safety-regulations accessed July 23, 2018.

Okoth-Okelloh, A., Onyango, R., Tonui, W., Okumu, W., Ouma, C., Occupational health and safety administration (OSHA) in the Morgues: Management and practice of the Universal Precautions in Morgues in Kenya, *Biosafety* (2015) Vol. 4, No. 1, pp. 1–12.

Perazzi, J., Merli, G., Paredes, L., The size of the informal economy in Venezuela, *El Norte: The Finnish Journal of Latin American Studies* (December 2010) Vol. 5. ISSN 1796-4539, pp. 1–12.

Pingle, S., Occupational safety and health in India: Now and the future, *Industrial Health* (2012) Vol. 50, pp. 167–171.

Sakurai, H., Occupational safety and health in Japan: Current situations and the future, *Industrial Health* (2012) Vol. 50, pp. 253–260.

Sanchez-Roman, F., Juarez-Perez, C., Madrid, G., Borja-Aburto, V., Haro-Garcia, L., Claudio, L., Occupational health in Mexico, *International Journal of Occupational and Environmental Health* (Oct/Dec 2006) Vol. 12, No. 4, pp. 345–354.

Santana, V., Workers' health in Brazil: Graduate research, *Revista de saude publica* (2006) Vol. 40, pp. 101–111. (N Esp) www.revistas.usp.br/rsp/article/view/32092/34158 accessed March 12, 2018.

Scholastic, Mexico: The economy (2018) www.scholastic.com/browse/subarticle.jsp?id=1107 accessed July 22, 2018.

Sese, A., Palmer, A., Cajal, B., Montano, J., Jimenez, R., Llorens, N., Occupational safety and health in Spain, *Journal of Safety Research* (2002) Vol. 33, pp. 511–525.

Shelley, T. (2007) *Exploited: Migrant Labour in the Global Economy*. London: Zed Press.

South Africa, Occupational Safety and Health Act (1993) www.labour.gov.za/DOL/downloads/legislation/acts/occupational-health-and-safety/a85-93.pdf accessed August 7, 2018.

Statista, India: total population (2018a) www.statista.com/statistics/263766/total-population-of-india/ accessed March 5, 2018.

Statista, Total population of China (2018b) www.statista.com/statistics/263765/total-population-of-china accessed July 23, 2018.

Swuste, P., Eijkemans, G. Occupational safety, health, and hygiene in the urban informal sector of Sub-Saharan Africa, an application of the Prevention and Control Exchange (PACE) Programme to the informal sector workers within Healthy City projects, *International Journal of Occupational and Environmental Health* (2002) Vol. 8, No. 2, pp. 113–118.

Tadesse, S., Israel, D., Occupational injuries among building construction workers in Addis Ababa, Ethiopia, *Journal of Occupational Medicine and Toxicology* (2016) Vol. 11, No. 6, p. 16.

Trading Economics, Spain unemployment rate (2018) https://tradingeconomics.com/spain/unemployment-rate accessed July 31, 2018.

U.S. Department of Labor, Finding on the worst form of child labour—Venezuela (2008) www.unhcr.org/refworld/docid/48caa499c.html accessed July 18, 2009.

USBLS, Unionization rates in 2015 (February 29, 2016) www.bls.gov/opub/ted/2016/government-utilities-and-transportation-and-warehousing-had-highest-unionization-rates-in-2015.htm accessed August 6, 2018.

WB, Doing business in Spain (2015) www.doingbusiness.org/~/media/WBG/DoingBusiness/Documents/Subnational-Reports/DB15-Spain-Overview.pdf accessed July 30, 2018.

World Population Review, South American Population (2018) http://worldpopulationreview.com/continents/south-america-population/ accessed March 12, 2018.

Worldatlas, Latina American countries (2018) www.worldatlas.com/articles/which-countries-make-up-latin-america.html accessed August 6, 2018.

Worldometers, China Population (July 22, 2018a) http://www.worldometers.info/world-population/china-population/ accessed July 23, 2018.

Worldometers, South American Population (2018b) www.worldometers.info/world-population/south-america-population/ accessed March 12, 2018.

Zhang, X., Wang, Z., Li, T., The current status of occupational health in China, *Environmental Health and Preventative Medicine* (2010) Vol. 15, pp. 263–270.

Zhi, S, Sheng, W., Levine, S. P. National occupational health service policies and programs for workers in small-scale industries in China. *AIHAJ* (2000) Vol. 61, pp. 842–849.

12 Global Issues and Aspects of the Disposal and Recycling of Electronics

Thomas P. Fuller
Illinois State University

CONTENTS

12.1 INTRODUCTION

E-waste, also known as electronic waste or electrical waste, has been defined as "Waste Electrical and Electronic Equipment (WEEE) that is dependent on electric currents or electromagnetic fields in order to function." This includes all components, subassemblies, and consumables that are part of the original equipment at the time of disposal. E-waste includes cell phones, fax machines, air conditioners, televisions, batteries, computers, laptops, washing machines, refrigerators, and any other major appliances and electrical or electronic equipment (EEE) (Brune, 2013).

The United States and China top the list of the major contributors of waste. The annual global production of e-waste reached 48.9 million metric tons in 2012 and is expected to exceed 50 million metric tons in 2018 (StEP, 2014; Baldé, 2015; Seeberger, 2016). The amount of e-waste globally is growing at an alarming rate with no foreseeable end. WEEE has been estimated to comprise up to 8% of all municipal waste (Widmer, 2005). Rapid technological innovation and changes in lifestyle drive the electronic industry to continually reveal new products to consumers (Annamalai, 2015). In addition to consumer sources, every office, manufacturing facility, hospital, and service business contribute to the EEE waste supply chain.

The composition of e-waste is varied and is based on the type of electronic device. Even within the same types of electronics, variations in composition can occur. E-waste can contain heavy metals (mercury, lead, cadmium), flame retardants, pentabromophenol, polybrominated diphenyl ethers(PBDEs), polychlorinated biphenyl(PCB), and other hazardous substances (Borthakur, 2016; NEWMOA, 2008; Sepúlveda, 2010). More than 1,000 different materials have been identified in WEEE (Cobbing, 2008). And these materials change and break down as the electronic device ages. Due to the chemical constituents and the potential for occupational and environmental negative consequences, e-waste is generally considered, and should be handled as, hazardous waste.

About 70% of the heavy metals and 40% of the lead in U.S. landfills come from electronic waste (Puckett, 2002). But with increasing limits of landfill disposal, e-waste from the United States is making its way to economically developing countries (EDCs) that will accept the environmental risks for the economic benefits that e-waste can provide. Besides being paid to take e-waste, the materials themselves can be harvested for the valuable materials contained within. Early personal computers could contain up to a gram of gold, and valuable copper comprises about 20% of total average e-waste mass (Widmer, 2005). It is easy to see the economic benefit to nations, organizations, and informal workers willing to harvest the materials and accept the risks.

E-waste has gone from a domestic to an interregional and global problem. In 2016, only 20% of e-waste was recycled (UN, 2017). And it is estimated that 75%–80% of the waste is shipped to Africa and Asia for disposal and recycling (Diaz-Barriga, 2013). The actual final destination of most e-waste is either unreported or unknown, and existing legislation is doing little to enforce and curb the illegal flow of e-waste around the world (Ongondo, 2011). WEEE interactions with the environment.

With the growing production of electronics, e-waste is becoming more of a global issue every day. The following section discusses various topics that are important for addressing the e-waste problem. This includes the disposal and recycling techniques of different regions throughout the world, the overall occupational and environmental hazards of e-waste, laws and regulations pertaining to the topic, and recommendations that can be put in place to properly handle the e-waste crisis.

12.2 REGIONAL AND NATIONAL DISPOSAL/ RECYCLING PROCESSES

12.2.1 West (North America)

North America typically uses three basic e-waste processes: disposal, recycling, and donation. It was found by Seeberger that the average household in the United States, in 2008, owned approximately 24 electronic devices (Seeberger, 2016). Eight years later, this number continues to grow. E-waste in North America amounted to

1.1 million tons in 2014. In Canada, e-waste produced amounted to 0.725 million tons in 2014 (Kumar, 2016).

The main method of handling e-waste in North America is disposal, more specifically landfilling. According to the U.S. EPA, of the 3.4 million metric tons of e-waste ready for disposal in 2012, 2.42 million tons (71%) ended up in landfills (Seeberger, 2016). While landfills are used to dispose of the majority of wastes due to its low cost, this disposal method does not recover any of the toxic materials contained. Metals such as lead and nickel can be released from buried electronics and cause harmful environmental effects. If these harmful materials could be removed from products before burial, or excluded from the components in electronics, negative environmental consequences could be avoided.

Recycling is the second most common method of electronics disposition in North America. The U.S. EPA estimates that about 29% of U.S.-generated e-waste is collected and sent to be recycled (Seeberger, 2016). This equates to approximately 1 million metric tons, compared to the 3.4 million metric tons sent to landfills. The recycling process constitutes three approaches to processing e-waste: taking the electronic device apart to reuse individual materials or components, reusing the product, and refurbishing the product.

According to the U.S. EPA's survey of seven recycling facilities throughout the United States in 2009, it was found that the average collection of consumer electronics was about 10,000 tons per year, of which 67% were recycled and 33% were reused or refurbished (U.S. EPA, 2011). The recycling processes vary between the different types of electronic products.

The last method in WEEE handling is often overlooked, or even ignored, during research of e-waste. This process stems off of the reuse and refurbish aspects of recycling donation (U.S. EPA, 2018). Overall, the method has nothing to do with the dismembering of e-waste but can help slow down the timeline of e-waste from a consumer to the dump. Instead of throwing an electronic device away, it could be donated or sold to another consumer, thus extending the product life cycle, reducing demand for new products, and ultimately slowing down the flow in the e-waste chain. Donation does not deal with obsolete e-waste or the products that have been passed around and then eventually become obsolete. This type of e-waste will still rely on the other methods for processing.

12.2.2 East (Asia, Middle East, Australia)

In Eastern Asia, the main method for e-waste processing is recycling. Recycling in Asia consists of primarily dismantling certain electronic products for metals. This is routinely accomplished by small-scale operations, in the informal sector, often on the side of the street, mainly throughout China, India, and Pakistan.

In the town of Guiyu, China, approximately 100,000 people are employed in small informal e-waste operations that preform hazardous recycling practices such as toner sweeping, open burning, cathode ray tube (CRT) cracking and dumping, and circuit board recycling. The wastes recycled at these small operations are received from Japan, South Korean, and Europe, but mainly from North America (Puckett, 2002). Other methods of handling e-waste in China include open burning to "cook" off plastics to retrieve metals, de-soldering printed circuit boards to release computer chips, acid baths, open dumping of recycling byproducts, and physical dismantling by force (Jan, 2013).

In India, the conditions and recycling processes are also shown to be hazardous. Workers employ crude methods for the extraction of metals and are often exposed to associated hazardous materials including PCBs and other heavy metals. There is seldom adequate ventilation or proper handling, and much of the work is done by women and children, often in the home. In India, more than 90% of e-waste recycling is done in the informal sector by primitive means, including burning in open air. Unfortunately, the elementary methods used to extract the valuable materials only yield 10%–20% of the precious metals contained in the electronics (Chatterjee, 2016). So even with the hazardous occupational and environmental exposures, there remain large lost opportunities for further economic benefit.

Open burning to remove plastic and retrieve copper, and open acid extraction to recover gold, platinum, palladium, and silver, is common in India, Vietnam, China, India, Pakistan, and the Philippines (Leung, 2006). Dealers look to buy copper, gold, and other metals that have value. This forces the workers to use any technique, often inherently hazardous, to destroy electronic equipment in search of these metals.

12.2.3 Africa, Australia, and Europe

Australia is similar to the United States in the handling and disposition of e-waste. Only 10% of Australian computers are currently recycled (TEC, 2008). Processing e-waste in Australia is primarily done through two different methods: disposal (landfilling) and recycling. Relying heavily on landfilling, it was found that this disposal method is causing serious problems in Australia. Studies have shown groundwater contamination from landfills that store e-waste. According to Seeberger (2016), "one site operated since 2005 with a capacity of 200,000 ton/year. It had the highest groundwater lead (Pb) levels: up to 38 μg/L (almost four times higher than the local drinking water guideline of 10 μg/L)." This highlights the potential environmental hazards of landfilling WEEE. It also demonstrates the amount of time it can take for hazardous materials in electronic components to break down and reach the environment, and then remain in the groundwater.

In Europe, e-waste handling methods are similar to those used in North America. In Europe, e-waste is commonly either disposed of in landfills or recycled. Recently however, European countries have begun to examine the health and environmental consequences of e-waste disposal.

Similar processes are used to handle and dispose of e-waste by most African countries. In Africa, the Bamako Convention created in 1991 bans the importation of hazardous wastes into member nations (OAU, 1991). This ban includes WEEE. Despite the ban on wastes entering the continent, violations occur, and there is still the issue of handling WEEE created within African nations themselves.

Overall, the disposal/recycling methods that are used in the treatment of e-waste tend to be relatively the same globally. Some areas of the world rely on certain techniques over others for various reasons. The provided methods are the most common not including the numerous countries that export waste to others.

12.3 PROBLEMS AND ISSUES WITH EXISTING SYSTEMS OF E-WASTE HANDLING AND DISPOSAL

12.3.1 Environmental Hazards

The massive amounts of e-waste created, and the lack of comprehensive environmental or occupational regulations regarding disposal, have led to significant exposures of hazardous materials to both the environment and workers. Unfortunately, these operations by unorganized workers often use rudimentary methods to extract the valuable metals. Physical methods include smashing, pulverizing, and cutting components to pieces. These operations by themselves expose workers to hazardous conditions, but they also release potentially hazardous dust particles into the work environment and surrounding neighborhoods. (Huang, 2016; Tsydenova, 2011) Other hazardous processes used include open burning or melting of components, acid baths, and other operations that lead to toxic vapors and fumes being released to the workplace and then the environment (Annamalai, 2015; Steiner, 2004).

Studies have shown that approximately 80% of the e-waste generated globally is recycled in informal settings in developing countries such as Nigeria, Ghana, China, and India (Perkins, 2014). Common techniques used by unskilled workers in makeshift settings include gold recovery from computer circuit boards with cyanide salt leaching or nitric acid and mercury amalgamation (Keller, 2006; Torre, 2006).

Mercury has been identified as a significant component of many types of e-waste including electric ovens and ranges, home appliances, space heaters, circuit boards, lighting, liquid crystal displays, cameras, copiers, flat screen televisions, batteries, and fax machines (NEWMOA, 2008). Circuit boards have also been shown to contain cadmium, lead, antimony, and beryllium (Cui, 2008; Guo, 2009). As burning, chipping, and physically dismantling the electronic components are commonly used to release the valuable metals, these hazardous materials are also released to the general surrounding environment.

The main releases from WEEE tend to be the original materials used in the product, auxiliary substances used in the recycling or dismantling of the product, and by-products formed by the transformation of the original materials (Sepúlveda, 2010). The main constituents and pathways from either recycling or disposal of WEEE are as follows:

- Leachates from ground dumping
- Particulate matter from dismantling
- Ashes from open burning
- Fumes of metals and plastics from melting and burning
- Liquid effluents from leaching (Sepúlveda, 2010)

A study of metals in surface dust in soil surrounding e-waste facilities in the Philippines showed increased levels of silver, arsenic, cadmium, cobalt, copper, iron, indium, magnesium, nickel, lead, and zinc (Fujimori, 2012). Workers themselves were shown to be exposed to these metals, in addition to children who played and lived in the surrounding neighborhood.

Water pollution can occur due to runoff or air dispersion from e-waste handling sites. It can also result from leaching at dumpsites and landfills where processed and unprocessed wastes have been disposed. Acids and other materials used in hydrometallurgical processes can also enter water and soil and result in the contamination of aquatic systems (Robinson, 2009). A basic diagram depicting WEEE interactions with the environment is shown in Figure 12.1.

In a systematic review by Grant (2013), e-waste community exposure of hazardous e-waste substances through inappropriate and unsafe handling and disposal was determined to lead to several adverse consequences. Negative health effects in populations exposed to e-waste included altered cellular function and expression, decreased lung function, changes in thyroid function, and altered temperament and behavior (Grant, 2013). These studies also showed that communities located near e-waste recycling or handling facilities suffered higher levels of spontaneous abortions, stillbirths, premature births, and reduced birth weights and sizes.

In studies of people living in proximity to recycling center operations in China, the exposed groups were found to have statistically higher levels of dioxin (Zhang, 2010; Ma, 2011; Chan, 2007). Other environmental exposures in China have identified high body burdens of polychlorinated dibenzo-*p*-dioxins and dibenzofurans in people living near recycling centers, and great potential for negative health implications in future generations (Chan, 2013).

Besides human exposure and health effects, flora and biota can be impacted negatively from these environmental toxins. Contamination associated with e-waste has been associated with environmental degradation in developing countries and the health of people living around e-waste handling facilities (Robinson, 2009). The primary issue with the e-waste toxins is on the contamination of surface and groundwater. This is a cause for concern due to the fact that many countries rely heavily on landfills and their primary disposal method for e-waste.

The benefits to developing countries from exporting their e-waste are the avoidance of strict environmental regulations and high labor costs. Unfortunately, this transfer introduces serious problems in the receiving communities associated with the handling and

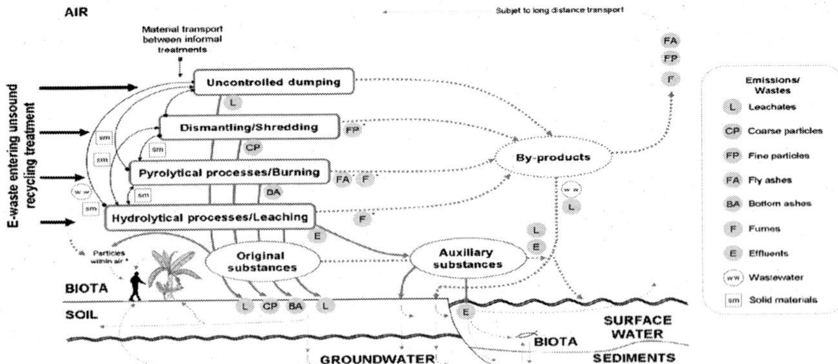

FIGURE 12.1 Sample of e-waste interactions with the surrounding environment. (From Sepúlveda, 2010.)

disposal of e-waste that lead to environmental contamination and occupational exposures to hazardous chemicals. Despite the economic benefits from the extraction of valuable materials and metals from e-waste, many developing countries lack the infrastructure, technology, methods, and facilities to safely dispose of and recycle e-waste. Without a concerted effort to curb the expansion and growth of the e-waste industry in informal and unregulated sections of developing countries, the related environmental degradation can be expected to continue and increase in the coming years (Shamim, 2015).

12.3.2 Occupational Hazards

Many of the chemicals emitted during e-waste dismantling or recycling are known to be hazardous to human health. Toxic effects can include skin diseases, nervous system damage, kidney malfunction, respiratory problems, adverse pregnancy and birth outcomes, and endocrine disruption, and underdevelopment of the brain in children (Lundgren, 2012; Wang, 2012; Frazzoli, 2010; Kristen, 2013). In a Swedish study of workers in WEEE, recycling plants used air monitoring and biological exposure indices to show that they were exposed to significantly increased levels of multiple toxic metals (Julander, 2014).

Much of e-waste recycling is done by workers in the informal sector, not by formal enterprises or registered businesses. In comparison with other workers, these informal sector e-waste workers are much less knowledgeable about their workplace hazards and risks of exposure to toxic chemicals (Ohajinwa, 2017). The work is very menial, and WEEE is often physically dismantled using hammers and chisels. Devices are sometimes heated/melted/burned to separate plastics and metals without ventilation necessary to reduce worker exposures or environmental releases. Gold and other valuable metals are often extruded and extracted from electronics in open-pit acid baths without the benefit of protective ventilation systems or appropriatepersonal protective equipment (PPE) (Chi, 2011).

Many of the workers in the informal sector handling hazardous e-waste include marginalized people, who are less educated than the general populations where they live. E-waste provides economic opportunities for many with no other options. Disadvantaged people, women, and children represent a substantial portion of e-waste workers (Oteng-Ababio, 2012; Chi, 2011). E-waste workers in developing countries may make as little $1.50 per day (Puckett, 2002).

In a study of e-waste workers in Thailand, it was observed that most informal e-waste operations were run by families and included support from women and children. People tended to dismantle equipment using their bare hands and primitive tools without any consideration of exposure to toxic chemicals. Even when these workers used open burning or hazardous chemicals to extract valuable metals such as gold, silver, copper, and nickel, they did not consider their activities hazardous to themselves, or the local environment (Pookkasorn, 2016).

In Nigeria, 57% of e-waste workers were not aware of any PPE that could be used to reduce exposures to hazardous chemicals (Ohajinwa, 2017). And only 18% of Nigerian e-waste workers ever wore any sort of PPE. This was mostly due to the lack of awareness and underestimation of potential hazards from associated chemical exposures (Ohajinwa, 2017). In a similar study conducted in Bangkok, it was determined that the majority of handlers and workers did not have a clear understanding or knowledge about good e-waste management including effects from e-waste releases on their health or the environment (Pookkasorn, 2016).

Children make good e-waste workers because their small dexterous hands can more easily dismantle small electronics parts. It is also a very low skill task that requires little knowledge or training. But the large number of children working in this industry is particularly problematic because the children are more likely to be exposed to toxic agents and they are more physiologically vulnerable to the hazardous health effects of the associated toxic chemicals. Children spend a significant amount of time outdoors and in contact with the ground and soil. They exhibit higher rates of hand-to-mouth contact than adults, and their lack of experience and undeveloped perceptions of risk place them at higher levels of toxic agent exposure than adults. Exposures of toxic agents to children can be longer than the routinely assumed 40 years for adults, and the exposures to toxic agents will have the opportunity to become evident much earlier in the life of a person exposed as a child. Physiological characteristics of children such as breathing rates, and food and water consumption, are significantly higher in proportion to their height and weight than those of adults (WHO, 2006). Children also have larger skin-to-body weight ratios and more absorbent skin than adults for most chemicals. So children tend to absorb proportionately more toxic chemicals and maintain higher doses than adults (WHO, 2006). Metabolic defenses to toxic agents are much less developed in children, and they are therefore more sensitive to damage. These physiological characteristics combined with the higher levels of exposure discussed above place children at a much higher level of risk of injury and illness compared with adults (Perkins, 2014).

Related to this is the problem of surrounding communities and exposure to vulnerable populations (children and elders). As stated by Heacock, "Chemicals can accumulate in children's bodies because their immature systems are unable to process and excrete some toxic materials effectually." (Heacock, 2016). The exposure can be the result of indirect or direct exposure and can cause many health and developmental issues. Some health effects are found in Table 12.1.

TABLE 12.1
Common E-Waste Components and Their Potential Health Effects

Toxicant	Health Effects (Children)
Lead	Impaired cognitive function, behavioral disturbances, attention deficits, hyperactivity
Mercury	Deficits in motor function, attention disorders
Cadmium	Deficits in cognition learning, behavior, and neuromotor skills
Hexavalent chromium	Unknown but is classified as a carcinogen
PCBs	General cognition, visual–spatial function, memory, attention, motor function

Source: Adapted from Chen (2011).

12.4 LAWS AND REGULATIONS

The Basel Convention on the Control of Transboundary Movements of Hazardous Wastes and their Disposal was adopted in 1989 and put into force in 1992. The objective of the convention is to protect the human health and the environment against the adverse effects of hazardous wastes. The convention proscribes requirements regarding transboundary movements of hazardous waste. Due to the hazardous properties such as mercury, flame retardants, and lead, e-waste is considered a hazardous waste and falls within the framework of the convention. The convention is the most comprehensive global environmental agreement on hazardous wastes and other waste, and 186 nations are parties. The United States remains one of the only developed countries in the world that has not ratified the Basel Convention and permits, and depends upon, transport of toxic e-waste internationally (UNEP, 1992).

The Basel Convention does not have set regulations specifying how countries manage, how to pack, or how to ship waste. The main focus of the convention is the enforcement of responsibility when a spill or improper disposal occurs. The convention enforces safe shipment by making the exporting party take liability for remediation and relocation of hazardous waste. So the international convention is more of a reactive regulation, primarily due to the lack of authority over agreement nations.

The United States, one of the largest e-waste producers, has no federal legislation that mandates the recycling of e-waste. While there are no federal regulations, "there are many states that have begun to mandate environmental responsibility by banning the disposal of e-waste into landfills" (E-waste, 2013). Currently, there are 18 states that have banned disposal into landfills. But the United States does not ban the transport of e-waste internationally.

Other regulations or laws that affect e-waste are bilateral agreements. This involves two or more parties (countries) that have come to an agreement on issues such as e-waste and radioactive waste. For example, the United States has five bilateral agreements that focus on hazardous waste. Agreements with Canada and Mexico allow for the export and import of hazardous materials between borders for recycling and disposal. The other three agreements with Costa Rica, Malaysia, and the Philippines allow hazardous waste shipment to the United States for recycling and disposal, but do not allow export of waste from the United States to those countries (EPA, 2018).

In EU, Waste Electrical and Electronic Directive, manufactures, and importers of within member states are required to take back their products from consumers and ensure safe disposal at the end of life (EU, 2012). Other regulations regarding WEEE are being developed or considered in the EU.

The Bamako Convention created by the Organization of African Unity in 1991 bans the importation of hazardous wastes into member nations. It is meant to control the transboundary movement of hazardous wastes within Africa and to curb transport of these materials onto the continent. The convention also requires member states to ensure safe management of hazardous wastes within their jurisdiction, and to take steps necessary to prevent waste pollution from negatively affecting human health or the environment (OAU, 1991). The Bamako Convention is much stricter than the Basel Convention and prohibits all import of hazardous wastes into African nation member states.

12.5 CONCLUSIONS/RECOMMENDATIONS

Solutions to the e-waste problem can be many pronged. Production of electronics products is expected to continue to rise to support global consumer demand and corporate profits for product manufacturers and distributors. With little regulatory impetus coming out in the major producer countries, the waste will continue to flow to EDCs when they reach their end of life. The fundamental interactions that support the generation and flow of e-waste are shown in Figure 12.2.

Greenpeace is challenging manufacturers to take action on their own to reduce the global impact of the waste of their products through two main approaches. One is to embrace and expand the principle of "individual producer responsibility" where each manufacturer develops programs to take financial responsibility for the disposal or recovery of their products. Besides ensuring the products are handled appropriately, it builds in incentives to design electronics that are more easily recoverable or less hazardous to the environment (Cobbing, 2008). The second Greenpeace recommendation is to design out toxicants from electronics products, eliminate hazardous substances, and replace harmful ingredients with safer alternatives.

Besides Greenpeace, other nongovernmental organizations (NGOs) need to become more active in e-waste management and analysis. NGOs can help provide oversight on corporate and national e-waste production and control. NGOs should conduct more local research, interviews, and surveys of social climate and activities to identify which methods and systems will be most beneficial in particular regions of interest (Orlins, 2016). These NGOs can then help build capacity and provide training in the management of e-waste locally to reduce occupational and environmental impacts. This can be especially useful in addressing the issues and concerns of activities in the informal sector.

E-waste has been a growing problem globally for many years and is on the rise. While many countries have disposal and recycling methods in place, these methods should be evaluated and improved to allow for safer disposal. Many problems have come to be due to e-waste and the challenges it provides for disposal. Overall, more

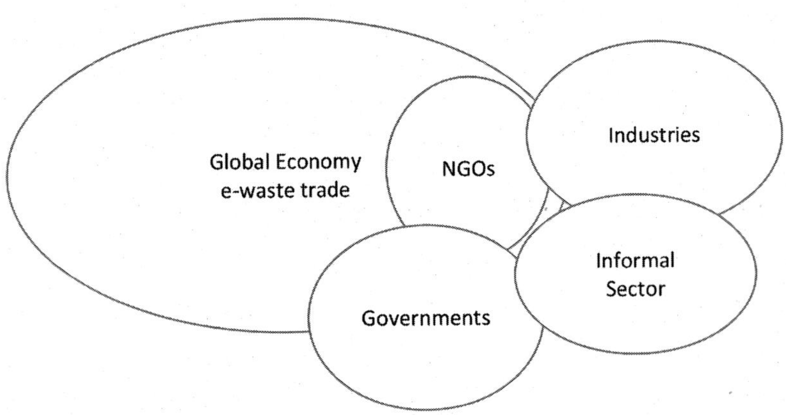

FIGURE 12.2 Intersecting actors in e-waste generation and flow. (Adapted from Orlins, 2016.)

action needs to take place to allow for safer disposal method to be put in place, regulated, and enforced. Environmentally sound management for recycling facilities should include (1) a systematic management approach, (2) risk assessment, (3) risk control and prevention, (4) effective communications and awareness training, (5) adherence to requirements and standards, (6) program review and corrective action, and (7) transparency (UNEP, 2011).

E-waste management programs must closely consider and include social, political, and economic conditions of the regions where they are to be put in practice. Interventions must consider and be able to address the economic and social conditions that impact the value of e-waste handling and recycling (Lundgren, 2012). A broad tripartite approach involving international organizations, governments, and research institutions is needed to address the complex interrelated issues surrounding e-waste creation and handling.

Sustainable management of raw materials and hazardous waste requires a more compatible system of production that considers the flows of associated wastes downstream. There need to be business models that include collaboration between EEE manufactures and downstream collection of recyclable and waste materials (Cucchiella, 2015). Rather than sending electronics to waste handling and allowing the transfer to developing countries, better methods to remove waste materials should be developed to allow safer and more efficient reprocessing. A sustainable economy is a recycling or recovering economy.

Although epidemiological studies to date have not definitively identified a causal relationship between e-waste exposures, the observation and evidence of human exposure to chemicals associated with e-waste warrant a precautionary approach to control and minimize e-waste (Grant, 2013). Research programs should be expanded to increase understanding and the body of knowledge regarding health effects to people and the environment. Policies, educational programs, and public health interventions should be expanded to ensure human health is not affected. Ecological and environmental studies should be completed to evaluate the impacts of e-waste on biota and fauna.

Regulation and legislation of e-waste need to be put in place to better regulate what is happening with the e-waste produced. This will not completely solve the problem but will be a stepping stone for the future. Extended producer responsibility legislation can place requirements on manufacturers to design and sell more environmentally friendly products using fewer hazardous materials and more recyclable components to reduce the volumes and risks associated with WEEE (Olds, 2012). Additional international standards need to be developed to fill existing gaps that still allow developed countries to transfer hazardous wastes externally (Renckens, 2008).

In 2011, the U.S. EPA and the Environmental Protection Administration of Taiwan created a collaborative International E-Waste Management Network. The groups are working to build global capacity for sound e-waste management and to exchange information and best practices (EPA, 2011).

The relative value of discarded electronics components to EDCs leads to the transfer of this waste from developed and financially secure countries to the poorer countries. Weak laws regarding e-waste disposal promote the growth of disposal and dismantlement in the informal sector.

Nearly two decades after the Basel Convention was created as a means to protect human health and the environment against the adverse effects, and restrict transboundary movements of hazardous wastes, the international mechanisms that were created to provide environmental justice and equity to all countries remained inadequate (Sonak, 2008). Numerous accounts presented in this chapter show that the same is true for the shipment and systems in place for hazardous electronic waste.

The continued economic exploitation of EDCs by more advanced developed nations is a root cause of the ongoing transfer of hazardous waste and hazardous operations to countries desperate for economic opportunities despite the occupational and environmental costs (Castleman, 2016). The transfer of hazardous electronic waste is no exception. Despite the creation of corporate responsibility indices used to demonstrate sound environmental management, little is being done by corporations to curb the creation, or provide a solution, to the growing problem of e-waste. Collaboration between governments needs to be increased to develop better ways to restrict the illegal transfer of hazardous e-waste to countries that are ill prepared to handle it in a safe and efficient manner.

REFERENCES

Annamalai, J., Occupational health hazards related to informal recycling of e-waste in India: An overview, *Indian Journal of Occupational & Environmental Medicine* (2015) Vol. 19, No. 1, pp. 61–65. Academic Search Complete. Web. 11 Dec. 2016.

Baldé, C. P., Wang, F., Kuehr, R., Huisman, J. (2015). *The Global e-waste Monitor 2014: Quantities, Flows and Resources.* Bonn, Germany: A Report; United Nations University, UNU-IAS institute for the Advanced study sustainability, pp. 1–74.

Borthakur, A., Health and environmental hazards of electronic waste in India, *Journal of Environmental Health* (2016) Vol. 78, No. 8, pp. 18. MasterFILE Premier. Web. 11 Dec. 2016.

Brune, M., Goldizen, F., Neira, M., Van Den Berg, M., Lewis, N., King, M., Suk, W. A., Carpenter, D. O., Arnold, R. G., Sly, P. D., Health effects of exposure to e-waste, *Lancet Global Health* (2013) Vol. 1, p. e70.

Castleman, B., The export of hazardous industries in 2015, *Environmental Health* (2016) Vol. 15, p. 8.

Chan, J., Wong, M., A review of environmental fate, body butdens, and human health risk assessment of PCDD/Fs at two typical electronic waste recycling sites in China, *Science of the Total Environment* (2013) Vol. 463–464, pp. 1111–1123.

Chan, J., Xing, G., Xu, Y., Liang, Y., Chen, L., Wu, S., Wong, C. K., Leung, C. K., Wong, M. H., Body loadings and health risk assessment of polychlorinated dibenzo-p-dioxins and dibenzofurans at an intensive electronic waste recycling site in China, *Environ Science and Technology* (2007) Vol. 41, pp. 7668–7674.

Chatterjee, S., E-waste recycling in India, presentation made to Department of Electronics and Information Technology (May12, 2016) www.oecd.org/environment/waste/Session_2-Part_1-EPR-Role-of-Informal-Sector-Sandip_Chatterjee.pdf accessed July 9, 2018, from Organisation for Economic Co-operation and Development (OECD) website.

Chen, A., Dietrich, K. N., Huo, X., Ho, S. M., Developmental neurotoxicants in e-waste: An emerging health concern, *Environmental Health Perspectives* (2011) Vol. 119, No. 4, pp. 431–438. Health Source: Consumer Edition. Web. 8 Dec. 2016.

Chi, X., Streicher-Porte, M., Wang, M., Reuter, M., Informal electronic waste recycling: A sector review with special focus on China, *Waste Management* (2011) Vol. 31, pp. 731–741.

Cobbing, M. (February 2008). *Toxic Tech: Not in Our Backyard-Uncovering the Hidden Flows of e-waste.* Amsterdam, the Netherlands: Greenpeace.

Cucchiella, F., D'Adamo, I., Koh, S., Rosa, P., Recycling of WEEs: An economic assessment of present and future e-waste streams, *Renewable and Sustainable Energy Reviews* (2015) Vol. 51, pp. 263–271.

Cui, J., Zhang, L., Metallurgical recovery of metals from electronic waste: A review, *Journal of Hazardous Materials* (2008) Vol. 158, No. 2–3, pp. 228–256.

Diaz-Barriga, F., Evidence-based intervention programs to reduce children's exposure to chemicals in e-waste sites (2013), *Discussion Paper for WHO Working Meeting on e-waste and Children's Health*, [cited 2014 Sept 17], p. 1e90.

ECycle, E-Waste Recycling Laws in the United States and Stopping the Export of E-Waste by Utilizing E-Stewards Certified Recyclers (09 August 2013), ECycle. N.p. Web. 06 Dec. 2016.

EPA, International E-Waste Management Network (IEMN) (2011) www.epa.gov/international-cooperation/international-e-waste-management-network-iemn accessed July 9, 2018.

EPA, International Agreements on Transboundary Shipments of Hazardous Waste (2018) www.epa.gov/hwgenerators/international-agreements-transboundary-shipments-hazardous-waste accessed May 15, 2018.

EU, Directive 2012/19/EU of the European Parliament and the Council on waste electrical and electronic equipment (WEEE) (2012) https://eur-lex.europa.eu/legal-content/EN/TXT/?uri=uriserv:OJ.L_.2012.197.01.0038.01.ENG accessed May 31, 2018.

Frazzoli, C., Orisakwe, O. E., Dragone, R., Mantovani, A., Diagnostic health risk assessment of electronic waste on the general population in developing countries' scenarios, *Environmental Impact Assessment Review* (2010) Vol. 30, pp. 388–399.

Fujimori, T., Takigami, H., Agusa, T., Eguchi, A., Bekki, K., Yoshida, A., Terazono, A., Ballesteros, F., Impact of metals in surface matrices from formal and informal electronic-waste recycling around Metro Manila, the Philippines, and intra-Asian comparison, *Journal of Hazardous Materials* (June 2012) Vol. 221–222, pp. 139–146.

Grant, K., Goldizen, F., Sly, P., Brune, M., Van Den Berg, M., Norman, R., Health consequences of exposure to e-waste: A systematic review, *The Lancet* (December 2013) Vol. 1, pp. 350–360.

Guo, J., Guo, J., Xu, Z., Recycling of non-metallic fractions from waste printed circuit boards: A review, *Journal of Hazardous Materials* (2009) Vol. 168, No. 2–3, pp. 567–590.

Heacock, M., Kelly, C., Asante, K., Birnbaum, L., Bergman, A., Bruné, M., Buka, I., Carpenter, D., Chen, A., Huo, X., Kamel, M., Landrigan, P., Magalini, F., Diaz-Barriga, F., Neira, M., Omar, M., Pascale, A., Ruchirawat, M., Sly, L. D., Sly, P., E-waste and harm to vulnerable populations: A growing global problem, *Environmental Health Perspectives* (2016) Vol. 124, No. 5, pp. 550–555. Health Source: Consumer Edition. Web. 7 Dec. 2016.

Huang, C., Bao, L., Luo, P., Wang, Z., Li, S., Zeng, E., Potential health risk for residents around a typical e-waste recycling zone via inhalation of size-fractionated particle-bound heavy metals, *Journal of Hazardous Materials* (2016) Vol. 317, pp. 449–456. Edselp. Web. 8 Dec. 2016.

Jan, J., Wong, W., A review of environmental fate, body burdens, and human health risk assessment of PCDD/Fs at two typical electronic waste recycling sites in China, *Science of the Total Environment* (2013) Vol. 463–464, pp. 1111–1123.

Julander, A., Lundgren, L., Skare, L., Grandér, M., Palm, B., Vahter, M., Lidén, C., Formal recycling of e-waste leads to increased exposure to toxic metals: An occupational exposure study from Sweden, *Environment International* (2014) Vol. 73, pp. 243–251.

Keller, M., Assessment of Gold recovery processes in Bangalore, India and evaluation of an alternative recycling path for Printed Wiring Boards. A case study (2006), *Diploma Thesis*, Institute for Spatial and Landscape Planning, Regional Resource Management, ETH. Zurich, 115 pp.

Kristen, G., Goldizen, F., Sly, P., Brune, M., Neira, M., Van Den Berg, M., Norman, R., Health consequences of exposure to e-waste: A systematic review, *The Lancet Global Health* (2013) Vol. 1, pp. e350–e361.

Kumar, A., Holuszko, M., Electronic waste and existing processing routes: A Canadian perspective, *Resources* (2016) Vol. 5, No. 35, pp. 1–19.

Leung, A., Cai, Z., Wong, M., Environmental contamination from electronic waste recycling at Guiyu, southeast China, *Journal of Material Cycles and Waste Management* (March 2006) Vol. 8, No. 1, pp. 21–23.

Lundgren, K. (2012). *The Global impact of e-waste: Addressing the Challenge.* Geneva, Switzerland: International Labour Organization, ISBN 978-92-2-126898-7 pdf.

Ma, J., Cheng, J., Wang, W., Kunisue, T., Wu, M., Kannan, K., Elevated concentrations of polychlorinated dibenzo-p-dioxins and polychlorinated dibenzofurans and polybrominated diphenyl ethers in hair from workers at an electronic waste recycling facility in Eastern China, *Journal of Hazardous Materials* (2011) Vol. 186, pp. 1966–1971.

NEWMOA, Northeast Waste Management Officials' Association: IMERC fact sheets on mercury uses (2008) www.newmoa.org/prevention/mercury/imerc/factsheets/ accessed July 20, 2018.

OAU, Organization of African Unity, Bamako Convention on the Ban of the Import into Africa and the Control of Transboundary Movement and Management of Hazardous Wastes within Africa by the Organisation of African Unity (1991) www.jus.uio.no/lm/hazardous.waste.ban.afrian.import.bamako.convention.1991/portrait.pdf accessed May 31, 2018.

Ohajinwa, C., Van Bodegom, P., Vijver, M., Peijnenburg, W., Health risks awareness of electronic waste workers in the informal sector in Nigeria, *International Journal of Environmental Health Research and Public Health* (August 2017) Vol. 14, No. 911, pp. 1–16.

Olds, L., Curb your e-waste: Why the United States should control its electronic waste exports, *Cardozo Journal of International & Comparative Law* (2012) Vol. 20, No. 3, pp. 827–873. Academic Search Complete. Web. 8 Dec. 2016.

Ongondo, F., Williams, I., Cherrett, T., How are WEEE doing? A global review of the management of electrical and electronic wastes, *Waste Management* (April, 2011) Vol. 31, No. 4, pp. 714–730.

Orlins, S., Guan, D. J., China's toxic informal e-waste recycling: Local approaches to a global environmental problem, *Journal Cleaner Production* (2016) Vol. 114, pp. 71–80.

Oteng-Ababio, M., When necessity begets ingenuity: E-waste scavenging as a livelihood strategy in Accra, Ghana, *African Studies Quarterly* (2012) Vol. 13, pp. 1–21.

Perkins, D., Drisse, M., Nxele, T., Sly, P., E-waste: A global hazard, *Annals of Global Health* (2014) Vol. 80, pp. 286–295.

Pookkasorn, S., Sharp, A., The management of waste from electrical and electronic equipment (WEEE) in Bangkok, Thailand (2016), *6th Int. Conf. on Biological & Environmental Sciences (BCES-2016)*, August 8–9, Pattaya, Thailand.

Puckett, J., Bystar, L., Westervelt, S., Gutierrez, R., Davis, S., Hussain, A., Dutta, M., (February 25, 2002). *Exporting Harm-the High-Tech Trashing of Asia.* Basel Action Network, Silicon Valley Toxics Coalition, San Francisco, California.

Renckens, S., Yes, we will! Voluntarism in US e-waste governance, *Review of European Community & International Environmental Law* (2008) Vol. 17, No. 3, pp. 286–299. Business Source Alumni Edition. Web. 8 Dec. 2016.

Robinson, B. H., E-waste: An assessment of global production and environmental impacts, *Science of the Total Environment* (2009) Vol. 408, No. (2), pp. 183–191.

Seeberger, J., Grandhi, R., Kim, S., Mase, W., Reponen, T., Ho, S., Chen, A., E-waste management in the United States and public health implications, *Journal of Environmental Health* (2016) Vol. 79, No. 3, p. 8. MasterFILE Premier. Web. 6 Dec. 2016.

Sepúlveda, A., Schuluep, M., Renaud, F., Streicher, M., Kuehr, R., Hageluken, C., Gerecke, A., A review of the environmental fate and effects of hazardous substances released from electrical and electronic equipments during recycling: Examples from China and India, *Environmental Impact Assessment Review* (2010) Vol. 30, pp. 28–41.

Shamim, A., Mursheda, A., Rafiq, I., E-waste trading impact on public health and ecosystem services in developing countries, *International Journal of Waste Resources* (2015) Vol. 5, No. 188, p. 2.

Solving the E-waste Problem, E-waste world map (2014) http://step-initiative.org/WorldMap.html.

Sonak, S., Sonak, M., Giriyan, A., Shipping hazardous waste: Implications for economically developing countries, *International Environmental Agreements: Politics, Law and Economics* (2008) Vol. 8, pp. 143–159.

Steiner, S. (2004) *Risk Assessment of e-waste Burning in Delhi, India*. Zurich: Federal Institute of Technology ETH, http://exigorecycling.com/pdf/E_waste_Delhi.pdf accessed June 1, 2018.

Tsydenova, O., Bengtsson, M., Chemical hazards associated with treatment of waste electrical and electronic equipment, *Waste Management* (2011) Vol. 31, No. (1), pp. 45–58.

Torre, M., Bachiller, D., Rendueles, M., Menéndez, C., Díaz, M., Cyanide recovery from gold extraction process waste effluents by ion exchange I. Equilibrium and kinetics, *Solvent Extraction and Ion Exchange* (2006) Vol. 24, pp. 99–117.

Total Environment Centre, Tipping point: Australia's e-waste crisis (2008) www.tec.org.au/images/e-waste%20report%20updated.pdf.

UN, Electronic waste poses 'growing risk' to environment, human health, United Nations News (December 13, 2017) https://news.un.org/en/story/2017/12/639312-electronic-waste-poses-growing-risk-environment-human-health-un-report-warns accessed May 29, 2018.

UNEP, Parties to the Basel Convention on the Control of Transboundary Movements of Hazardous Wastes and their Disposal (1992) www.basel.int/Countries/StatusofRatifications/PartiesSignatories/tabid/4499/Default.aspx#enote1 accessed May 30, 2018.

UNEP, Guideline on environmentally sound testing, refurbishment and repair of used computing equipment. Conference of the Parties to the Basel Convention on the Control of Transboundary Movements of Hazardous Wastes and Their Disposal (February 17, 2011) http://archive.basel.int/industry/compartnership/docdevpart/ppg11DraftGuidelineFinal-2011-03-15.pdf accessed July 17, 2018.

U.S. Environmental Protection Agency, Electronics waste management in the United States through 2009. (Publication No. 530-R-11-002) (May 2011) https://nepis.epa.gov/Exe/ZyPDF.cgi/P100BKKL.PDF?Dockey=P100BKKL.PDF accessed July 7, 2018.

U.S. EPA, Electronics donation and recycling (November 2, 2018). EPA. Environmental Protection Agency, https://www.epa.gov/recycle/electronics-donation-and-recycling accessed December 27, 2018.

Wang, X., Miller, G., Ding, G., Lou, X., Cai, D., Chen, Z., Meng, J., Tang, J., Chu, C., Mo, Z., Han, J. Health risk assessment of lead for children in tinfoil manufacturing and e-waste recycling areas of Zhejiang Province, China, *Science of the Total Environment* (2012) Vol. 426, pp. 106–112.

WHO, Environmental health criteria 237: Principles for evaluating health risks in children associated with exposure to chemicals (2006) www.who.int/ipcs/publications/ehc/ehc237.pdf accessed May 30, 2018.

Widmer, R., Oswald-Krapf, H., Sinha-Khetriwal, D., Schnellmann, M., Böni, H., Global perspectives on e-waste, *Environmental Impact Assessment Review* (2005) Vol. 25, pp. 436–458.

Zhang, J., Jiang, Y., Zhou, J., Wu, B., Liang, Y., Peng, Z., et al. Elevated body burdens of PBDEs, dioxins, and PCBs on thyroid hormone homeostasis at an electronic waste recycling site in China, *Environmental Science & Technology* (2010) Vol. 44, pp. 3956–3962.

13 Global Burden and Aspects of Occupational Cancer

Thomas P. Fuller
Illinois State University

CONTENTS

13.1 INTRODUCTION

The global burden of cancer is a major source of morbidity and mortality. In 2012, there were 14 million new cases of cancer and 8 million cancer-related deaths globally (IARC, 2014). By the year 2030, the incidence of cancer cases is globally expected to increase to 22 million (Bray, 2015). In general, the highest cancer incidence rates occur in the high-income countries of North America and Western Europe, in addition to Japan, Korea, and Australia (IARC, 2018a). The percentage of cancer cases attributable to workplace exposures in these developed nations is expected to decline in the coming years due to occupational exposure restrictions that have been in place now for several decades (Boffetta, 1999; IARC, 2018b). Overall age-adjusted death rates have been falling dramatically in the United States for lung, bronchus, colon, prostate (men), and breast (women) cancers for the past 20 years (ACS, 2018). According to International Agency for Research on Cancer (IARC), based on population sizes, more than 60% of all cancer cases occur in Africa, Asia, and Central and South America. These regions also account for 70% of all cancer fatalities (IARC, 2014). In addition, the burden of disease is expected to increase greatly in these lower economic countries in the next two decades due to aging populations, industrial growth, and other environmental and social risk factors (Stewart, 2016; Torre, 2015).

The broad differences in types of cancers in various world regions lend information about the causes and preventions that could be taken to lower the number of cases. Many of the cancers in the industrialized countries are associated with diet and lifestyle. In these regions, cancers of the breast, colorectum, and prostate are common. However, cancers of the liver, stomach, and esophagus are more common in low-income countries. Decreased survival rates in developing countries are likely to be the result of reduced availability of advanced clinical care or early diagnosis.

In economically advanced countries, efforts have been taken to prevent occupational cancer by reducing the workplace exposure to carcinogens (Blair, 2011). Workers in developed countries tend to be more educated and aware of the hazards and health consequences of exposure to carcinogens, and they are more reluctant to be exposed. Workers in economically developing countries (EDCs) that are becoming more industrialized will likely be at increasingly greater risk of exposure to carcinogens in the workplace. As industrialized countries move towards service-oriented businesses, there are fewer workers exposed to hazardous working conditions, and there are fewer workers exposed to carcinogens overall.

Many of the "dirty" industries (manufacturing, mining, chemical, and petroleum production) that have left the advanced countries are moving rapidly to the less developed nations, and as a result, more workers in the developing countries will be exposed to carcinogens in the future. Sixty-three percent of all cancers occur in low- and middle-income countries (Espina, 2013). Dirty industries are also less likely to be adequately regulated in these less developed nations. Workers in EDCs are less aware and often less concerned about the risks associated with the exposure to carcinogens. These workers also have fewer options for employment and may be willing to accept risks for that reason. Workers in these underdeveloped countries may also be more susceptible to harmful effects of carcinogens due to their poor nutritional levels and other physiological and environmental factors (Hashim, 2014). Technologies that might be used in advanced countries to protect workers from hazardous exposures are often not available in EDCs. And neither companies nor countries are fully aware of the long-term economic benefits of keeping the workforce healthy.

Cancer comes with high socioeconomic costs. Cancer costs the U.S. economy more than $243 billion in 2009. Ninety-nine billion was due to medical costs, and more than $144 billion was from lost productivity due to illness or death (Reuben, 2010). If occupational cancers were assumed to make up only 8% of the total number of cancers, this would imply that occupational cancer costs the U.S. $243 billion × 0.08 = $19.44 billion (Nurminen, 2001; Steenland, 2003; Rushton, 2012).

Studies have shown that more than one-third of all cancers are preventable, including those that arise from occupational exposures (Danaei, 2005). Occupational studies of nasal cancer in furniture workers exposed to wood dust as early as 1940 showed a diminution of the disease after exposure was reduced (Hayes, 1986). Numerous opportunities for occupational prevention of exposure to carcinogenic agents or conditions continue to exist today. Collaborative strategies for research and intervention policies and education are needed to fully reduce the incidence of occupational cancer worldwide (Espina, 2013).

13.2 OCCUPATIONAL CANCER DEFINED

Ramazzini was an Italian physician who associated several diseases with various occupational exposures in the 17th century. He related exposures to irritating chemicals, dusts, and metals to health effects in his treatise "Diseases of Workers" (Ramazzini, 1718). Ramazzini related diseases to various workplace exposures in 54 different occupations. He noticed that nuns had a lower incidence of cervical cancer and higher-than-expected incidence of breast cancer. In the 18th century, Percival Pott found links between prolonged exposure to soot, and scrotal cancer among chimney sweeps in London (Pott, 1775).

Since those early years, cancer has been associated with numerous other occupations including paraffin workers, shale oil workers, and in the textile industry. In addition, lung cancer cases were noted in miners and bladder cancer in dye makers. In the first part of the 20th century, cancer clusters were found in varied but discrete occupational settings such as nickel refineries and the manufacture of asbestos products, among others (Siemiatycki, 2006). In most of these discoveries, cancer was related solely to the associated occupation, rather than particular causative agents, through primitive retrospective cohort studies. Cancer itself was not related to a particular known agent until the 1950s, when cigarette smoking was associated with lung cancer.

Today, much of what is known about cancer still comes from occupational exposures. Research on occupational exposures to carcinogens forms the basis of much of the understanding about relative risks, prioritization of controls, and the development of the means for cancer prevention. And as understanding of occupational cancer increases, the consequences of public exposures to the same carcinogens become evident. For example, the awareness that occupational exposures to diesel engine emissions are associated with cancer provided useful information about the relative risks of general public exposures to diesel exhaust, and the need to reduce environmental exposures from vehicle emissions.

As many agents can cause cancer in both environmental and occupational settings, and individuals can be exposed to carcinogens in both the workplace and other public settings, or the home, it can be difficult to define an "occupational exposure." There is no simple definition; however, we can consider an occupational carcinogen to be one where the primary exposure to the agent occurs in the workplace and the primary epidemiological evidence comes from increased cancer risks associated with occupational settings at discrete times and places. Some agents are particularly difficult to define as "occupational" as they are commonly encountered in the everyday environment; sunlight, tobacco smoke, and diesel exhaust may be considered to be in this category. However, outdoor workers exposed to sunlight for extended periods, bartenders or waiters exposed to secondhand smoke, or traffic patrol personnel standing in heavy levels of exhaust all day may be considered occupational exposures.

Originally, carcinogens were loosely identified when cancer clusters occurred in a particular industry work group. As epidemiological studies increased in sophistication, they could more closely relate specific worker exposures to cancer. In addition, over time, direct evidence of carcinogenicity was derived from experimental animal studies. Today, toxicological studies of carcinogenesis can include complex analyses of chemical and metabolic interactions of agents with cells and tissues within living organisms. Absorption and metabolism of toxic agents within the body can be followed and analyzed to observe the changes in the organism including toxic, mutagenic, and carcinogenic effects. In the near future, it may be possible to determine carcinogenicity by comparing a chemical molecule with those of other chemicals to identify similar characteristics or traits that equate the likeliness that the new agent is also cancer causing in the absence of time-consuming and expensive epidemiologic or animal studies.

Despite what we have learned about the cancer-causing properties of numerous chemicals and hazardous agents, occupational exposures to them are still widespread. Tens of thousands of workers are still routinely exposed to well-recognized carcinogens such as heavy metals, diesel engine emissions, asbestos, polycyclic aromatic hydrocarbons, and silica. And although there have been increasing controls over occupational carcinogens in developed countries, there is growing evidence that the use of hazardous processes and materials is shifting to EDCs at an increasing pace (Hashim, 2014; Pearce, 1994). The use of carcinogenic agents is expanding in countries where workers tend to be less educated about, and less aware of, the hazardous or carcinogenic properties of chemicals, where there are fewer government regulations for the handling and training on toxic chemicals, and many of the workers are children. Along with this shift in the use of carcinogenic agents in the workplace, a surge in occupational cancers can be expected to occur in developing countries in the decades to come.

Studies have estimated that between 4% and 8% of all cancers in developed countries are caused by occupational exposures (Nurminen, 2001; Steenland, 2003; Rushton, 2012). Attributable fractions for different cancers vary by agent and by regions. In U.S. studies, workplace exposures accounted for 6%–17% of all lung cancers in men (Driscoll, 2005; Steenland, 2003). A similar study conducted in Finland showed that 29% of male lung cancer cases were due to occupational factors (Nurminen, 2001). In Europe, up to 10% of all bladder cancers were attributable to occupational exposures (Boffetta, 1999). In a recent report of a large occupational cancer study in France, it was shown that 42% of occupational cancers were caused by exposure to asbestos (ANSES, 2018). In the United States, cancer of the trachea, bronchus, and lungs made up 12% of all workplace fatalities (Nelson, 2005).

In one global study, cancer was identified as the top killer of workers, surpassing both workplace diseases and accidents in the number of deaths worldwide (Hamalainen, 2007). In a recent analysis of work-related illnesses, cancer was attributed to 26% of all occupationally related fatalities (Hamalainen, 2017).

13.3 GENERAL METHODS USED TO EVALUATE CARCINOGENICITY

In recent years, advances in molecular biology and cancer research have demonstrated that cancer develops from a complex multifactorial web of causes. The most relatable evidence for establishing a relationship between an agent and a carcinogenic outcome in humans historically, however, has been through epidemiological studies. Diseases are identified in a human population, and then the factors associated with its occurrence are identified (Hill, 1971). Some of the first formal epidemiological studies that identified carcinogenic effects were in workers exposed to asbestos, benzene, and vinyl chloride (Fontham, 2009). As it remains today, most agents identified to be carcinogenic to the larger population have first been identified in occupational populations. Other useful epidemiological information comes from studies of human lifestyle (e.g., tobacco and alcohol consumption) and patients receiving various medical therapies or treatments.

Limitations in epidemiological evaluations include the existence of confounders that hide or confuse the understanding of relationships, inaccurate estimates or measures of true exposure or dose, and other study biases that hide the true effects of the agent. Scrupulous study design and mathematical manipulation aid in assuring validity and accuracy of cancer studies. Often, an agent is evaluated by more than one study before a conclusion is drawn. And studies included in the evaluation pass the strict guidelines for the strengths of cancer association between exposure duration and level, consistency with other studies, and the timing of exposure to the development of disease.

In developed countries, the studies of occupational exposures to carcinogens are expensive and time consuming, but they are performed when possible and especially when there are explicit concerns regarding the agent or exposure. These types of exposure studies are grossly lacking in workers exposed in developing countries, and the resulting epidemiological analyses are not possible.

The second most useful method for determining potential carcinogenicity of an agent is the long-term bioassay in experimental animals. Animal testing for toxic effects of chemicals began as early as 1900. Since that time, animal studies have been used to identify approximately 30% of carcinogens known to humans (Huff, 1993, 1999; Fung, 1995; Maronpot, 2004). Due to anatomical, metabolic, genetic, and physiological differences between animals, they are not the perfect means of predicting carcinogenicity of humans to all agents. However, the usefulness of the information they provide has been demonstrated for numerous agents. In recent years, however, the use of animals for toxicity evaluation for humans has been curtailed for ethical reasons.

In order to replace the use of animal studies, new toxicokinetic and mechanistic studies are expanding drastically. These methods use what is already known about the mutational signatures that comprise tumors and cancer cells and compare them with other similar chemicals that would be expected or likely to result in similar outcomes. It becomes a systematic method to evaluate mechanistic data and identify likely toxicological outcomes for various chemicals by comparing them to the biological outcomes of other known carcinogens. Some of the mechanistic signals of carcinogenesis have included such characteristics as transformation of metabolites that can damage DNA, alteration of gene expression, disruption of the immune system, and interference with molecular communication. Results of these types of studies, combined with sophisticated mathematical and computational manipulations, can lead to the evaluation of significantly more potential carcinogens in much less time than previously possible. Future advances in these types of studies will continually expand to include toxicological predictions based on physical and chemical properties of molecules, genomic responses of biological samples, cancer pathway and network analyses, and even clinical studies of molecular changes in tissues of exposed humans (Cote, 2016; NTP, 1999; EPA, 2017).

13.4 CANCER RESEARCH ORGANIZATIONS

Several internationally respected organizations conduct research and publish reports on cancer-causing agents and environmental conditions. Each organization identifies the methods used to study and evaluate carcinogens. Some of the methods and assumptions used between agencies overlap, whereas some do not. Each organization identifies different levels or types of cancers and creates a rating scale of severity where different chemicals or agents can be categorized. Although some calls for greater harmonization in cancer research methodology and classification, differences between organizations' methods and levels remain for a variety of complex reasons. In most cases, it is useful to evaluate the results from two or more of the cancer research organizations to obtain a more valid and useful perspective of the occupational applicability to the cancer research information available. Some of the more important and significant organizations that provide current information on carcinogens are described in the following sections.

13.4.1 U.S. National Toxicology Program

In the United States, the National Toxicology Program (NTP) prepares a Report on Carcinogens (RoC) for the Secretary of Health and Human Services through a series of review mechanisms, and publishes reports periodically. Monographs on carcinogens are prepared according to the strict review mechanisms and processes (NTP 2016a, b). The NTP makes two carcinogen classifications: agents are classified as "known to be human carcinogens" when there is sufficient evidence of carcinogenicity from studies in humans, which indicates a causal relationship between exposure and disease, and agents classified as " reasonably anticipated to be human carcinogen" are those that are indicated by limited evidence of carcinogenicity in humans where a causal link is credible but alternative explanations such as bias and confounders cannot be excluded. Evidence of possible human carcinogenicity from animal studies or from convincing information about mechanistic processes that indicate an agent may be carcinogenic to humans can also be used to classify an agent in this category.

In the 14th NTP ROC published in 2016, there are 62 agents listed as known human carcinogens and 186 listed as reasonably anticipated to be human carcinogen. In the latest listing, there are six new agents known to be human carcinogens:

- Epstein–Barr virus
- Human immunodeficiency virus type 1 (HIV-1)
- Human T-cell lymphotropic virus type 1
- Kaposi sarcoma-associated herpesvirus
- Merkel cell polyomavirus
- Trichloroethylene

Five of the six new listings are linked to viruses for which there are no vaccines available. In total, the viruses are linked to more than 20 different types of cancers. These viruses and subsequent cancers are more likely to develop in people with weakened immune systems. These new biological agents, in addition to other infectious diseases previously identified to cause cancer, such as hepatitis B virus (HBV) and hepatitis C virus (HCV), are especially alarming since it is estimated that 16,000 HCV, 66,000 HBV, and 1,000 HIV health-care worker infections occur globally each year due to sharps injuries. The fraction of infections of HCV, HBV, and HIV attributable to occupational exposures represents up to 39%, 37%, and 4.4%, respectively, of all transmissions identified globally (Pruss-Ustun, 2005). In addition, these numbers are most likely underestimated due to the lack of sharps exposure reporting.

In addition to the six new carcinogens listed in the 2016 NTP report, cobalt and cobalt compounds that release cobalt ions in vivo were added to the list as reasonably anticipated to be a human carcinogen in an additional report published by the Department of Health and Human Services in 2016 (NTP, 2016c).

13.4.2 International Labor Organization

The International Labor Organization (ILO) established the first list of occupational diseases in 1925 (Kim, 2013). It was an important tool for the harmonization of international policies and regulations to protect workers from occupational diseases. In 1934, the list was updated with Convention number 42 to include primary epitheliomatous cancer of the skin. In 1980, the list was updated again to include lung cancer and mesothelioma caused by asbestos.

Today, the ILO recognizes cancer as a leading cause of worker illness and fatality. The ILO Convention C139 was created in 1974 to curtail worker exposures to carcinogens in the workplace. It requires each ratifying member country to work to prohibit or reduce the use of carcinogens, and to take action to minimize worker exposures to carcinogenic substances and agents in the workplace. It also requires the establishment of medical surveillance for workers exposed to carcinogens during and after employment, and the establishment of appropriate associated record-keeping systems (ILO, 1974a). At the same period, the ILO created a recommendation number 147 on how to prevent occupational hazards from carcinogens, (ILO, 1974b).

In 1977, the ILO created a comprehensive guideline on the control and prevention of occupational cancer. Topics discussed in this document included suggested methods for the determination of safe exposure levels for workers exposed to carcinogens through the use of epidemiologic and animal studies. It also proposed various preventive measures to be taken to minimize worker exposures. Other subjects in the manual included workplace and biological monitoring, administrative controls, and medical surveillance of workers (ILO, 1977).

The ILO document Occupational Safety and Health Series 74 published in 2010 by the ILO provides a list of 21 agents currently believed to cause cancer (ILO, 2010). Occupational carcinogens listed in this report included the following:

- Asbestos
- Benzidine and its salts
- Bis(chloromethyl) ether
- Chromium VI and chromium VI compounds
- Coal tars, coal tar pitches, or soots
- Beta-naphthylamine
- Vinyl chloride
- Benzene
- Toxic nitro- and amino-derivatives of benzene or its homologues
- Ionizing radiations
- Tar, pitch, bitumen, mineral oil, anthracene, or the compounds, products, or residues of these substances
- Coke oven emissions
- Compounds of nickel
- Wood dust
- Arsenic and its compounds
- Beryllium and its compounds
- Cadmium and its compounds
- Erionite
- Ethylene oxide
- Formaldehyde
- HBV and HCV

13.4.3 International Agency for Research on Cancer

Over the past several decades, IARC has been evaluating and identifying workplace carcinogens. The IARC is an international agency within the World Health Organization that performs and publishes critical reviews of epidemiological studies and experimental data on the carcinogenicity of chemicals, agents, mixtures, processes, and biological materials to which humans are exposed. Collective bodies of experts review the available information about human exposures and potential carcinogenic outcomes and provide a cancer rating for each subject. The IARC ratings are provided in Table 13.1 (IARC, 2014).

The IARC has identified 32 carcinogenic occupational agents. In addition, they have identified 11 exposure circumstances that are carcinogenic to humans. They have also identified 27 agents and 6 exposure circumstances that are probably carcinogenic to humans. A partial listing of the IARC carcinogenic ratings for occupational agents, occupations, and industries is provided in Table 13.2 (IARC, 2014).

TABLE 13.1
Current IARC Cancer Groups

Group 1: Carcinogenic to humans
Group 2A: Probably carcinogenic to humans
Group 2B: Possibly carcinogenic to humans
Group 3: Unclassifiable as to carcinogenicity in humans
Group 4: Probably not carcinogenic to humans

TABLE 13.2
Partial Listing of IARC Group 1 and 2A

	Group 1: Carcinogenic to Humans	
Agent, Occupation, or Industry	**Cancer Site/Cancer**	**Main Industry or Use**
Acid mists, strong inorganic	Larynx	Chemical
Arsenic	Lung, skin, bladder	Glass, metals, pesticides
Benzidine	Bladder	Pigments
Leather dust	Nasal cavity	Shoe manufacture and repair
Mineral oils	Skin	Lubricant
Nickel compounds	Nasal cavity, lung	Metal alloy
Silica dust	Lung	Construction, mining
Trichloroethylene	Kidney	Solvent, dry cleaning
Vinyl chloride	Liver	Plastics
Wood dust	Nasal cavity	Wood
Aluminum production	Lung, bladder	
Coal-tar distillation	Skin	
Coke production	Lung	
Painter	Bladder, lung, mesothelioma	
Rubber manufacture	Stomach, lung, bladder, leukemia	
	Group 2A Carcinogens	
Acrylamide	—	Plastics
Bitumens	Lung	Roofing
Indium phosphide	—	Semiconductor
Polychlorinated biphenyls	—	Electrical components
Vinyl bromide	—	Plastics, textiles
Art glass workers	Lung, stomach	
High temperature food frying	—	
Hairdressers and barbers	Bladder, lung	
Shiftwork with circadian disruption	Breast	Nursing and others

Source: Adapted from IARC (2014).
Carcinogens with Cancer Site and Industry.

13.4.4 European Union

International classifications for carcinogens have been harmonized at the community level by the European Union in Regulation (EC) No 1272/2008. Cancer-causing agents were originally established by the Commission on Cancer in 1963 and identified in a Cancer Liaison Program (CLP) (ACS, 2018). The classifications are as follows:

- CLP class 1A: known carcinogenic potential for humans
- CLP class 1B: presumed carcinogenic potential for humans
- CLP class 2: suspected human carcinogens

13.5 OCCUPATIONAL EXPOSURE LIMITS FOR CARCINOGENIC AGENTS

Occupational exposure limits (OELs) have been identified for thousands of chemicals and toxic agents by numerous regulatory and standards development organizations globally. Typically, OELs represent the maximum allowed concentration of a given substance in the air in the workplace. They are generally based on a time-weighted average over an 8-h period. Levels can be set when risks can be reasonably established and safe levels of exposure can be set. Similarly, for non-genotoxic agents (substances that do not directly damage DNA), safe levels of occupational exposure can also be determined. However, for most genotoxic carcinogenic agents, a threshold effect cannot be identified and it becomes extremely difficult to derive a safe level of occupational exposure. Infectious disease carcinogens are considered not to have thresholds.

At this time, there are no international OELs that are consistently recognized or agreed upon. Many occupational carcinogens do not have OELs in many countries. In many countries, OELs have not been established because safe levels of occupational exposure have not been agreed upon. Some governments have characterized risk quantitatively and compared exposures to commonly accepted social and economic criteria in order to set OELs. Even where OELs for carcinogens have been established at the national level, broad scientific and policy differences lead to greatly differing OELs between different countries in many instances (Ding, 2014). In some countries, the as low as reasonably achievable approach used to minimize worker exposure to ionizing radiation has been applied to no-threshold carcinogens.

13.5.1 The European Union

In the European Union (EU), indicative OEL values (IOELVs) for carcinogenic substances are determined by a commission of independent scientific experts, called the European Scientific Committee on Occupational Exposure Limits (SCOEL). Determinations are based on the types of carcinogenic effects, mechanisms of action, and whether the substance is genotoxic. Carcinogenic substances are categorized into the following groups:

Group A: Non-threshold genotoxic carcinogens identified by linear non-threshold (LNT) models of extrapolation test results from high-dose animal studies and low-dose human studies

Group B: Genotoxic carcinogens where studies are not sufficient to demonstrate the LNT model and a dose threshold is not identified

Group C: Genotoxic carcinogens where a practical threshold can be determined from available data

Group D: Non-genotoxic carcinogens and non-DNA-reactive carcinogens where a threshold can be set based on a no observable adverse effect level

OELs have been set by SCOEL for carcinogens in groups C and D.

In 2004, the EU promulgated Directive 2004/37/EC for the protection of workers from the risks related to exposure to carcinogens or mutagens. These regulations require employers to access worker risks of exposure to carcinogens. Specific hygienic workplace practices such as contamination control and personal protective equipment (PPE) are mandated. These laws also delineated requirements for worker information and training (EU, 2004).

13.5.2.1 Registration, Evaluation, Authorisation and Restriction of Chemicals

Registration, Evaluation, Authorisation and Restriction of Chemicals (REACH) is the European Community intergovernmental body responsible for registration, evaluation, authorization, and restriction of chemical substances. REACH limits on carcinogenic substances are based on two different effect levels. Derived no-effect levels (DNELs) are set for carcinogens that have a threshold (REACH Regulation Article 119). For carcinogens without a threshold, REACH sets a derived minimal effect level (DMEL) of cancer risk levels of either 10^{-5} or 10^{-6} as indicative tolerable risk levels for workers. Methodologies for DNEL and DMEL are determined by REACH and followed by registrants that manufacture/import/or use over 10 tons per year (ECHA, 2010, 2011).

13.5.2.2 The Netherlands

In the Netherlands, OELs are recommended by the Dutch Health Council. OELs for carcinogenic and mutagenic substances are based on one of the two substance risk levels. A prohibitive risk level limits the additional risk of cancer to less than 10^{-4} per year (4×10^{-3} for a 40-year working lifetime). An additional target risk level allows an additional risk of up to 10^{-6} per year (4×10^{-5} during a 40-year working lifetime).

13.5.2.3 France

In France, OELs are derived for chemicals with and without a threshold. For carcinogens without a threshold, low-risk occupational exposure levels are set at three different levels to control excess risk of developing cancer (10^{-4}, 10^{-5}, or 10^{-6}) INRS (2008). Methods to be used to measure and assess pollutants in the workplace are developed by the OEL committee of the National Agency for Safety and Hygiene in Food, the Environment and work (ANSES, 2017).

13.5.2.4 Germany

Germany sets OELs for carcinogenic substances for which there is evidence of a threshold. For carcinogens without a threshold, risk-based target values are used to limit worker exposures. An acceptable risk target level is currently set at 4×10^{-4}, but moves to 4×10^{-5} in 2018. An additional tolerable risk target level is 4×10^{-3} if certain specified safety and control measures are in place BAuA (2011).

13.6 CONTROLS AVAILABLE TO REDUCE OCCUPATIONAL EXPOSURES TO CARCINOGENS

Despite what is known about the industries and agents that cause cancer, workers continue to be exposed to carcinogenic agents. In a French study from 2005, it was shown that more than 13% of workers are still being exposed to substances known to cause cancer (DARES, 2005). Numerous and significant opportunities to reduce workplace exposures remain untapped. The costs to implement controls and reduce worker exposures remain to be misunderstood in terms of the payback in worker health and other direct and indirect costs. The effectiveness of the most sophisticated means of control is not understood and is not fully used even in the most developed countries. Further yet, undeveloped nations may not only fully understand the consequences of worker exposures to carcinogens but also not be aware of the means to curtail exposures to workers. The most relevant methods to reduce occupational exposure to carcinogenic agents involve four main approaches: (1) elimination or substitution, (2) engineering controls, (3) administrative controls, and (4) PPE.

13.6.1 Elimination and Substitution

Attributable risks to workers from carcinogens can be reduced either by reducing the number of carcinogens to which workers are exposed or by reducing the exposure levels to the carcinogens. The first method that can be used to reduce exposures to carcinogens is often via elimination of the agent in the workplace. In some cases, it is as simple as excluding the material from the production or fabrication process. An example here is the use of alternative fixatives in pathology and histology to replace carcinogenic formaldehyde mixtures. In the 1970s, Sweden was one of the first countries to ban the use of asbestos in products and by the 1990s had seen a reduction in the incidence of pleural mesothelioma in workers (Hemminki, 2008). In other cases, the carcinogenic process can be altered in some way to reduce worker contact or proximity to carcinogens. Substitution of one type of carcinogenic raw material used to produce a product with one that is less toxic and not carcinogenic can be a useful means to reduce occupational exposures (IARC, 2014). Engineering controls such as ventilation or enclosures can be used to reduce worker exposures too. Examples here could be as simple as using lids to keep carcinogenic vapors from escaping into the workers' breathing zone. A process change such as implementing wet hygienic methods to keep dust levels from grinding processes down and prevent the spread of contaminates could also be used to reduce airborne levels of dust. Wet spray methods could also be implemented to control the generation of airborne dust in stone-cutting operations where silica is a significant component. These alternative production methods can reduce the amounts of silica dust workers who would inhale considerably.

13.6.2 Engineering Controls

The next method to minimize worker exposures to carcinogens is the use of engineering controls. Engineering controls are defined as any system or device that separates the worker from the hazardous agent. This could include a safer needle device to keep nurses from being stuck with a contaminated patient needle to reduce the transmission of bloodborne pathogens such as HIV or HBV. Engineering controls can also include ventilation systems that draw carcinogenic aerosols away from the workers' breathing zones. Ventilation engineering controls can include sophisticated fume hoods and withdrawal systems, or simple designs such as natural airflow or wind patterns in a factory or hospital to carry carcinogenic aerosols away from the workers.

13.6.3 Administrative Controls

Administrative controls to reduce exposure to carcinogens can include written programs, policies, and procedures that help workers understand carcinogenic hazards and actions needed to be taken to reduce exposures. Worker education, training, and record keeping are all considered forms of administrative controls.

13.6.4 Personal Protective Equipment

Lastly, the use of adequate and effective PPE can lead to significant reductions in worker exposure to carcinogens. Impermeable gloves, aprons, goggles, and face shields are a last line of defense to protect the skin or eyes from coming in direct contact with carcinogens. Respirators can be an effective means to reduce worker exposure to carcinogens through the airborne pathway. In order for any PPE method to be effective, however, workers need adequate levels of training on PPE use, including donning and doffing.

Unfortunately, many of the engineering controls and even PPE used, and even taken for granted, in developed countries are not always available in economically challenged countries. Informal sector workers exposed to airborne carcinogenic dust or vapors may not have access to local exhaust ventilation not only due to the cost of such systems but also because there is no electricity. Appropriate PPE for many jobs is often unavailable in developing countries. Even adequate gloves or simple respiratory protection devices such as an N95 respirator may not be available in sufficient quantities.

With the greater understanding about the relationships between infectious agents and cancer, new opportunities for primary prevention arise. The need to protect health-care and other workers at risk from infectious agents including HIV, herpes virus, Epstein–Barr virus, and HBV becomes even more urgent when the additional risk of cancer is added to the already-inherent biological risks. Many EDCs already burdened with catastrophic numbers of HIV are now also at an elevated risk of cancer in addition. Health worker training in the effective use of PPE to protect from bloodborne pathogens, the use of impervious gloves, safe needle devices, immediate prophylactic treatment of workers exposed to known infectious agents, the diligent use of medical surveillance of workers, and the use of available vaccines will not only be expected to reduce the transmission of infectious agents to workers but also reduce the subsequent development of infectious agent-related cancers in the population (Vineis, 2014). It is essential that these protective devices and systems be made broadly and consistently available to workers in these developing countries in order to curtail occupational cancer rates.

13.6.5 Regulation

Another strategy that can be used to mitigate occupational cancer is the development of regulation governing the potential for worker exposures to carcinogens. Regulations on the use of carcinogens and limits on the levels and durations of worker exposures can have significant effects on global outcomes and prevalence of disease (Landrigan, 2011). National development and expansion of existing cancer and infectious disease reporting mechanisms would also aid in reducing cancer rates. Development of vaccines for infectious diseases can also be a tool to reduce the associated subsequent occupational cancers. These societal benefits should be included in decisions regarding funding for research and vaccine development.

13.6.6 Communications and Education

Communications and public health education programs that are expanded to reach greater populations and provide more information regarding hazardous agents, pathways of exposure, and the relations between infectious diseases and cancer will also benefit in the reduction of occupational cancer burden (Landrigan, 2011). Professional organizations can contribute to the awareness and understanding of occupational carcinogens and how they can be controlled to reduce worker exposures and negative outcomes. Conferences, seminars, and workshops can be held on specific occupational carcinogen topics with specific work groups. Health-care workers can be targeted for training on the recent findings regarding carcinogenicity related to infectious agents and the importance for improved worker protections.

13.7 OCCUPATIONAL CANCER RESEARCH

Despite the growing shift in occupational cancers to developing countries, associated research is lagging behind. Most research on occupational cancer has been in the formal sector of developed countries. Very little research on carcinogens in the informal and precarious workforce sector and in developing countries has been completed (Santana, 2009). These informal workers typically work outside of formal industry, in homes or informal settings. They do not have occupational safety and health training or support. They are particularly precarious in that they typically do not fall under any social support system and workers do not typically have any health-care benefits, including emergency medical care if they get hurt while they are working.

A 2009 report produced by the U.S. Department of Health and Human Services on reducing environmental cancer risks included a section on occupational cancers. It highlighted four general areas where research on occupational cancers should be enhanced: identification of occupational carcinogens, epidemiologic research, risk assessment, and prevention (Reuben, 2010).

Identification of occupational carcinogens can be improved through advanced employee surveillance systems and more accurate workplace exposure assessments. Better strategies for predicting adverse effects of working conditions and mixtures of hazardous agents in combination with advanced computational analyses will also improve the understanding of carcinogens.

Epidemiological studies must be better designed to identify exposed populations and routes of exposure. Cancers in women and minority workers need to be evaluated more broadly. And the relationships between maternal and paternal exposures and genetic effects in offspring need to be evaluated (Ward, 2003).

Increased use of biomarkers associated with occupational carcinogens is seen as a way to improve knowledge about risk. With better understanding of occupational cancer and risks, better primary and secondary prevention and communication strategies can be developed and implemented.

Improved research on the control and prevention of occupational cancer includes greater emphasis on the front end of industrial process design to minimize the potential for worker exposure to carcinogens. Means to identify and protect workers at particularly high risk should be prioritized and implemented (Reuben, 2010).

13.8 CONCLUSIONS/RECOMMENDATIONS

Occupational exposure to carcinogenic agents and working conditions accounts for a significant portion of the global burden of disease and death. And the impacts of these exposures reach far beyond the injured worker to affect families, organizations, and nations.

Due to the lack of reporting, the true occupational contribution to the rates of certain cancers remains unknown. New evidence of the relations between infectious diseases and cancer is particularly alarming. In addition, many occupational illnesses, injuries, and fatalities go unreported for large segments of the informal and contingent workforces. In low-income countries, the lack of efficient health and statistical data collection makes it difficult to accurately assess the burden of cancer on the population (IARC, 2017). Informal workers such as people working from home, children, and migrants are often not included in health statistics databases (Nelson, 2005). Therefore, the true risk of occupational cancer is expected to be higher than what has been reported.

It is estimated that 30%–50% of all cancers could be prevented if public health strategies were put in place to counter known risk factors (Stewart, 2016). This includes a reduction in the contribution from occupational cancers through the use of elimination, substitution, engineering controls, administrative controls, and PPE.

In a 2014 review of occupational cancer epidemiology research, it was determined that there has been a substantial decrease in the number of studies published between 1991 and 2001 (Raj, 2014). This same study demonstrated a lack of data from industrializing countries such as Russia, China, and India, and there are nearly no studies from the Middle East or most countries in Africa. In addition, few studies include female workers.

In not knowing what the true nature of the cancer hazard is or what exposure levels are, many workers go unprotected from toxic or fatal exposures. Much remains to be done in terms of worker protections, ranging from accurate exposure assessment and risk characterization, exposure evaluation, and development and implementation of controls to minimize exposures. Workers exposed to carcinogens in developing countries where sophisticated controls such as ventilation or respiratory protection may not be available are in need of research for alternative methods of protection such as the use of natural ventilation or readymade PPE, and the use of housekeeping and hygienic methods. As the numbers of exposures and illnesses are expected to increase in the next decades, even small changes and improvements can have major impacts globally.

REFERENCES

ACS, American Cancer Society. (2018). *Cancer Facts & Figures*. Atlanta: American Cancer Society, pp. 1–76.

ANSES, Occupational Exposure Limits for Chemical Agents (2017) www.anses.fr/en/content/occupational-exposure-limits-chemical-agents accessed October 23, 2017.

ANSES, Réseau national de vigilance et de prevention des pathologies professionnelles (rnv3p) (June 6, 2018) Presented at the 35th Congress on Occupational Medicine and Workplace Health, Marseille, June 5–8, 2018 www.anses.fr/fr/content/r%C3%A9seau-national-de-vigilance-et-de-pr%C3%A9vention-des-pathologies-professionnelles-publication?utm_source=lettre-information-INRS-juillet-aout-2018&utm_medium=email&utm_campaign=newsletter-INRS accessed July 20, 2018.

BAuA (2011) *Announcement 910- Risk Figures and Exposure-Risk Relationships in Activities Involving Carcinogenic Hazardous Substances (in German only)*. Dortmund, Germany: Federal Institute for Occupational Safety and Health (BAuA), GMBl 2011 S. 194 [Nr. 10].

Blair, A., Marrett, L., Freeman, L., Occupational cancer in developed countries, *Environmental Health* (2011) Vol. 10, No. Suppl 1, p. S9.

Boffetta, P., Kogevinas, M., Introduction: Epidemiologic research and prevention of occupational cancer in Europe, *Environmental Health Perspectives* (May 1999) Vol. 107, No. Supp. 2, pp. 229–231.

Bray, F., Jemal, A., Torre, L., Forman, D., Vineis, P., Long-term realism and cost-effectiveness: Primary prevention in combatting cancer and associated inequalities worldwide, *Journal of the National Cancer Institute* (2015) Vol. 107, No. (2): djv273. 8 pp.

Cote, I., Andersen, M., Ankley, G., Barone, S., Birnbaum, L., Boekelheide, K., Bois, F. Y., Burgoon, L. D., Chiu, W. A., Crawford-Brown, D., Crofton, K. M., The next generation of risk assessment multiyear study—Highlights of findings, applications to risk assessment and future directions, *Environmental Health Perspectives* (2016) Vol. 124, No. 11, pp. 1671–1682. [epub posted 4/19/16].

Danaei, G., Vander Hoorn, S., Lopez, A., Murray, C., Ezzati, M., Comparative risk assessment collaborating group (cancers) (2005) Causes of cancer in the world: Comparative risk assessment of nine behavioural and environmental risk factors, *Lancet* (2005) Vol. 366, No. 9499, pp. 1784–1793.

DARES (2005) Les expositions aux produits cancerogne, Premiere Syntheses Information, DARES, No. 28.1 2005.

Ding, Q., Schenk, L., Hansson, S., Setting risk-based occupational exposure limits for no-threshold carcinogens, *Human and Ecological Risk Assessment Journal* (2014) Vol. 20, No. 5, pp. 1329–1344.

Driscoll, T., Nelson, D., Steenland, K., Leigh, J., Concha-Barrientos, M., Fingerhut, M., Prüss-Üstün, A., The global burden of disease due to occupational carcinogens. *American Journal of Industrial Medicine* (2005) Vol. 48, pp. 419–431. http://dx.doi.org/ 10.1002/ajim.20209 PMID:16299703.

EC, Regulation (EC) No 1272/2008 of the European Parliament and of the Council of 16 December 2008 on classification, labelling and packaging of substances and mixtures, Annex VI (2008) Official Journal of the European Union Updated list from European chemical substances information system (ESIS): https://eur-lex.europa.eu/legal-content/EN/TXT/?uri=CELEX:32008R1272 accessed August 2, 2018.

ECHA, Guidance on information requirements and chemical safety assessment. Chapter R.8: Characterisation of dose [concentration]-response for human health. European Chemical Agency (ECHA) Version 2 (2010) https://echa.europa.eu/documents/10162/13632/information_requirements_r8_en.pdf accessed August 2, 2018.

ECHA, Registered substances. Chemical Substance Search. European Chemicals Agency (ECHA) (2011) https://echa.europa.eu/information-on-chemicals, accessed August 2, 2018.

EPA, Toxicity Forecasting: Advancing the Next Generation of Chemical Evaluation. The U.S. Environmental Protection Agency (2017) www.epa.gov/chemical-research/toxicity-forecasting accessed October 16, 2017.

EU, Directive 2004/37/EC Directive 2004/37/EC of the European Parliament and of the Council of 29 April 2004 on the protection of workers from the risks related to exposure to carcinogens or mutagens at work (2004) Sixth individual Directive within the meaning of Article 16(1) of Council Directive 89/391/EEC.

Espina, C., Porta, M., Schuz, J., Aguado, I., Percival, R., Dora, C., Slevin, T., Guzman, J., Meredith, T., Landrigan, P., Neira, M., Environmental and occupational interventions for primary prevention of cancer: A cross-sectional policy framework, *Environmental Health Perspectives* (2013) Vol. 121, No. 4, pp. 420–426.

Fontham, E., Thun, M., Ward, E., Balch, A., Delancey, J., Samet, J., American cancer society perspectives on environmental factors and cancer. *CA: Cancer Journal for Clinicians* (2009) Vol. 59, No. 6, pp. 343–351.

Fung, V. A., Barrett, J. C., Hu, J. E., The carcinogenesis bioassay in perspective: Application in identifying human cancer hazards. *Environmental Health Perspectives* (1995) Vol. 103, No. 7–8, pp. 680–683.

Hamalainen, P., Takala, J., Leena, K., Saarela, L., Global estimates of fatal work related diseases, *American Journal of Industrial Medicine* (2007) Vol. 50, pp. 28–41.

Hamalainen, P., Takala, J., Kiat, T. (2017). *Global Estimates of Occupational Accidents and Work-Related Illnesses 2017.* Workplace Safety and Health Institute, ISBN: 9789811148446.

Hashim, D., Boffetta, P., Occupational and environmental exposures and cancers in developing countries, *Annals of Global Health* (2014) Vol. 80, pp. 393–411.

Hayes, R., Gerin, M., Raatgever, J., de Bruyn, A., Wood-related occupations, wood dust exposure, and sinonasal cancer, *American Journal of Epidemiology* (1986) Vol. 124, No. 4, pp. 569–577.

Hemminki, K, Hussain, S., Mesothelioma incidence has leveled off in Sweden, *International Journal of Cancer* (2008) Vol. 122, No. 5, pp. 1200–1201.

Hill, A. (1971). *Principles of medical statistics* (9th ed.). New York: Oxford University Press.

Huff, J., Chemicals and cancer in humans: First evidence in experimental animals. *Environmental Health Perspectives* (1993) Vol. 100, pp. 201–210.

Huff, J. (1999). Value, validity, and historical development of carcinogenesis studies for predicting and confirming carcinogenic risks to humans. In K. T. Kitchin (Ed.), *Carcinogenicity Testing, Predicting, and Interpreting Chemical Effects* (pp. 21–123). New York: Marcel Dekker.

IARC, Forman, D., Ferlay, J. (2014). The global burden and regional burden of cancer. In B. Stewart, C. Wild (Eds). *World Cancer Report 2014* (pp. 15–53), International Agency for Research on Cancer, Lyon, France, ISBN 978-92-832-0443-5.

IARC, Press Release No. 251, Latest data show a global increase of 13% in childhood cancer incidence over two decades, World Health Organization (April 11, 2017) www.iarc.fr/en/media-centre/pr/2017/pdfs/pr251_E.pdf accessed August 2, 2018.

IARC, Fact Sheet (2018a) http://gco.iarc.fr/today/fact-sheets- cancers?cancer=29&type=0&-sex=0 accessed August 2, 2018.

IARC, Cancer Incidence, Mortality and Prevalence Worldwide GOBOCAN 2012 (2018b) http://globocan.iarc.fr/old/burden.asp?selection_pop=224900&Text-p=World&selection_cancer=290&Text-c=All+cancers+excl.+non-melanoma+skin+cancer&pYear=13&type=0&window=1&submit=%C2%A0Execute accessed August 2, 2018.

ILO, C139-Occupational Cancer Convention (1974a) www.ilo.org/dyn/normlex/en/f?p=NORMLEXPUB:12100:0::NO::P12100_INSTRUMENT_ID:312284 accessed October 10, 2017.

ILO, Recommendation 147, Concerning prevention and control of occupational hazards caused by carcinogenic substances and agents (1974b) www.ilo.org/dyn/normlex/en/f?p=NORMLEXPUB:12100:0::NO::P12100_ILO_CODE:R147 accessed August 2, 2018.

ILO. (1977). *Occupational Safety and Health Series No. 39.* Occupational Cancer Prevention and Control. ISBN 92-2-101827-X.

ILO. (2010). *List of Occupational Diseases-Occupational Health Series 74.* ILO. ISBN 978-92-2-123795-2.

INRS. (2008). *Occupational Exposure Limit Values for Occupational Exposure to Chemical Agents in France.* Paris, France: Institut National de Recherche et de Sécurité (INRS), Edition ED984.

Kim, E., Kang, S., Historical review of the list of occupational diseases recommended by the international labour organization (ILO), *Annals of Occupational and Environmental Medicine* (2013) Vol. 25, p. 14.

Landrigan, P., Espina, C., Neira, M., Global prevention of environmental and occupational cancer - editorial, *Environmental Health Perspectives* (July 2011) Vol. 119, No. 7, p. a280.

Maronpot, R., Flake, G., Huff, J., Relevance of animal carcinogenesis findings to human cancer predictions and prevention. *Toxicologic Pathology* (2004) Vol. 32, No. Suppl 1, pp. 40–48.

Nelson, D., Concha-Barrientos, M., Driscoll, T., Steenland, K., Fingerhut, M., Punnett, L., Pruss-Ustun, A., Leigh, J., Corvalan, C., The global burden of selected occupational diseases and injury risks: Methodology and summary, *American Journal of Industrial Medicine* (2005) Vol. 48, pp. 400–418.

National Toxicology Program (NTP). Carcinogens Report, Tenth Edition—Substances, mixtures and exposure circumstances for listing or delisting, *Federal Register* (1999) Vol. 64, No. 74, pp. 19188–19189.

National Toxicology Program (NTP), History of the Report on Carcinogens (RoC). The National Toxicology Program (2016a) http://ntp.niehs.nih.gov/pubhealth/roc/history/index.html accessed July 12, 2016.

National Toxicology Program (NTP), Tox 21: Toxicology in the 21st Century. The National Toxicology Program (2016b) http://ntp.niehs.nih.gov/results/tox21/index.html accessed March 21, 2016.

National Toxicology Program (NTP), 14th Report on Carcinogens 2016, US Department of Health and Human Services (2016c) https://ntp.niehs.nih.gov/pubhealth/roc/index-1.html accessed August 2, 2018.

Nurminen, M., Karjalainen, A., Epidemiologic estimate of the proportion of fatalities related to occupational factors in Finland. *Scandinavian Journal of Work Environment & Health* (2001) Vol. 27, pp. 161–213. http://dx.doi.org/10.5271/sjweh. 605 PMID:11444413.

Pearce, N., Matos, E., Vainio, H., Boffetta, P., Kogevinas, M. (1994). *Occupational Cancer in Developing Countries* (IARC Scientific Publication No. 129). Lyon, France: International Agency for Research on Cancer, p. 181.

Pott, P. (1775). *Chirurgical Observations Relative to the Cataract, the Polypus of the Nose, the Cancer of the Scrotum, the Different Kinds of Ruptures, and the Mortification of the Toes and Feet.* London: Printed by T.J. Carnegy, for London Hawse, W. Clark and R. Collins.

Pruss-Ustun, A., Rapiti, E., Hutin, Y., Estimation of the global burden of disease attributable to contaminated sharps injuries among health-care workers, *American Journal of Industrial Medicine* (2005) Vol. 48, pp. 482–490.

Raj, P., Hohenadel, K., Demers, P., Zahm, S., Blair, A., Recent trends in published occupational cancer epidemiology research: Results from a comprehensive review of literature, *American Journal of Industrial Medicine* (2014) Vol. 57, pp. 259–264.

Ramazzini, B. (1718). De Morbis Artificum. Padua Italy: J.B. Conzattum. In W. Wright (Ed.), *Diseases of Workers [English Translation].* London: Hafner, 1964.

Reuben, S., Reducing environmental cancer risk: What can we do now. 2008–2009 Annual Report of the President's Cancer Panel. National Cancer Institute. US Department of Health and Human Services, 240 pp (2010) https://deainfo.nci.nih.gov/advisory/pcp/annualreports/pcp08-09rpt/pcp_report_08-09_508.pdf accessed August 2, 2018.

Rushton, L., Hutchings, S., Fortunato, L., Young, C., Evans, G. S., Brown, T., Bevan, R., Slack, R., Holmes, P., Bagga, S., Cherrie, J. W., Van Tongeren, M., Great Britain, *British Journal of Cancer* (2012) Vol. 107, No. Suppl 1, pp. S3–S7. PMID:22710676.

Santana, V., Riberiro, F., Occupational cancer burden in developing countries and the problem of informal workers, *Environmental Health* (2011) Vol. 10, No. Suppl 1, p. S10.

Siemiatycki, J., Richardson, L., Boffetta, P. (2006). Occupation. In: D. Schottenfeld, J. F. Fraumeni, Jr., (Eds). *Cancer Epidemiology and Prevention* (3rd ed.). New York: Oxford University Press, pp. 322–354.

Steenland, K., Burnett, C., Lalich, N., Ward, E., Hurrell, J., Dying for work: The magnitude of US mortality from selected causes of death associated with occupation. *American Journal of Industrial Medicine* (2003) Vol. 43, pp. 461–482. http://dx.doi.org/10.1002/ajim. 10216 PMID:12704620.

Stewart, B., Bray, F., Forman, D., Ohgaki, H., Ulrich, A., Wiod, C., Cancer prevention as part of precision medicine: Plenty to be done, *Carcinogenisis* (2016) Vol. 37, No. (1), pp. 2–9.

Torre, L., Bray, F., Siegel, R., Ferlay, J., Lortet-Tieulent, J., Jemal, A., Global cancer statistics, 2012, *CA: Cancer Journal for Clinicians* (2015) Vol. 65, pp. 87–108.

Vineis, P., Wild, C., Global cancer patterns: Causes and prevention. The cancer wars series, *The Lancet* (February 8, 2014) Vol. 383, pp. 549–557.

Ward, E., Schulte, P., Bayard, S., Blair, A., Brandt-Rauf, P., Butler, M., Dankovic, D., Hubbs, A., Jones, C., Karstadt, M., Kedderis, G., Melnick, R., Redlich, C., Rothman, N., Savage, R., Sprinker, M., Toraason, M., Weston, A., Priorities for development of research methods in occupational cancer, *Environmental Health Perspectives* (January 2003) Vol. 111, No. 1, pp. 1–12.

14 Migrant Worker Occupational Safety and Health

Thomas P. Fuller
Illinois State University

CONTENTS

14.1 INTRODUCTION

As the world becomes increasingly globalized, people, institutions, governments, and financial systems become more entwined and interdependent. In the past two decades, the world has become drastically more connected through the Internet, improvements in air travel and other forms of transportation, and expansion of global business endeavors. As political interconnections have expanded, and people become more aware of opportunities in other regions, migration increases to take advantage of better political conditions, economics, and social systems. Migration can also benefit governments, businesses, and communities. In many cases, both origin and destination countries work together to create opportunities for migration.

Migration is a global phenomenon of gigantic importance that affects all aspects of society in some capacity. Although most migration occurs legally, in recent years there has been a drastic increase in migration due to poverty, famine, conflict, persecution, and environmental disaster. As a result, migrants end up in regions that do not have necessary political, social, or economic capabilities to absorb and adjust to them. These increased numbers of unqualified or unneeded migrants place additional economic and social burdens on the host countries, leading to further regional instability and erosion. Migrants who enter these regions or countries that do not want them, or do not have infrastructure to take care of them, live in relatively insecure and precarious conditions. Living conditions for these migrants are often substandard and do not meet normal public health standards. They often have limited or no access to health-care systems.

Despite the existence for hundreds of years, the complexities of the politics, economics, and cultural aspects of today's migration make it difficult to understand and sometimes politically charged. It is important for educated professionals to better understand how migration impacts their areas of endeavor and communities. For example, health-care providers and public health agencies need to study migrant groups to see how services can be better provided and how migrants impact communities. It is equally important for occupational safety and health (OSH) professionals to expand their awareness of how migrants assimilate into the work environment and how to best protect them from injury, illness, or death. Migrants comprise a special class of workers that need specialized assessment and protection controls, and OSH professionals need to know how to address those needs.

14.2 MIGRANT POPULATIONS

There were 258 million international migrants in the world in 2017, comprising 3.4% of the global population (UN DESA, 2017c). And according to the International Labor Organization (ILO) migration is likely to intensify in the coming years due to increasing work deficits in large geographic regions (ILO, 2017). People in search of work will be migrating in increasing numbers, particularly from Africa and Latin America. Younger workers between the ages of 15–24 are more likely to migrate. Numbers of global migrants over the past several years are shown in Table 14.1.

Separate from migration numbers that represent more or less permanent movements, there were more than 40 million internally displaced persons and 22.5 million refugees in 2016 (IDMC, 2017; UNHCR, 2017). With ongoing political conflict in the Middle East and other regions throughout the world, these numbers of people moving across national boundaries are expected to increase in the near term. The relative numbers of migrants in various regions are shown in Figure 14.1.

Work is the major reason for international migration. Migrant workers comprise about 70% of all migrants. And about 75% of those go to high-income countries, with 23% going to middle-income countries (IOM, 2017 migrant report). Approximately 55.7% of migrants are male, whereas 44.3% female. Industries and sectors where migrants work are shown in Table 14.2.

14.3 GEOGRAPHIC REGIONS AND MIGRANT MOVEMENTS

The United States is the leading destination country for international immigration, followed by Germany, Russia, and Saudi Arabia. In Saudi Arabia, migrants comprise

TABLE 14.1
Number of International Migrants as a Percentage of the World's Population

Year	Number of Migrants	Migrants as a % of World's Population
1970	84,460,125	2.3
1975	90,368,010	2.2
1980	101,983,149	2.3
1985	113,206,691	2.3
1990	152,563,212	2.9
1995	160,801,752	2.8
2000	172,703,309	2.8
2005	191,269,100	2.9
2010	221,714,243	3.2
2015	243,700,236	3.3

Source: UN DESA (2015).
Note: The number of entities (such as states, territories, and administrative regions) for which data were made available in the 2015.
UN DESA *Revision of International Migrant Stock* was 213. In 1970, the number of entities was 135.

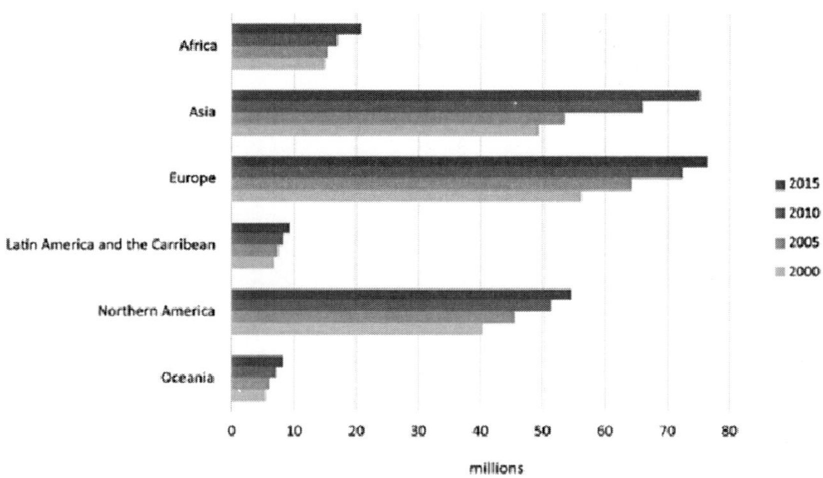

FIGURE 14.1 International migrants, by major region of residence, 2000–2015 (million). (IOM 2017 migrant report, p. 16.)

88% of the total population (UN DESA, 2017a, 2017b). The leading origin countries of migration are India, Mexico, Russia, and China. A graphical representation of destination and origin countries is provided in Figure 14.2.

According to the World Bank, two-thirds of international immigrants reside in high-income economies (the United States, Canada, Australia, Western Europe).

TABLE 14.2
Common Occupational Sectors for Migrants

Sector	Percentage
Services other than domestic	64.1
Domestic services	7.0
Manufacturing and construction	17.8
Agriculture	11.1

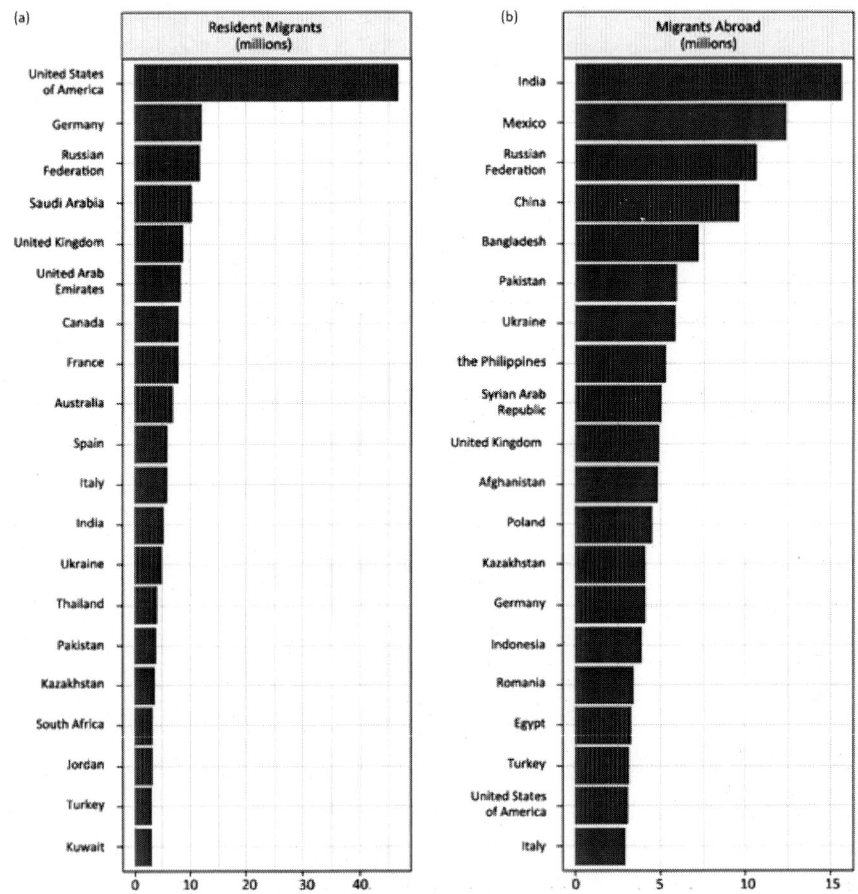

FIGURE 14.2 Top 20 destinations (a) and origins (b) of international migrants in 2015 (million).

Just less than a third reside in middle-income countries, with the remainder in low-income countries (World Bank, 2018; Docquier, 2011).

These estimates are inherently difficult to document, however, for a variety of reasons. Individual immigration status is often fluid and difficult to capture in time.

A temporary visit visa may change to an unauthorized overextension. These "irregular" migrant populations become difficult to count, and different national systems for identifying migrants add another dimension making accurate comparisons between countries difficult (Gordon, 2009). Another problem in creating accurate measures of migrants involves the common confusing bureaucracy surrounding governmental recognition of migrant status and the clandestine nature of migrant activity (Koser, 2010). Measuring migratory movement is a difficult endeavor and requires extensive resources and infrastructure that are often beyond those available to many governments.

14.4 BENEFITS OF MIGRATION

Despite much of the negative political rhetoric regarding migration, there are many social and economic benefits to both origin and destination countries. One recent OECD study found that the taxes migrants pay host governments minus the benefits they receive tend to be positive (OECD, 2013). Migrants can have a positive effect on labor productivity and gross domestic product per head. They can have a positive effect on the labor market, especially in markets where there are particular shortages of specific workers. In one Norwegian study, it was shown that a 10% increase in migrant employment in the construction industry led to a 0.6% decrease in wages of construction workers, but that these wage and cost reductions were passed along to consumers (Bratsberg, 2012).

Migration can have positive impacts on the countries of origin. It can reduce unemployment and poverty. It can lessen social burdens on already weak and unstable social systems. Money sent by migrants back to the families in their home countries can have a considerable positive financial impact on the economies of those countries. They can represent a relatively stable source of income over time. In 2016, the flow of remittances worldwide reached US$429 billion (World Bank, 2017).

Migrants, and their families who remain behind in their country of origin, can benefit immensely. Wages of migrants abroad can be many times higher than from their same job at home (Clemens, 2009). The largest relative income gains occur in the least skilled and most restricted migrants (Gibson, 2011). Other benefits to migrants generally include better health than in their native countries, higher school enrollment rates, and large reductions in infant mortality (World Bank, 2016).

14.5 MIGRATION GOVERNANCE

International migration governance primarily stems from United Nations member state agreements to cooperate on various aspects of migration including the development of laws and norms, and to create governance institutions and mechanisms. The benefits of enhanced global governance of migrations include improved control and awareness of the movements of people, better use of resources for monitoring through coordinated transit points, and control of illicit activities including smuggling, human trafficking, organized crime, and terrorist activities. Better migration control and coordination also improves awareness of financial remittances and flow of monies across borders.

Unfortunately, international controls and agreements for migration policy lag behind those developed for other social areas, such as human rights, trade, and even environmental impacts. A more effective global governance system for migration could create beneficial opportunities for participating states. Guiding principles, rules, and norms, whether binding or not, can create benchmarks of state behavior and guidance for the development of programs and policies where there are currently none.

While the UN convention on refugees has been widely ratified, and a specific UN agency for refugees has been created, the UN High Commissioner for Refugees (UNHCR), instruments to support migrants have received far less support. International treaties that exist for human trafficking and migrant smuggling have been ratified, yet protections for migrants, sometimes in extremely vulnerable situations, generally do not exist.

There are a variety of reasons that international migration policy is difficult to develop. Migration encompasses such a broad range of technical topics across international boundaries the complexities are difficult to understand. Economics, demographics, religion, cultural values, and political stability can all be influenced by migration. The development of policies that change the dynamics in one country over those in another is difficult to agree upon. The sovereignty of countries and the right of States to self-determination often outweigh the common good of the international community. Due to the complexity of migration, it is often misunderstood by the general population, and it becomes a politicized topic of discussion in public immigration debates. The lack of information about how migration will affect a country's economy, culture, and social systems makes it difficult to rationally debate the issues and form consensus. Often when clear data are available, it is disregarded for cultural or political reasons.

CASE

There is often concern that an influx of migrants into a region will negatively affect employment and the wages of existing national workers. However, in a comprehensive study in Denmark, it was shown that the new migrant workers tended to take the lower paid jobs, but the existing Danish workers in those jobs tended to move on to other higher paying positions in the region. Overall employment in the area increased, along with Danish worker wages.

(Foged, 2015).

One of the main reasons international migration policy development is so difficult is due to the imbalance of power between the origin and destination countries. Migrants have little bargaining power when they arrive at their destination countries. They tend to, and need to, accept what civil privileges and freedoms they are given, and they have little recourse to object to their conditions. In that regard, it suggests even greater need for international cooperation for the development of benchmarks and standards for all countries to follow in terms of migrant rights.

14.6 INTERNATIONAL ORGANIZATION FOR MIGRATION

The International Organization for Migration (IOM) was founded in 1951 as an independent regional organization. Over the years, it became a much more global organization, and in 2016, it joined the United Nations as a related organization. The IOM has convened an annual International Dialogue on Migration since 2001 to bring stakeholders together to discuss relevant and emerging migration topics. In 2015, the IOM created a Migration Governance Framework that outlines the essential elements necessary for safe, orderly, and responsible state migration policies. This governance framework advocates three main principles: adherence to international standards and fulfillment of migrant rights, development of comprehensive and coordinated state policies and programs, and international collaboration and coordination.

14.7 GOVERNING CONSENSUS OF MIGRANT LABOR

Although there has been some development of international cooperation on such issues as refugee persecution and human trafficking, less has been implemented on the rights of migrants. The ILO has done the most to address health and safety working conditions of migrants. On the global level, there are the 1949 Migration for Employment Convention (Revised) (ILO Convention No. 97), the 1975 Convention Concerning Migrations in Abusive Conditions and the Promotion of Equality of Opportunity and Treatment of Migrant Workers (ILO Convention No. 143), and the 1990 Convention on the Protection of the Rights of All Migrant Workers and Members of their Families.

Topics addressed in ILO Convention No. 97 include policies regarding migrant recruitment, remuneration, work age, employment taxes, worker training, and membership in trade unions (ILO, 1949). Convention 143 expands responsibilities for monitoring and control of migration by States and provides for human rights of migrants and expands their equalities of opportunity and treatment in host countries (ILO, 1975).

The only place working conditions are discussed in either of these two conventions is in Article 12 of Convention 143. It requires members to "guarantee equality of treatment, with regard to working conditions, for all migrant workers who perform the same activity whatever might be the particular conditions of their employment." Note that this does not imply equal working conditions with native workers in the host country. In fact, except for this clause, governance guidelines or standards for working conditions in international migration governance are nearly nonexistent. A word search for "occupational" of the 2018 World Migration Report had no hits. Occupational health and safety issues were not mentioned in the entire 347-page report (IOM, 2017).

The European Union (EU) recently drafted new rules that would require employers in one EU country to provide the same working conditions for legal EU workers from other countries working in their nations. The policy will basically continue the EU Directive 96/71/EC from 1996, which requires equal working conditions, including health and safety, for posted workers. However, this new revision will still not protect working conditions of migrants coming from outside the EU. And the extent that the agreement will affect worker safety overall is presently unclear.

The EU Strategic Framework on Health and Safety at Work 2014–2020 describes future commitments to worker rights. The document addresses emerging risks and aging workers specifically, but does not mention migrant or immigrant workers and rights (EU, 2014). Nor are there any explicit U.S. OSHA regulations that specifically address migrants, although there are mentions sporadically in OSHA alerts and other informational materials. And language deficiencies are implicitly covered in the Global Harmonization Standard for hazardous material postings and labels. In Canadian OSH legislation, significant gaps have been identified in the working conditions and occupational safety applied to migrant farm and agricultural workers (Otero, 2010; McLaughlin, 2014).

14.8 ECONOMIC AND SOCIAL DISPARITY OF MIGRANTS

Migrants have consistently been shown to suffer economic discrepancies due to a variety of factors. Migrant wages have been shown to be up to 13% less than natives working in the same firms (Bartolucci, 2014). It is well known that migration across national borders, even between developed countries, causes a loss in the intrinsic value of human capital. Different schools and educational systems, loss of credentialing, and differences in equipment, modalities, and methods can cause migrants to be less valuable, even if only for perceived differences. Losses are significantly influenced by cultural differences between the origin and destination countries. Migrant human capital value is significantly higher for migrants from highly developed countries, with similar language and cultural values, than for migrants who come from economically developing countries with different languages and cultures (Sanroma, 2015). The more different they are, the greater the losses in relative value of human capital (Cheswick, 1979; from Carneiro, 2012). As migrants find themselves unqualified for similar positions they had held in their countries of origin, they begin to accept offers at lower skilled occupations, amplifying the loss in wage phenomenon. Migrant language skills of the destination country have been shown to be important determinants of wages (Cheswick, 2002).

Employers in destination countries rank migrant qualifications lower than natives, and it accounts for two-thirds of wage differentials (Carneiro, 2012; Kampelmann, 2016; Chan, 2014). In one study, managers were shown to be more likely to hire workers with similar national origins. And since only 1.5% of the managers were foreign-born, it significantly negatively impacted immigrant wages (Aslund, 2014). The lack of managers from other countries then tends to hold migrant job placement and wages back further.

Migrants searching for positions may intentionally downgrade their qualifications to increase the search pool and find better matches (Weiss, 2003). They often cluster into ethnic enclaves, and seek work in the same areas, which also has a downward impact on wages (Carneiro, 2012; Aydemir, 2008). Migrant wages eventually begin to rise with work experience in the destination country and by obtaining local training (Cohen, 2002).

14.9 EDUCATION AND LANGUAGES OF MIGRANTS

Some studies have indicated that migrants with lower levels of education and poor language skills may accept more hazardous jobs because they do not fully understand the risks. And a study by Orrenius showed that workers that spoke no English faced greater injury and fatality rates than other workers (Orrenius, 2009). Approximately 39% of foreign-born adults in the United States do not have a high school diploma, compared with 11% of natives. In a study by Grieco (2003), 35% of immigrants reported speaking English poorly, and 12% did not speak it at all. The Center for Disease Control and Prevention reported that worker deaths among Hispanic immigrants were greater due to poor English language and literacy abilities, partly related to the inability to understand safety training and documents (CDC, 2008).

In addition to reduced levels of training for migrants due to language and educational deficiencies, migrants are often hired through contractors or manpower agencies (sometimes located overseas). It is often assumed that these contractors or agencies will assume the health and safety training requirements internally, but these companies are often unfamiliar with specific hazards of a migrant's job assignments, and equally unprepared to offer these workers adequate training. These workers are often temporary, and extensive safety training is not deemed worth the time for the short period the workers will be on-the-job (Belin, 2011). The lack of inexperience of these workers often then exacerbates the training failures further (HSE, 2010).

14.10 OCCUPATIONAL HEALTH RISK FACTORS FOR MIGRANTS

Although a considerable amount has been written about the economic and social consequences of migration, much less is known about the OSH conditions of migrants compared with native workers. In a study in China, migrant workers accounted for a disproportionate burden of occupational morbidity and mortality (Fitzgerald, 2013). Other studies have also shown that migrant workers tend to have higher injury rates, and be employed in more dangerous industries, than native-born workers (Schenker, 2010; Ahonen, 2007; Reid, 2010).

A 2009 study by Orrenius showed that during the period 1992–2005, while the overall U.S. worker fatality rate declined, the percentage of fatalities by foreign-born workers rose from 3.5% to 18% of the total (Orrenius, 2009). This same study concluded that immigrants do tend to work in more dangerous industries and occupations.

Immigrants may take riskier jobs for a variety of reasons. One may be similar to taking a job for lower wages; by accepting the riskier job, the likelihood of being hired may increase. Another reason may be cultural, in different levels of acceptance of risk. Or to put it another way, the value of life, or limb. Immigrant workers may be used to more dangerous working conditions that were accepted in their origin country, and they may not perceive the working conditions in the destination country as particularly dangerous, again pointing towards cultural differences in risk acceptance. Migrants also tend to work high levels of overtime, and at accelerated paces, adding to the risk of an injury- or work-related illness (HSE, 2010; Otero, 2010; Hennebry, 2015). And they have been shown to take additional work risks in order to please their supervisors (Vartia-Väänänen, 2007). In fact, part of the reason employers like to hire migrants is their willingness to work long hours and do work that natives are not inclined to perform (Basok, 2002).

Migrants also tend to be younger than most native populations (EC, 2010). This would tend to make them more likely to accept more physical challenges and risks than older workers with more experience. Migrants have also been shown to be assigned more hazardous work by their supervisors than native workers in the same job categories (Eurofound, 2007).

The social and economic precariousness of migrant work can also play a role in determining the risk factors associated with safety on the job. Workers in more precarious jobs, such as migrants, have been shown to suffer disparities in their general health compared to workers in more stable jobs (Puig I Barrachina, 2013). Employees that report high job insecurity exhibit higher levels of workplace injuries and illnesses (Probst, 2001). These workers may show lower levels of motivation towards compliance with safety rules.

14.11 OCCUPATIONAL HEALTH OUTCOMES

In a study by Chan (2006), 21 of 29 construction fatalities in New York City during a 12-month period were immigrants with poor language skills. In a review of the United States Bureau of Labor and Statistics data for the period 1992–2005, Orrenius found a dramatic increase in the percentage of immigrant to overall fatalities per 100,000 workers shown in Figure 14.3.

The job design and type of business where migrants are employed can have an impact on the number and level of injuries or illnesses they might incur. Barling reported that job characteristics associated with high-performance and high-quality work systems led to fewer occupational injuries (Barling, 2003). When workers have a high perception of job insecurity, they tend to show less motivation to follow safety rules, which results in higher numbers of accidents and injuries (Probst, 2001).

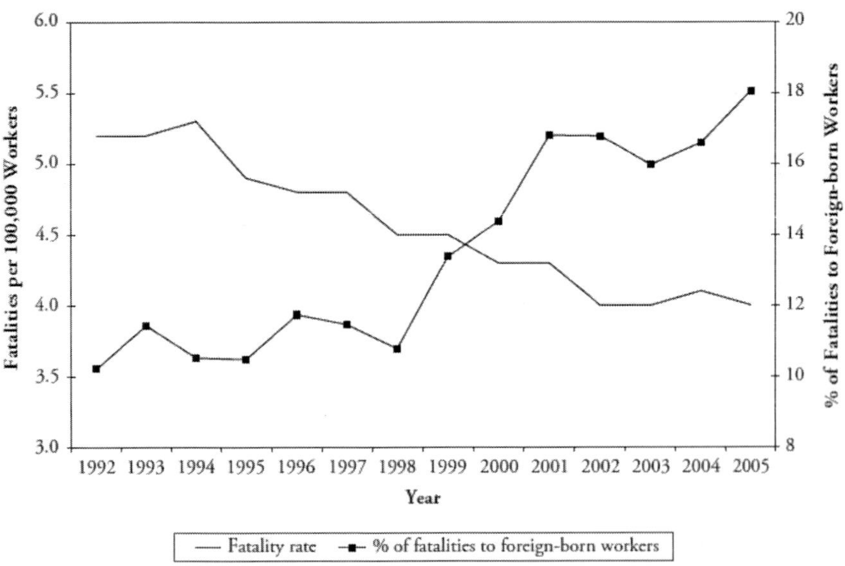

FIGURE 14.3 Overall fatality rate and fatalities to foreign-born workers (Orrenius, 2009).

Companies operating in highly developed countries with well-trained workers engaged in job design through their knowledge and preexisting skill set will tend to have lower injury rates. Employees participate in decision-making, including safety decisions, and thus tend to be more proactively involved, with positive results. Simard reported that workers who participated in management decisions were much more proactively involved in their own safety, and not just for the sake of regulatory compliance (Simard, 1997).

Conversely, it follows then that in countries, work sectors, and jobs with much lower level of work sophistication, performing simpler tasks, the workers would be less involved in the design and control of their workplaces, with adverse outcomes in terms of workplace health and safety. In addition, it would follow that in countries with cultural and managerial styles that discourage worker involvement in decision-making, workers would be less interested in safety. Workers performing high-performance work tend to be more satisfied (Berg, 1999).

Turnover rates for workers performing higher skilled jobs also tend to be lower (Guthrie, 2001). Both of these factors would play a major role in the participation in safety training and the likelihood to apply the information gained to the workplace. High turnover rates make the cost of training go up and overall effectiveness of safety training go down. This is again a particular problem for migrants who are often moving and changing jobs.

In a paper by Barrachina, it was noted that precarious workers, namely, women, those in nonsupervisory positions, and those working in Eastern and Southern Europe, were more likely to suffer job dissatisfaction and poor work-related health (Barrachina, 2013). Certain types of contract workers are less likely to report unsafe working conditions and then suffer the consequences in terms of injury and illness (Van Aerden, 2015).

Studies have shown that contract workers suffer more frequent and severe injuries than regular full-time workers (Blank, 1995; Quinlan, 1999). These workers often have less experience or formal safety training, and they tend to work longer hours and in more hazardous jobs (Wright, 1998). They also tend to work in more informal work settings and structures where workplace risks are higher (Glodstein, 2001).

Negative health outcomes in migrant workers are often closely associated with the types of jobs and sectors they work in (Vartia-Väänänen, 2007). The increased proportions of migrants working in more hazardous and physically demanding sectors such as construction, agriculture, and fishing include significant hazardous tasks and exposures and increase their likelihood of workplace-associated injury or illness (Eurofound, 2007). Migrants are commonly exposed to excessive vibration, noise, heat, pesticides, and other airborne contaminants. Immigrants to the United States from Mexico have been shown to have higher work-related fatalities (Loh, 2004). These higher fatality rates at least in part arise because of their disproportionate employment in the hazardous industries of construction, and agriculture (Richardson, 2003).

Many immigrants work in service industries such as retail shops or driving taxicabs. Many own their own businesses. But working with the public in service industries can also be hazardous. Between 1992 and 2005, more than 3,000 foreign-born were murdered on the job in the United States, making homicide the leading cause of immigrant workplace fatality (Franklin, 2006; Sincavage, 2005).

14.12 POTENTIAL UNDERREPORTING

There is reason to speculate that immigrant injury and illness rates are underreported. Immigrant workers may be less likely to report their injuries or illnesses for fear of reprisal, or missing work without compensation. Partly because of their reduced language skills, and partly because of their lack of understanding of social systems, migrants do not partake in available health-care benefits when they are available to the extent of natives (Rust, 1990).

Contract workers are less likely to report workplace injuries or seek associated medical treatment due to fears of being dismissed. In small to midsized enterprises, contract workers are fearful of reporting unsafe conditions to authorities. Often when they do, they either lose their jobs or their employers "go out of business" altogether rather than pay penalties or fix the unsafe conditions (Underhill, 2011).

When migrants use health-care services, it tends to be at very late stages of illness or disease (Vartia-Väänänen, 2007). Migrants tend to work in more precarious and informal industries and in seasonal jobs (Gravel, 2014). These smaller employers would be less likely to partake in OSHA required reporting mechanisms, and would be less likely to report injuries and illnesses in these workers. Employers may be less likely to report a fatality as work related in migrant cases to avoid scrutiny of working conditions or possible inspections. Undocumented workers would be less likely to report injuries or illnesses for fear of drawing attention to themselves with the authorities. They are often also unaware of how to access occupational health services or even basic medical treatment. And if there was not going to be either medical assistance or workers' compensation, then there are few benefits to reporting an injury or illness. So as a result, the injury, illness, and fatality rates for immigrants are probably even greater than what is reported (Orrenius, 2009).

14.13 RECOMMENDATIONS FOR IMPROVEMENT

The practical implications of companies or countries using contract or migrant workers are clear. Additional accidents and injuries will occur with organizations set up to take advantage of a flexible workforce. When using these workers, additional controls may be warranted to confront the increased and new risks. Injuries and illnesses may be reduced through additional hiring screening, on-the-job training, better communication of safety practices, and a culture of open communications and continuous improvement. Some workers may need special training on safety practices, including such topics as the usefulness of personal protective equipment. They may need special encouragement to report injuries, exposures to hazardous agents, or unsafe conditions.

In a study by Amuedo-Dorantes (2002), it was shown that additional worker safety training for temporary workers can significantly reduce the likelihood of injuries. Considering the high numbers of immigrants who do not speak English, and the high injury and fatality rates for immigrants, it is clear that safety training and written materials, including labels and postings, need to be available to the workers in their native languages.

One way to increase the number of migrants in a workforce and to increase wages would be to promote more immigrants into managerial positions (Aslund, 2014). It might also follow then that an increase in foreign-born managers might also improve the health and safety working conditions at the location.

Research on migrant working conditions and occupational health and safety outcomes should be expanded in order to gain better understanding of the issues and to find possible solutions. Injury and illness reporting deficiencies should be rectified by pointed research into the topic to better assess conditions and outcomes.

Professionals in both OSH and other areas should be informed of migrant issues and particular vulnerabilities in their workplaces. OSH professionals who understand the issues and concerns will be in a better position to do something about them. This could mean better communications with managers and supervisors about migrants, increased emphasis on cultural differences and language deficiencies, and involvement in corporate hiring and promotional practices for immigrant workers. Social service and health-care providers in a community should be aware of special migrant health conditions, cultural differences, and potential workplace hazards.

Migrants in unsafe working conditions around the world could benefit from increased world governance that clarified national responsibilities to protect all workers from hazardous working conditions and agents. The decoupling of normal health and safety regulations from people with few protections is unethical from a professional standpoint for practicing OSH professionals. Governments should work to create controls and protection standards and regulations oriented directly and specifically at the protection of migrant worker rights within individual nations and across international boundaries. The ILO has provided numerous reports and guidance documents that can be used to guide management principles and development of state of federal regulations. Pressure from practicing professionals who become informed on the issue of migrant workplace safety can begin the dialogue towards the development of regulations and policies necessary to protect these vulnerable workers.

REFERENCES

Ahonen, E., Benavides, F., Benach, J., Immigrant populations, work and health—A systematic literature review, *Scandinavian Journal of Work Environment & Health* (2007) Vol. 33, pp. 96–104.

Amuedo-Dorantes, C., Work safety in the context of temporary employment: The Spanish experience, *Industrial and Labor Relations Review* (2002) Vol. 55, No. 2, pp. 262–285.

Aslund, O., Hensvik, L., Skans, O., Seeing similarity: How immigrants and natives manage in the labor market, *Journal of Labor Economics* (2014) Vol. 32, No. 3, pp. 405–441.

Aydemir, A., Skuterud, M., The immigrant wage differential within and across establishments, *Industrial and Labor Relations Review* (April 2008) Vol. 61, No. 3, pp. 334–352.

Barling, J., Iverson, R., Kelloway, E., High-quality work, job satisfaction, and occupational injuries, *Journal of Applied Psychology* (2003) Vol. 88, No. 2, pp. 276–283.

Barrachina, P. (2013). *Monitoring Employment-Related Health Inequalities in Europe: The Case of Unemployment and Precarious Employment.* Barcelona: Universitat Pompeu Fabra, p. 339.

Bartolucci, C., Understanding the native-immigrant wage gap using matched employer data: Evidence from Germany, *ILR Review* (October 2014) Vol. 67, No. 4, pp. 1166–1202.

Basok, T. (2002). *Tortillas and Tomatoes. Mexican Transmigrant Harvesters in Canada.* Montréal: McGill-Queen's University Press.

Belin, A., Zamparutti, T., Tull, K., Hernandez, G., Graveling, R. (August 2011). *Occupational Health and Safety Risks for the Most Vulnerable Workers.* Brussels: European Parliament.

Berg, P., The effects of high performance work practices on job satisfaction in the United States steel industry, *Industrial Relations* (1999) Vol. 54, pp. 111–134.

Blank, V., Andersson, R., Linden, A., Nilsson, B., Hidden accident rates and patterns in the Swedish mining industry due to involvement of contractor workers, *Safety Science* (1995) Vol. 21, No. 1, pp. 23–35.

Bratsberg, B., Raaum, O., Immigration and wages: Evidence from construction, *The Economic Journal* (December 2012), Vol. 122, No. 565, pp. 1177–1205.

Carneiro, A., Fortuna, N., Varejao, J., Immigrants at new destinations: How they fare and why, *Journal of Population Economics* (2012) Vol. 25, pp. 1165–1185. As a result, migrants have been shown to suffer wage discrimination by employers.

Centers for Disease Control and Prevention (CDC). Work-related injury deaths among Hispanics—United States, 1992–2006, *Mortality and Morbidity Weekly Report* (2008) Vol. 57, pp. 597–600.

Chan, S. (November 22, 2006). *Fatal Construction accidents in the City Rise Sharply Over 12 Months.* New York: New York Times, p. C13.

Chan, J., Assimilation of Hong Kong immigrants in Canada, *Pacific Economic Review* (2014) Vol. 19, No. 4, pp. 439–465.

Cheswick, B. (1979). The economic progress of immigrants: some apparently universal patterns. In W. Fellner, (Ed.), *Contemporary Economic Problems.* Washington, DC: American Enterprise Institute, pp. 357–399. (from Carneiro, 2012).

Cheswick, B., Miller, P., Immigrant earnings: Language skills, linguistic concentrations and the business cycle, *Journal of Population Economics* (2002) Vol. 15, No. 1, pp. 31–57.

Clemens, M., Montenegro, C., Pritchett, L., (2009). *The Place Premium: Wage Differences for Identical Workers across the U.S. Border.* Washington, DC: Center for Global Development.

Cohen, S., Eckstein, Z., Labor mobility of immigrants: Training, experience, language and opportunities, *International Economic Review* (2002) Vol. 49, No. 3, pp. 837–872.

Docquier, F., Ozden, C., Peri, G., The Wage Effects of Immigration and Emigration. Policy Research Working Paper 5556 World Bank. (February 2011) http://documents.worldbank.org/curated/en/675801468331816158/pdf/WPS5556.pdf accessed January 14, 2018.

EU (June 6, 2014). European Commission-Communication from the Commission to the European Parliament, the Council, the European Economic and Social Committee and the Committee of the Regions on an EU Strategic Framework on Health and Safety at Work 2014–2020, Brussels.

Eurofound (May 2007). *Employment and Working Conditions of Migrant Workers.* Dublin: EWCO, p. 37.

European Commission, Demography Report (2010). Commission Staff Working Document, Directorate-General for Employment, Social Affairs and Inclusion and Eurostat, the Statistical Office of the European Union, p. 44.

Fitzgerald, S., Chen, X., Qu, H., Sheff, M., Occupational injury among migrant workers in China: A systematic review, *Injury Prevention* (October 2013) Vol. 19, No. 5, pp. 348–354.

Foged, M., Peri, G., Immigrants' effect on Native workers: New analysis on longitudinal data, *American Economic Journal: Applied Economics* (2015) Vol. 8, No. 2, pp. 1–14.

Franklin, S., Little, D. (October 8, 2006). *Cabbies, Clerks Put Lives on the Line: Immigrants' Top Cause of Death on the Job: Homicide.* Chicago: Chicago Tribune, p. 1.

Gibson, J., McKenzie, D. (2011). *Australia's Pacific Seasonal Worker Pilot Scheme: Development Impacts in the First Two Years*. New Zealand: University of Waikato.

Glodstein, G., Helmer, R., Fingerhut, M., Mobilizing to protect worker's health: The WHO global strategy on occupational health and safety, *African Newsletter on Occupational Health and Safety* (2001) Vol. 11, pp. 56–60.

Gordon, L., Scanlon, K., Travers, T., Whitehead, C. (May 2009). *Economic Impact on London and the UK of an Earned Regularization of Irregular Migrants in the UK*. London: Greater London Authority. ISBN 978-1-84781-258-2 Available from www.london.gov.uk/sites/default/files/gla_migrate_files_destination/irregular-migrants-report.pdf accessed January 14, 2018.

Gravel, S., Premji, S. (2014). Travailleur migrants: une histoire sans fin de cumul des précarités de statut, d'emploi et de conditions de santé et de sécurité au travail, Perspectives interdisciplinaires sur le travail et la santé, 16–21 Travailleurs immigrants et santé et sécuritié du travail.

Grieco, E. (2003). *English Abilities of the US Foreign-Born Population*. Washington, DC: Migration Information, Migration Policy Institute. www.migrationpolicy.org/article/english-abilities-us-foreign-born-population/ accessed January 8, 2018.

Guthrie, J., High involvement work practices, turnover and productivity: Evidence from New Zealand, *Academy of Management Journal* (2001) Vol. 44, pp. 180–190.

Hennebry, J., McLaughlin, J., Preibisch, K. (September 18, 2015). Changing Workplaces Review, Ministry of Labour. Addressing agricultural migrant worker protection.

HSE (February, 2010). Protecting migrant workers, Health and Safety Executive www. hse. gov.uk/migrantworkers/employer/ protecting.pdf accessed December 12, 2018.

Internal Displacement Monitoring Center (IDMC) (2017). *Global Report on Internal Displacement*. Geneva: IDMC. www.internal-displacement org/global-report/grid 2017/pdfs/2017-GRID.pdf accessed January 14, 2018.

ILO, International Labor Organization, C097—Migration for Employment Convention (Revised), (No. 97) (Entry into force: 22 Jan 1952) (1949) www.ilo.org/dyn/normlex/en/f?p=NORMLEXPUB:12100:0::NO::P12100_INSTRUMENT_ID:312242 accessed January 7, 2018.

ILO, International Labor Organization, C143—Migrant Workers (Supplementary Provisions) Convention, 1975, Convention concerning Migrations in Abusive Conditions and the Promotion of Equality of Opportunity and Treatment (1975) www.ilo.org/dyn/normlex/en/f?p=NORMLEXPUB:12100:0::NO::P12100_ILO_CODE:C143 accessed January 7, 2018.

ILO, Future of Work, Major Trends (2017) www.ilo.org/global/topics/future-of-work/trends/WCMS_545683/lang--en/index.htm accessed August 24, 2017.

IOM (2017). *World Migration Report 2018*. Geneva: International Organization for Migration, ISBN 978-92-9068-742-9.

Kampelmann, S., Rycx, F., Wage discrimination against immigrants: Measurement with firm-level productivity, *IZA Journal of Migration* (2016) Vol. 5, p. 15. Migrants have been shown to suffer significant negative wage gaps compared with natives in the same jobs.

Koser, K., Dimensions and dynamics of irregular migration, *Population, Space and Place* (2010) Vol. 16, No. 3, pp. 181–193.

Loh, K., Richardson, S. Foreign-born workers: Trends in fatal occupational injuries, 1996–2001, *Monthly Labor Review* (2004) Vol. 127, No. 6, pp. 42–53.

McLaughlin, J., Hennebry, J., Haines, T., Paper versus practice: Occupational health and safety protections and realities for temporary foreign agricultural workers in Ontario, *Perspectives Interdisciplinaire sur le travail et la santé* (2014) Vol. 16-2, pp. 2–17.

Organisation for Economic Co-operation and Development (OECD) (2013). The fiscal impact of immigration in OECD countries. In T. Liebig & J. Mo (Eds). *International Migration Outlook*. Paris: OECD.

Orrenius, P., Zavodny, M., Do immigrants work in riskier jobs? *Demography* (August 2009) Vol. 46, No. 3, pp. 535–551.

Otero, G., Preibisch, K. (August, 2010). *Farmworker Health and Safety: Challenges for British Columbia*. Richmond: Worksafe BC, RS2006-0G11 www.sfu.ca/~otero/docs/Otero-and-Preibisch-Final-Nov-2010.pdf accessed December 12, 2018.

Probst, T., Brubaker, T., The effects of job insecurity on employee safety outcomes: Cross-sectional and longitudinal explorations, *Journal of Occupational Health Psychology* (2001) Vol. 6, No. 2, pp. 139–159.

Puig I Barranhina, V., Monitoring employment-related health inequalities in Europe: the case of unemployment and precarious employment, (2013), *Thesis Dissertation*, Universitat Pompeu Fabra.

Quinlan, M., The implications of labour market restructuring in industrialized societies for occupational health and safety, *Economic and Industrial Democracy* (1999) Vol. 20, No. 3, pp. 427–460.

Reid, A., Migrant workers, *Ethnicity & Health* (2010) Vol. 15, pp. 436–437.

Richardson, S., Ruser, J., Suarez, P. (2003). Hispanic Workers in the United States: An Analysis of Employment Distributions. In *Fatal Occupational Injuries, and Non-Fatal Occupational Injuries and Illnesses* (pp. 43–82) Washington, DC: National Academies Press, in Safety Is Seguridad.

Rust, G., Health status of migrant farmworkers: A literature review and commentary, *AJPH* (October 1990) Vol. 80, No. 10, pp. 1213–1217.

Sanroma, E., Ramos, R., Simón, H., Portability of human capital and immigrant overeducation in Spain, *Population Research and Policy Review* (2015) Vol. 34, pp. 323–241.

Schenker, M., A global perspective of migration and occupational health, *American Journal of Industrial Medicine* (2010) Vol. 53, pp. 329–337.

Simard, M., Marchand, A., Work groups' propensity to comply with safety rules: The influence of micro-macro organisational factors, *Ergonomics* (1997) Vol. 40, pp. 172–188.

Sincavage, J. R., Fatal occupational injuries among Asian workers, *Monthly Labor Review* (2005) Vol. 128, No. 10, pp. 49–55.

UN DESA, Department of Economic and Social Affairs, Trends in International Migrant Stock: The 2015 Revision (2015).

UN DESA, Department of Economic and Social Affairs, International Migration Report 2017 (December 18, 2017a) www.un.org/development/desa/en/news/population/international-migration-report-2017.html accessed June 1, 2018.

UN DESA, International Migrant Stock: The 2017 revision (2017b) www.un.org/en/development/desa/population/migration/data/estimates2/estimates17.shtml accessed May 1, 2018.

UN DESA, Department of Economic and Social Affairs, Population Division, Population Facts (December, 2017)(2017c) No. 2017/5.

Underhill, E., Quinlan, M., How precarious employment affects health and safety at work: The case of temporary agency workers, *Relations Industrial Relations* (2011) Vol. 66, No. 3, pp. 397–421.

United Nations High Commissioner for Refugees (UNHCR) (2017). *Global Trends Forced Displacement in 2016*. Geneva: UNHCR. www.unhcr.org/5943e8a34.pdf. http://www.unhcr.org/5943e8a34.pdf accessed January 14, 2018.

Van Aerden, K., Moors, G., Levecque, K., Vanroelen, C., The relationship between employment quality and work-related well-being in the European Labor Force, *Journal of Vocational Behavior* (2015) Vol. 86, pp. 66–76.

Vartia-Väänänen, M., Pahkin, K. (2007). *Literature Study on Migrants.* Luxembourg: European Agency for Safety and Health at Work, European Risk Observatory.

Weiss, Y., Sauer, R., Gotlibovski, M., Immigration, search and loss of skill, *Journal of Labor Economics* (2003) Vol. 21, No. 3, pp. 221–225.

World Bank (2016). *Migration and Development. A Role for the World Bank Group.* Washington, DC: World Bank, http://documents.worldbank.org/curated/en/690381472677671445/Migration-and-development-arole for-the-World-Bank-Group.

World Bank (April 2017). *Migration and Remittances: Recent Developments and Outlook. Migration and Development Brief 27.* Washington, DC: World Bank, http://pubdocs.worldbank.org/en/992371492706371662/MigrationandDevelopmentBrief27.pdf accessed January 14, 2018.

World Bank, World Bank Country Income Group Classifications, in World Bank Country and Lending Groups (2018) http://datahelpdesk worldbank.org/knowledgebase/articles/906519-world-bank-country-and-lending-groups (World Bank, n.d.a) accessed January 14, 2018.

Wright, C., Lund, J., Under the clock: Trade union responses to computerized control in U.S. and Australian grocery warehousing, *New Technology, Work and Employment* (1998) Vol. 13, No. 1, pp. 3–15.

15 Child Labor

Thomas P. Fuller
Illinois State University

CONTENTS

15.1 INTRODUCTION

Children that work present special problems to occupational safety and health (OSH) professionals with the responsibility of protecting workers from hazardous conditions. This issue also represents a problem to the larger OSH professional community, because of the responsibility to protect all workers on a public health level, and not just from an organizational or business standpoint.

A job that may be relatively safe for an adult to perform, such as lifting an object, by nature of their physical characteristics, could be hazardous to a child. An occupational exposure to a recommended occupational limit value based on risk evaluations for an adult working 40 h per week for 40 years may not be applicable to a child beginning their working life well before 18, often working longer than 40 h per week, and with greatly different physiological and metabolic responses to toxic agents. Children are also at greater risk than adults of being injured by hazardous machinery or operations, sometimes due to their size and sometimes due to their inexperience and lack of training.

Some work performed by children is considered good in that they learn social skills of responsibility and gain experience. It is often the case that the work begins to interfere with normal schooling and life experiences and eventually hinders their social development, in addition to causing physical harm. The resultant lack of schooling leads to a young adult with poor social, reading, and math skills who must then endeavor to find useful and meaningful work with these disadvantages that last a lifetime. This is in addition to the potential physical ailments resulting from childhood work exposures.

Contrary to how many people think, child labor is not a third world problem, but occurs on a global scale. Many children in the United States work in hazardous conditions and are exposed to toxic chemicals on the job. And many advanced nation corporations manufacture products with parts and materials produced abroad with child labor infused into the supply chain. In addition to the moral reasons for eliminating child labor, there are business reasons. Child labor is used because it is typically cheaper than adult labor. Companies or countries that use child labor in their processes, or in their supply chains, gain an unfair advantage over their competitors. If for no other reasons, nations and businesses should have an interest in policing child labor for the sake of business value.

Lastly, the social consequences of child labor are seen in the precariousness of the eventual post-child labor adult working class. Uneducated, unskilled, injured, or ill, this group is not the shining model for a working social class. They live on the edge of poverty, often in the informal sector with few medical or social services available to them. They seldom pay taxes, and when they have children, they will likely be raised in similar environments to how they were raised, working, and not being educated. For developed countries that provide aid to many of the less developed countries, this is a revolving door where financial support can never solve the social problems unless child labor (which causes many of the social problems) is eliminated.

This chapter describes some of the definitions of child labor in use by different organizations, and explains the consequences to children and society. It then tries to provide some realistic alternatives and potential solutions to the child labor problem and provide resources for more information and guidance on the topic.

15.2 CHILD LABOR OF THE PAST

Various forms of child labor have existed through time. Child labor, indentured servitude, and child slavery existed in America since its first beginnings. Starting in agriculture, children quickly moved into urban jobs and factory work with the growth of cities and industrialization. Children were often preferred workers because they were seen as easier to manage, less expensive, and less likely to form labor unions.

The use of child labor peaked in the early 20th century in the United States, France, and the United Kingdom. But the growth of worker advocacy groups like the National Consumers League and Working Women's Societies led to increasing demands for social reform in workplaces, which included regulations for the limitation of child labor. In 1904, the National Child Labor Committee was created to fight for children's protections and rights to education. Finally, in 1938, the Fair Labor Standards Act was passed that included federal standards and limitations for the use of child labor (UI, 2018).

15.3 CHILD LABOR TODAY

According to the U.S. Census Bureau, the world's population is nearly 7.5 billion (USCB, 2017). Children make up about 27% of the world's population at approximately 1.9 billion (Gapminder, 2011).

Globally, approximately 218 million children between the ages of 5 and 17 work. Of these, 151.6 million are considered to be child laborers. These are children whose work interferes with their normal educational and other development. It can include long working hours, or working during normal school hours. The definition of child labor also includes children who work in hazardous conditions, or with toxic agents. A total of 72.5 million children are considered to be employed in hazardous activities. Hazardous activities include conditions of physical, psychological, or sexual abuse, and working underground, under water, at dangerous heights, and in confined spaces. Hazardous work also includes working with dangerous machinery or heavy loads. Exposure to hazardous chemical or physical agents, extreme noise, temperatures, and vibration is also considered hazardous. Working long durations, and shift or night work, is also considered hazardous by the International Labor Organization (ILO) (ILO, 2017a).

About 70.9% of child laborers work in agriculture (farming, hunting, forestry, and fishing). This is followed by work in the informal economy with 17.2% of children in service activities (retail, restaurants, hotels, transport, storage) and 11.9% of child labor in industry (construction, mining, utilities, manufacturing) (ILO, 2017a). Forty-four percent of child laborers are between the ages of 5 and 11 (ILO, 2013). Seventy percent of those workers meeting the definition of child laborers are unpaid family members (FAO, 2018). In terms of prevalence, Africa has the highest rates of child labor. Relative percentages of child labor for other regions are shown in Figure 15.1.

In agriculture (including hunting, forestry, and fishing), children face long hours in extreme temperatures, are exposed to pesticides, and work in conditions without adequate sanitation, food, or water. They often work for little or no pay, but for food and a place to sleep. Children have been documented harvesting bananas in Ecuador, cotton in Egypt, tea in Argentina, and cut flowers in Columbia (UI, 2018). A study of U.S. teenagers working in agriculture showed that this age group was highly unlikely to wear assigned appropriate personnel protective equipment (PPE) such as goggles, helmets, and earplugs (McCurdy, 2012).

In manufacturing, children have been documented making carpets in India, soccer balls in Pakistan, bricks in Nepal and India, and even fireworks in Guatemala and Peru. In mining, children have been reported as young as 6- or 7-year-old breaking

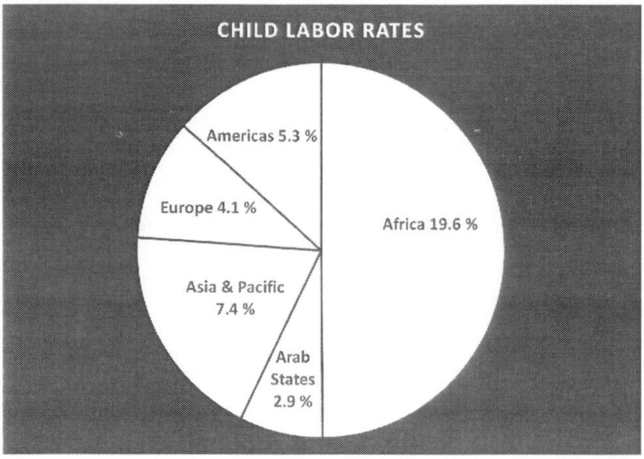

FIGURE 15.1 Relative rates of child labor globally. (Based on ILO, 2018.)

TABLE 15.1
Child Labor Distribution by Branch of Economic Activity

Sector	Share (%)
Agriculture	58.6
Industry	7.2
Services	32.3
Domestic work	6.9

Source: Adapted from ILO (2013).

and washing rocks. Nine-year-olds were being used to set underground explosives. Other activities included mining for gold in Columbia, chrome in Zimbabwe, and coal in Mongolia. Although poverty increases the likelihood of child labor and it comprises up to 22% of the labor force in low-income countries, it is not the only factor. Upper middle-income countries have as much as 6.2% of the workforce from children (ILO, 2013 "Making Progress"). Child labor distribution by branch of economic activity is shown in Table 15.1.

15.4 CHILD LABOR TERMINOLOGY

The United Nations Convention on the Rights of a Child, Article 1 defines a child as a "human being under the age of 18" (UN, 2018). The United Nations defines child labor as "work that is inappropriate for a child's age, affects children's education, or is likely to harm their health, safety or morals" (FAO, 2018). The ILO defines child labor as children doing work that is dangerous or unsuitable for them (ILO, 2016). It can depend on the child's age, the hours and type of work, and the conditions in which the work is performed. Not all work performed by children under 18 is considered child labor.

Child work can take many forms and occur at many ages. Positive forms of child work include good experiences that educate and develop their capabilities. Work can create skills and awareness in children and adolescents that lead to increased feelings of self-worth and satisfaction. The understanding of the value and benefits of labor are positive outcomes that can remain with a child for a lifetime. Beneficial work can improve health and intelligence such as chores around the home, helping in a family business, and working outside of school hours—all are positive learning experiences and are generally regarded as good. These types of activities can also contribute to the welfare of their families and help prepare the child for a productive working life as an adult.

Negative forms of child labor are considered as those that are harmful to physical and mental development and well-being. This could include any work that interferes with their schooling, subjects them to harmful working conditions, or deprives them of normal childhood development and dignity. Interference with normal schooling would include any work that interfered with school attendance. This includes leaving school early or arriving late. It could also include excessively long and heavy work that interfered with a child's ability to complete assignments or adequately engage while at school.

Work that subjects children or adolescents to toxic chemicals, harmful physical agents such as excessive noise or radiation, and musculoskeletal hazards such as heavy lifting or awkward postures are considered negative forms of child labor. In the absence of accurate scientific data regarding the negative effects of these harmful agents on children, a precautionary approach should be taken regarding their occupational exposures. Safe levels of exposure to a particular chemical for an adult cannot be considered "safe" for a child. Due to the differences in physiology and metabolism, children are typically considered to be more sensitive to harmful effects of chemical agents. Adult threshold limit values can be signals that exposure is hazardous to children, but it is generally assumed that children should not be exposed to the same levels as adults due to their developmental status. A safe lifting weight for an adult should not be considered safe for a child. Nor should safe working durations, rates, or postures for adults be considered safe for children. This includes excessive work such as shown in Figure 15.2. In addition to all the other "unknown variables" regarding hazardous exposures to children and potential negative outcomes, the exposure durations for child laborers are considerably longer than for adults.

Cumulatively negative exposures to excessive levels of noise to a child could be expected to have significantly higher likelihood of resulting in negative consequences in their lifetime, and sooner in their lifetime. Exposure to carcinogenic agents, which often take long periods before manifesting in an exposed worker, would be assumed to occur much sooner in the life of a person who is exposed as a child.

Extremely negative forms of child labor include those in which the child is deprived of all rights and protections normally provided by a society or government. These children may be working to support their families, or be separated from their families, or orphaned. They work for survival and perform all forms of hazardous work, do not attend school, and do not partake in normal childhood activities or

FIGURE 15.2 Child labor as identified by excessive work. (Courtesy of photographer Steve Thygerson.)

experiences. These children often exist and work outside of normal societal support systems such as medical benefits, welfare systems, or any form of social net or support. When these children are injured or sick, there is no support structure or access to health care.

The most extreme forms of child labor are defined in Article 3 of ILO Convention Number 182:

> (a) all forms of slavery or practices similar to slavery, such as the sale and trafficking of children, debt bondage and serfdom and forced or compulsory labour, including forced or compulsory recruitment of children for use in armed conflict;
> (b) the use, procuring or offering of a child for prostitution, for the production of pornography or for pornographic performances;
> (c) the use, procuring or offering of a child for illicit activities, in particular for the production and trafficking of drugs as defined in the relevant international treaties;
> (d) work which, by its nature or the circumstances in which it is carried out, is likely to harm the health, safety or morals of children.
>
> *(ILO, 1999).*

15.5 WHAT CHILD LABOR LOOKS LIKE

According to the ILO Recommendation Number 190, hazardous child labor is codified as work that exposes children to the following:

- Physical, psychological, or sexual abuse
- Work underground, under water, at dangerous heights, and in confined spaces
- Dangerous machinery, equipment, or tools
- Manual handling of heavy loads
- Hazardous chemicals, agents, processes, temperatures, noise, or vibration
- Particularly stressful working conditions including long hours, night shifts, or confinement

FIGURE 15.3 Children at work in the construction industry. (Courtesy of photographer Marianne Levitsky.)

Children at work in the construction industry are shown in Figure 15.3.

Child labor is not limited to developing countries, but some of the most advanced nations have considerable child labor rates. In the United States, more than two million children are exposed to occupational hazards each year (Castillo, 1999).

Child labor is very common in the United States, and in a 2003 study, it was shown that nearly 30% of farm-working children had driven tractors, all-terrain vehicles, or pickup trucks. Despite being considered hazardous work, many children under the age of 16 reported driving self-propelled combines, handling toxic fertilizers, and working in silos (Park, 2003).

According to the U.S. Department of Homeland Security, human trafficking might be identified by a long list of conditions that might be observable. Some of those conditions are listed as follows:

- The appearance that the person is disconnected from friends or family.
- The child is not attending school.
- Sudden and dramatic changes in behavior.
- Bruises or visible injuries.
- The person is confused, disoriented, fearful, timid, and submissive.
- They seem to have been coached or memorized what to say.
- The person is often in the company to whom they seem subservient or to whom they defer.

- The person does not seem to have many personal possessions, or basic items most people would have (e.g., a wallet, different and weather appropriate clothing, appropriate shoes).
- The child has little freedom of movement, appears malnourished or poorly clothed.

If human trafficking is suspected, the trafficker should never be confronted directly. Suspicious activities or conditions should be reported immediately to the U.S. Immigration and Customs Enforcement (ICE) Homeland Security Investigations (HIS) Tip Line 24h per day (USHS, 2018).

15.6 NEGATIVE CONSEQUENCES OF CHILD LABOR

15.6.1 Harmful Effects to the Child

Unintentional injuries are the leading cause of death of children between the ages of 10 and 19. In 2004, nearly a million children globally died due to injuries (WHO, 2008a). More than 95% of injury deaths to children occur in low- and middle-income countries. Child survival has been described as one of the most significant moral dilemmas of the 21st century (WHO, 2008b; Lee, 2003).

Injuries in young workers aged 15–24 is a public policy priority in many countries because this group consistently shows the highest rate of workplace injuries (Salminen, 2004; Santana, 2012). In one study, the fatality rates to young agricultural workers in the United States were shown to be three times higher than the death rates to adult private sector workers nationally (Hard, 1999). In addition, there is good reason to believe that the injury and illnesses reported by young people are actually even higher. Research has shown that up to 81% of workplace injuries go unreported (Pransky, 1999; Probst, 2008; Rosenman, 2006). Underreporting of workplace injuries is particularly prevalent in young workers (Tucker, 2014).

Due to anatomic and physiological differences in children, they can be much more susceptible to hazardous working conditions or exposures than normal healthy adults. Rapidly growing cells, tissues, and organs are more susceptible to harmful effects of toxic chemicals, and greater absorption rates in children, and are more likely to lead toxic health effects or cancer (Bruckner, 1999). Children have higher breathing rates, and aspirational absorption efficiency, than adults so children can be expected to receive a higher dose through the inhalation route than adults with the same exposure (Chung, 1993). Children have immature immune systems that put them at greater risk to some agents. They also absorb and retain heavy metals in the brain more easily than adults.

Children are often assigned some of the simplest and most routine tasks on a given jobsite. They may work long hours in awkward postures bent at their waist or knees. As their bodies are still developing, they are particularly vulnerable to the harmful effects of overexertion and repetitive motions. Due to the combination of the abovementioned factors, children are at an increased risk of musculoskeletal injuries.

Exposures to the same levels of noise can lead to greater risk of hearing loss to children due to the sensitivity of their hearing. The smaller size of children makes them more susceptible to harmful musculoskeletal injuries, and they require more food and rest than adults. Children do not have the same ability to assess risk as adults, and they tend to accept work in more hazardous environments. Young workers lack experiences and may be less familiar with safe operating procedures for sophisticated machines and equipment. Sixty-four percent of workplace fatalities of workers under 18 in the United States are from transportation incidents and contact with objects or equipment (NIOSH, 2003). Young workers may lack emotional and psychological maturity required for some tasks, and they may be unfamiliar with their basic rights as workers and be less willing to complain about adverse working conditions.

Due to these differences in children, they are much more likely to be injured in the workplace. In the United States, children between the ages of 15 and 17 were 30% (4.9/2.9) more likely to be injured at work than older workers (CDC, 2001). Child workers have also been shown to be less likely to report workplace hazards than adults (Breslin, 2007).

Growth deficiency is prevalent among children that work. Working children fall behind in height and weight compared to other children. Often exhausted and malnourished, working children often do not earn enough to feed themselves adequately (ILO/IPEC, 2017a).

In a recent study by Sturrock (2016), child labor was closely associated with poor mental health. Risk factors included work in domestic labor, working at a younger age, and increased intensity of work. Mental health problems most evident in child workers included low self-esteem and isolation. Working children were also demonstrated to have mood and anxiety disorders (Fekadu, 2006). Child workers have also been shown to have problems in sleeping (Kiran, 2007).

In some cases, the long hours of work performed by children directly impact their education, either by time away from schooling or by negatively influencing their ability to achieve their full potential. Adolescents working more than 20 h per week are more likely to drop out of school or complete fewer months of education. The worst forms of child labor including slavery, prostitution, soldering, and drug trafficking can have traumatic effects on the health and development of children (UI, 2018).

Child labor leads to long-term social and personal consequences related to the lack of education and social stigmatization. A child that does not receive primary or secondary education is at an extreme disadvantage socially. They may not be able to read or perform basic math skills necessary to function effectively in modern society. This leads to a lifetime of work in the most menial jobs and industries. Adults who worked as child laborers are also likely to have work-related negative health outcomes related to hazardous exposures as a child. These may include musculoskeletal disorders or outcomes of exposure to hazardous chemicals, such as pesticides. Exposure limits created to represent safe levels of exposure to adults for a period of 40 h a week for 40 years are often significantly exceeded in the child labor workforce.

15.6.2 Negative Impacts on Society

In free trade societies, the use of child labor provides unfair advantages. Businesses and governments that use child labor reap benefits in that labor costs are greatly reduced below typical market wages. Since workers are illegal, taxes are typically not paid for child laborers. And employers are not required to pay into work compensation, pension, vacation, and medical insurance benefits. Child laborers often do not fall within state run welfare of health-care systems, so nations allowing child labor may have unfair advantages in reduced operating costs.

The monetary benefits of child labor are never shared equally in society. An individual business may be more financially successful, but other members of society will end up paying more. In the long run, nations will end up with a large segment of the population that are disabled or incapable of effective and efficient work. Social costs of caring for large populations of injured or ill child workers will cost the state more in the long run. If these injured and ill children are not supported by the state, which is often the case, they are often cared for by other family members, or other philanthropic charitable or religious organizations.

15.7 CHILD LABOR IN AGRICULTURE

Agriculture is the sector where most child labor is found. Many children begin working as early as 5 years old (ILO, 2006). And it has been shown to be the most hazardous work for children. Agriculture, forestry, and fishing make up approximately 43% of all workplace fatalities for young workers in the United States (NIOSH, 2003). For young agricultural workers, the risk of fatality is four times that of young workers in other industries (BLS, 2000). In the Philippines, children working in agriculture were at five times the risk of injury than children working in other sectors (Castro, 2013).

Child laborers in agriculture are exposed to hazardous machinery, confined spaces, work at elevations, excessive work durations, exposure to dangerous tools and equipment, ultraviolet radiation, livestock, excessive levels of noise, and hazardous chemicals/pesticides.

In 2014, it was estimated that approximately 720,000 youth under the age of 20 worked on farms in the United States. It is common for children working in agriculture to drive tractors or use heavy machinery. Tractor overturns are extremely common and were the leading work-related cause of death to farmers in 2015, with 401 fatalities (NIOSH, 2017). Between the years of 1995 and 2002, an average of 113 youth die each year in the United States working in agriculture. The leading sources of youth deaths were attributed to heavy machinery (23%), motor vehicles (19%), and drowning (16%).

Many children working in agriculture work with and are exposed to increased levels of pesticides (Calvert, 2003). Regulations that limit access to pesticides vary by country. In the United States, children under the age of 16 cannot handle category I or II pesticides. But when in the proximity of spraying and other applications, it is easy for these same children to become exposed. These exposures can lead to both immediate and long-term effects that include rash, nausea, vomiting, headaches, and dizziness. Large doses can cause loss of consciousness, coma, and death (Alarcon, 2005). Long-term exposures to pesticides at even low levels have been associated with chronic health problems including cancer, neurologic disorders, and infertility (Alavanja, 2004). Children seldom receive adequate training on the use or hazards associated with pesticides (HRW, 2010).

15.8 CHILD LABOR IN RETAIL TRADE

The retail trade is associated with the second highest level of workplace fatalities in young workers in the United States, 63% of which were caused by homicide or violent acts (Windau, 1999). Particular hazards in retail fatalities include handling large sums of cash, working alone, and working late evening or early morning.

More than 60% of nonfatal injuries to young workers in retail trades occurred in eating and drinking establishments. The most common injuries were cuts, burns from hot surfaces of grease from fryers, and scrapes (Mardis, 2003).

15.9 CHILD LABOR IN CONSTRUCTION

Children working in construction are at serious risk from a wide range of conditions including falls, falling objects, electrocution, collapsing structures, and heavy machinery crashes. Other hazards include exposure to dangerous power tools, excessive noise, extreme thermal conditions, and toxic substances such as asbestos and silica, or solvents such as toluene or epoxy resins.

15.10 CHILDREN WORKING AT HEIGHTS

Several common child labor practices put children at significant risk of falls. Agriculture is the most common setting for nonfatal serious falls resulting in head and limb injuries. Particular dangers in this sector include climbing tall trees to pick crops, working on unprotected platforms, silos, unlit shafts and wells, and deep drainage ditches (ILO, 2006).

Many children work at heights on ladders in agriculture or on scaffolding in construction industries. Children often work as pickers in fruit harvesting. They may carry heavy bags and work in wet weather conditions.

15.11 LACK OF AND FAILURE TO USE PROTECTIVE EQUIPMENT

As stated above, 73 million children worldwide work in hazardous working conditions and with hazardous agents. Unlike their adult counterparts, children do not typically have access to properly fitting protective equipment. Protective gloves are not sized for the small hands of children. Protective helmets are seldom sized properly for children. Protective suits and respirators for use with hazardous chemicals or pesticides are never available in children's sizes. As a result, children often go without even this last line of defense from workplace hazards. Children are also less likely to fully understand the usefulness or rational for using PPE.

15.12 CAUSES OF AND CONDITIONS FOR CHILD LABOR

Child labor does not only occur in economically developing countries, but also occurs in developed countries that have their own problems with child labor. In the United States, more than 30% of all fatal injuries to young workers occurred in family businesses (NIOSH, 2003). Many of these occur in agriculture on family farms. In developing countries where parents and families depend on children to help support and provide for the existence, child labor can be easily understood. In developed countries, the use of children to perform hazardous and dangerous work is much more difficult to understand. It is presumed that parents clearly do not understand the danger they put their children in, and more needs to be done to educate them to the hazards and ramifications.

Causes of child labor have historically included poverty, lack of education, and repressions of worker rights in general. Limited regulatory prohibitions that condone or promote child labor also exacerbate the natural societal forces that cause children to work. More than one quarter of the world's population live in extreme poverty setting the stage for child laborers to assist in the support of the family for basic survival. Even in societies where child labor is banned, the laws may not be enforced or the use of child labor is well hidden. Governments and courts may have few resources to enforce regulations, even in countries where strong child protections exist on paper. In certain countries, broad exceptions have been made to allow child labor in certain industries such as agriculture.

CASE

A Safety Manager of a large dairy in the Midwest U.S. went to a nearby Safety program to give a class presentation. He spoke all about the activities and hazards associated with dairy operations, including working on a dairy farm. When a student asked if they had any policies on the use of children to work on the farms the Safety Manager said "No, the safety practices taken are up to each of the farmers in the regional dairy cooperative. If they dairy company asked farmers not to use their children in the operations of the farm, the farmers might decide to sell their milk to someone else." What would be some alternative solutions available to this Safety Manager of the dairy company buying milk from the cooperative of farms?

15.12.1 FAMINE AND DISEASE

Children often become orphans as the result of famine and disease. The HIV crisis in Africa has led to thousands of orphans who need to fend for themselves and their siblings.

15.12.2 POVERTY

Global spending and social support have a huge influence over the poverty levels of children throughout the world, and therefore their likeliness to work. Cuts in social spending force more children to work at earlier ages. Poverty-stricken families with several children are often led to "sell" or "rent" their children out as indentured servants. Children are often sent to do more hazardous jobs in place of their parents to reduce the likelihood of injury to the parent and possibility of the parent being out of work (ECLT, 2017).

Poverty and lack of decent work in general society tend to lower the working ages for children as they become more important to the financial survival of the family. Or in the worst cases, the family collapses and the children need to fend for themselves on the street. The lack of social protection systems for the poor, and particularly for children, forces children into labor just to provide for themselves. More than knowledgeable adults, children are not aware of social services that might be available to them.

Poor nations faced with large interest payments due on development loans to the World Bank and the International Monetary Fund are typically required to follow monetary and social policies according to the loan agreements. These policies require the developing nations to allow open trade and competition that drive prices of goods down, but other internal operating costs up. As a result, government spending on regulation and enforcement on occupational and public health topics, and education, are reduced. In regions with huge governmental debts, the percentage of child labor increases, and the number of children receiving and education goes down (UI, 2018).

15.12.3 Lack of Social Structure or Government

Weak governmental systems and political instability often lead to economic downturns and vulnerabilities. The poor and children are often most affected and are least protected from harsh economic and political conditions. Lack of access to legal protections is common for the poor, and without adult intervention, nearly impossible for children. Children do not typically pay taxes and cannot typically receive related social service benefits, or even basic medical care.

In the United States, child labor is permitted in agriculture when working on family-owned farms. Despite the documented increased risks to children who work with hazardous farm equipment, toxic chemicals used for pesticide and rodenticides, and hazardous levels of noise that exceed otherwise permissible limits, children in the United States continue to be injured and killed in agriculture and ranching.

15.12.4 War

War is another factor that contributes to child labor. Many children become orphans as the result of war, and these children need to work to survive to support themselves and sometimes their siblings. In a study from Bangladesh, injury was the primary cause of losing a parent (Rahman, 2005). These orphaned workers often work in street industries such as the collection of bottles for refunds or recycling, polishing shoes, washing cars, cleaning, dish washing, and collecting and selling wood. In a Cambodian study of child labor in the sugarcane industry, the consent to work in the formal sector on a plantation was given by the children themselves (ILO, 2015; Cambodia Ministry of Planning, 2013).

As devastating as war is in so many other social ways, it often leads to increases in the levels of child labor. Often it is because one or both of the child's parents are deceased as a result of the conflict. They need to provide for themselves and sometimes other younger siblings. In countries under civil war or social collapse, orphanages as we might think of do not exist. Children are left to live on the street and make a living, however, they can. Often at a very early age, these orphaned children are recruited to serve in the armies that defeated them and killed their families. In countries with rampant armed conflict, child labor is performed by 17% of all children (ILO, 2017b).

15.13 WAYS TO COMBAT CHILD LABOR

By looking at the long list of causes of child labor above, it is understood that regulations and laws alone cannot eliminate the practice. Strong international standards and conventions are a beginning and a way to educate policy makers and politicians about the problems that child labor brings to children and society as a whole.

Building upon the guidance of international guidelines and standards, independent nations can begin to draft regulations and laws that begin to restrict and limit child labor within and across their borders. Laws can govern minimum working ages and compulsory school attendance. But in order to be truly effective, the laws need to be followed up with programs for inspection and enforcement, which is typically lacking, particularly in developing countries where the fundamental reasons for child labor have not yet been addressed.

A variety of means have been identified to combat and eliminate child labor. The first step is to become informed about the national laws regarding child labor, and the current status and extent of child labor in your society. These national codes can be a starting point for any organization. In addition, management and employees should be educated on child labor definitions and issues so there is a better understanding of the issues.

Companies and organizations should next review requirements of international buyers and brands regarding child labor standards and expectations. Standardized international industry codes on the use of child labor have been developed. Companies and organizations should be sure they are aware of all expectations of buyers, local codes, and industry standards for the use of child labor to ensure they remain at the highest possible level of compliance. Corporate social responsibility programs that consider environmental and sustainability in enterprise supply chains, including internationally, must expand these programs to include OSH of their global workforce. Supply chain review must include the identification and elimination of inappropriate or illegal child workers, and child labor.

Employers should take positive actions to ensure the ages of their workers. ILO standards in the Minimum Age Convention (No. 138) set the minimum age for workers at the minimum school-leaving age of the country. It must not be less than 15 for any developed country, and the limit is 14 for developing countries. Companies should take the most stringent actions possible to ensure the ages of young workers, and document all findings.

Organizations and nations that review their supply chains to ensure worker health and safety should become familiar with the signs of child labor and be sure to eradicate it when identified. A comprehensive analysis of safe and health working conditions in supply chains should be able to note when children are working within that system. Policies, programs, and procedures should be clear that child labor is not permitted, and they should include reporting schemes and response protocols when it is identified. Companies should conduct onsite meetings and audits to engage with companies to discuss approaches to eliminate child labor within the supply chain (ILO, 2017b).

Another aspect of child labor directly related to supply chains is the expanding growth of informal labor that supply chains have become dependent upon. Working conditions in informal labor tend to be more hazardous than those in formal manufacturing industries. And informal industries that tend to be unregulated also tend to involve more child workers. By using formal manufacturers in official supply chains, it will reduce the use of informal sector workers and the use of unmonitored child labor (ILO, 2017b).

One approach to reducing the impacts to child workers is to reduce the hazard level of the working conditions. If workplace risks to children are adequately identified and addressed, then controls could be implemented directly to eliminate the hazardous characteristics of the work (ILO, 2014c).

Besides creating internal policies and programs to prevent child labor, companies and organizations can also create or join voluntary industry-wide collaborative initiatives against child labor. Eliminating child labor from the market eliminates unfair competitive advantages. By creating consensus and adhering to agreed upon standards, companies level the business playing field and create a holistic natural approach to eliminating the worst forms of child labor.

Cross-industry collaborations can go even further in reducing the use of child labor within and entire industry, or region. The ILO Global Compact Child Labor Platform is one example where industries can get involved and accelerate progress in reducing child labor. The labor platform provides companies and organizations with models for program and policy development and implementation (ILO, 2014a).

In a study regarding child labor in brick kilns in Nepal, it was determined that the most effective intervention strategies would include the following:

- Financial support to families to provide support for education and prevent financial bondage
- Onsite educational programs
- Additional research to determine which harm reduction strategies are most effective (Lamar, 2017)

Direct interventions in child labor have been separated into two categories: protective and preventative (Paruzzolo, 2009). Protective interventions tend to involve the direct removal of children already working in hazardous locations in the worst forms of child labor. But these types of preventions tend to not be lasting, with children often reengaging in the same labor after time (deGroot, 2007), and often leave the children in the same conditions that caused the need to work in the first place, poverty, and disadvantage (Bhukuth, 2006).

Protective approaches to child labor intervention tend to take a more holistic approach maintaining balance with the social and familial systems. Examples can be offsetting poverty by providing financial support structures, integrating educational services to match specific needs more closely, and targeting the initial causes of child labor (Lieten, 2010; Paruzzolo, 2009).

15.14 ORGANIZATIONS AND GUIDELINES

Various international conventions and guidelines have already been in existence for many years. The ILO is one of the more active organizations working to identify and curtail child labor globally. Some of the major ILO conventions and recommendations include the following documents:

- Convention concerning the Prohibition and Immediate Action for the Elimination of the Worst Forms of Child Labour, 1999 (No. 182), and Recommendation (No. 190)
- Convention concerning Minimum Age for Admission to Employment, 1973 (No. 138), and Recommendation (No. 146)
- Convention concerning Safety and Health in Agriculture, 2001 (No.184), and Recommendation (No.192)

ILO Convention 138 was created in 1973 and sets clear guidelines and limits on the use of child labor in agreement nations. ILO 138 child labor limits are shown in Table 15.2.

The ILO International Programme on the Elimination of Child Labour is source of information on child labor with many ideas and suggestions for the identification and eradication of child labor. The website includes links to guideline documents and online training programs (ILO, 2018).

The Global March Against Child Labor (GMACL) is a worldwide network of trade unions, teachers' and social organizations that work together towards the elimination and prevention of all forms of child labor. The organization supports local, regional, and national efforts through a variety of international instruments related to the protection of children's rights. They bring together a range of stakeholders to build and strengthen collaborative activities and programs (GMACL, 2018).

The Save the Children's Resource Center is an online library of information on topics of child protection, rights governance, health, nutrition, education, and poverty (Save the Children's Resource Center, 2018). The purpose for the center is to help build global capacity for the development and protection of children. The library is open to the public, and articles are free.

TABLE 15.2
ILO Convention 138 Child Labor Limits

ILO Convention 138	Minimum Age at Which Children Can Start Work	Possible Exceptions for Developing Countries
Hazardous work	18 (16 under strict conditions)	18 (16 under strict conditions)
Basic minimum age (should not be below the age for finishing compulsory schooling)	15	14
Light work that does not threaten health and safety or hinder education	13–15	12–14

15.15 RECOMMENDATIONS

As in other situations, workplace injuries to children are preventable. Childhood labor injuries can be combated on a variety of fronts. A public health approach to injury prevention involves data collection, risk factor analysis, development of interventions, evaluation, and widespread implementation of proven prevention methods.

15.15.1 Policy Development

The development of international conventions and recommendation of practice are a beginning point for nations and global corporations to begin to identify and address child labor, either within their borders or within their supply chains. National regulations and programs must also be created in order to further evaluate and enforce noncompliant formal companies, and to also begin to educate and inform those working in informal sectors. Government policies and regulations should promote basic rights of children, including access to education and other social support services.

Global organizations need to incorporate the evaluation of supply chains to ensure child labor is not part of the system at all manufacture, production, and distribution levels. Workers in these organizations, particularly working in the area of corporate social responsibility, need to be aware of international, national, and home country regulations and laws on the use of child labor. Programs and policies need to be developed and implemented on every level. These programs must include the evaluation of working conditions that may move a child from an "acceptable work" category in terms of working hours and conditions to a "child labor" category based on identified hazardous working conditions.

In order for developing countries to build capacities, funding must be provided for each of the steps. Occupational health and safety and policy professionals in developed countries need to be made aware of the extent and significance of child labor conditions and problems. The significance and impact of child labor is not always widely understood, and the rational for prevention and elimination is pervasively underappreciated by professionals and politicians in economically advanced countries. The lack of awareness and understanding undermines the organizational drive to extend energy and resources to research and solve the problem of child labor in developing countries.

15.15.2 Research

As most of the more significant and harmful forms of child labor are happening in developing countries, in order to begin public health prevention approaches, national capacities for data collection and record analysis must be improved. Training on research collection methodologies must be expanded to provide on-the-ground professionals to collect and analyze injury, fatality, and illness statistics. Effective intervention strategies must be created and implemented over time.

In one study of child labor in sugarcane in Cambodia, the need for additional research was identified in some general areas (ILO, 2015). It could be assumed that some or all of these basic research needs could also apply to other child labor industries and regions. The areas included the following:

- Macro-level supply chain analysis to provide information on social conditions that are the cause for child labor, existing industry production and distribution pathways. Supply chain analysis would also evaluate the roles of key actors including farmers, the children, owners and manufacturers, industry representatives, government officials, and end-user corporations.
- Deeper evaluation of social conditions and relations between children, parents, and the immediate environment.
- Comprehensive analysis and reporting of OSH risks to children in specific jobs and industries. This research should provide concise recommendations for risk reduction and provide easy to understand reasons why the changes are necessary for the protection of the child.
- The long-term impacts on child health and well-being in relation to physical, nutritional, educational, and social outcomes.

Research goals and needs have also been outlined in ILO documents in detail. Research needs in hazard assessment is one of the topic areas. This study also provides direction on how to formulate and conduct research regarding child work and labor (ILO, 2014c).

15.15.3 Dissemination of Information and Public Outreach

In order for the negative global consequences of child labor to be eliminated, what is known about the harmful personal and social impacts must be more broadly investigated and disseminated. The moral and ethical dilemmas for occupational health and safety professionals need to be discussed and debated at all levels in the government, professional organizations, and corporations. Once professionals understand the issues and need to act, effective and inexpensive interventions need to be identified and implemented.

Families, parents, and even children themselves need accessible and understandable information about the physical, mental, and social outcomes that result from the worst forms of child labor. Awareness about health and safety risks can be a gateway to further acceptance of more complex social and cultural changes necessary for sustainable change, including creation and implementation of new social systems and associated regulations (ILO, 2015).

A broader community of professionals including occupational hygienists, healthcare practitioners, and social workers need to be better informed about child labor issues, why it is an important global social issue, and why it should be important to them as professionals and individuals. Global corporations should be knowledgeable about this issue, and know their supply chains, for purely business reasons, but also for public relations purposes (ILO, 2014b). Lastly, professional organizations and nongovernmental organizations and their members should be aware of child labor conditions and issues, and do their part to eradicate the worst forms wherever possible through research, white papers and policy statements, and outreach and training programs.

Labor and professional organizations need to do more to inform their members about the social and personal ramifications of the expanding use of child labor. Child labor is a violation of basic human rights and most international treaties and standards. Professional organizations need to work with other organizations, businesses, governments, communities, families, and members to increase awareness the hazards and extent of child labor.

The U.S. Department of Homeland Security Blue Campaign provides a foundation for identifying and reporting human trafficking.

REFERENCES

Alarcon, W., Calvert, G., Blondell, J., Mehler, L., Sievert, J., Propeck, M., Tibbette, D., Becker, A., Lackovie, M., Soileau, S., Das, R., Beckman, J., Male, D., Thomas, C., Stanbury, M., Acute illness associated with pesticide exposures at schools, *Journal of the American Medical Association* (July 27, 2005) Vol. 294, No. 4, pp. 455–465.

Alavanja, M., Hoppin, J., Kamel, F., Health effects of chronic pesticide exposure: Cancer and Neurotoxicity, *Annual Review of Public Health* (2004) Vol. 25, pp. 155–197.

Bhukuth, A., Is child labor a substitute for adult labour?: A case study of brick kiln workers in Tamil Nadu, India, *International Journal of Social Economics* (2006) Vol. 33, No. 8, pp. 594–600.

Bureau of Labor Statistics (BLS). (2000). *Report on the Youth Labor Force.* Washington, DC: U.S. Department of Labor, Bureau of Labor Statistics, pp. 58–67.

Breslin, F., Polzer, J., MacEachen, E., Morrongiello, B., Shannon, H., Workplace injury or "part of my Job"?: Towards a gendered understanding of injuries and complaints among young workers, *Social Science & Medicine* (2007) Vol. 64, pp. 782–793.

Bruckner, J., Weil, W., Biological factors which may influence an older child's or adolescent's responses to toxic chemicals, *Regul Toxicol Pharmacol* (1999) Vol. 29, pp. 158–164.

Calvert, G., Acute pesticide-related illnesses among working youths, 1988–1999, *American Journal of Public Health* (April 2003) Vol. 93, No. 4, p. 609.

Cambodia Ministry of Planning, Household Survey by CIDS. (June 2013). *Cambodia Labour Force and Child Labour Survey, 2012.* Phnom Penh: International Labour Organization. Geneva, and National Institute of Statistics (NIS) of Cambodia www.ilo.org/wcmsp5/groups/public/---asia/---ro-bangkok/---sro-bangkok/documents/publication/wcms_230723.pdf accessed December 12, 2018.

Castillo, D.N.; Adekoya, N.; Myers, J.R., Fatal work-related injuries in the agricultural production and services sectors among youth in the United States, 1992–1996, *Journal of Agromedicine* (1999) Vol. 6, No. 3, pp. 27–41.

Castro, C., Hunting, K., Measuring hazardous work and identifying risk factors for non-fatal injuries among children working in Philippine agriculture, *American Journal of Industrial Medicine* (June 2013) Vol. 56, No. 6, pp 709–719.

Centers for Disease Control and Prevention (CDC), Nonfatal occupational injuries and illnesses treated in hospital emergency departments—United States, 1998, *MMWR* (2001) Vol. 50, No. 16, pp. 313–317.

Chung, I., Phalen, R., Dunn-Rankin, D., Predicted aerosol aspiration efficiency for infants, children, and adults, *Applied Occupational and Environmental Hygiene* (1993) Vol. 8, No. 7, pp. 639–44.

de Castro, A. (2012). Hazards at work: Child labor in agriculture. Institute for Labor Studies, Working Paper A.

deGroot, A. (2007). *Deprived Children and Education in Nepal.* Amsterdam, the Netherlands: International Research on Working Children (IREWOC). Web-ISBN: 978-90-79078-03-5, www.crin.org/en/docs/Nepal_Education.pdf accessed May 12, 2018.

Eliminating Child Labor in Tobacco Foundation (ECLT), Causes of Child Labor (2017) www.eclt.org/about/overview.html accessed December 29, 2017.

FAO, Food and Agriculture Organization of the United Nations, Child Labor in Agriculture (2018) www.fao.org/childlabouragriculture/en/ accessed March 10, 2018.

Fekadu, D., Alem, A., Hägglöf, B., The prevalence of mental health problems in Ethiopian child laborers, *Journal of Child Psychology and Psychiatry, and Allied Disciplines* (2006) Vol. 47, pp. 954–959. doi:10.1111/j.1469–7610.2006.01617.x.

Gapminder, The World Has Reached Peak Numbers of Children (2011) www.gapminder.org/news/world-peak-number-of-children-is-now/ accessed December 12, 2018.

GMACL, Global March Homepage (2018) www.globalmarch.org/ accessed May 9, 2018.

Hard, D., Myers, J., Snyder, K., Casini, V., Morton, L., Cianfrocco, R., Fields, J., Young workers at risk when working in agricultural production, *American Journal of Industrial Medicine Supplement* (1999) Vol. 1, pp. 31–33.

Human Rights Watch (HRW). (2010). *United States Fields of Peril Child Labor in US Agriculture.* New York: Human Rights Watch. ISBN 1-56432-628-4 www.hrw.org/sites/default/files/reports/crd0510webwcover_1.pdf accessed December 12, 2018..

ILO, Convention 182 Worst Forms of Child Labor Convention (1999) www.ilo.org/dyn/normlex/en/f?p=NORMLEXPUB:12100:0::NO::P12100_ILO_CODE:C182 accessed May 11, 2018.

ILO, (2006). *Tackling Hazardous Child Labour in Agriculture: Guidance on Policy and Practice User Guide.* Geneva: International Labor Organization. ISBN 92-2-118933-3.

ILO, (2013). *Making Progress Against Child Labor—Global Estimates and Trends 2000–2012.* Geneva. International Labor Organization. ISBN: 978-92-2-127182-6 (Web PDF).

ILO, (2014a). *Looking for Answers: Research on Hazardous Work of Children.* Geneva: International Labor Organization. ISBN: 978-92-2-129377-4 (Print); 978-92-2-129378-1 (Web PDF) accessed May 12, 2018.

ILO, Good Practice Notes—With Recommendations for Companies. Child Labor Platform (2014b) www.unglobalcompact.org/docs/issues_doc/human_rights/CRBP/CLP-Good-Practice-Notes-Nov2014.pdf accessed May 9, 2018.

ILO, (2014c). *Looking for Answers: Researching Hazardous Work of Children.* Geneva: International Labor Organization. ISBN: 978-92-2-129377-4 file:///C:/Users/tfulle2/Downloads/Looking_for_Answers_researching_hazardous_work_of_children_web_20150917.pdf accessed December 12, 2018.

ILO, (2015). *Rapid Assessment on Child Labour in the Sugarcane Sector in Selected Areas in Cambodia: A Comparative Analysis.* Geneva: International Labor Organization. ISBN: 9789221298724 (web pdf), www.ilo.org/wcmsp5/groups/public/---asia/---ro-bangkok/---sro-bangkok/documents/publication/wcms_391960.pdf accessed May 11, 2018.

ILO, (2016). *Checkpoints for Companies—Eliminating and Preventing Child Labour.* Geneva: International Labor Organization.

ILO, (2017a). *Global Estimates of Child Labor—Results and Trends, 2012–2016.* Geneva: International Labor Organization. ISBN: 978-92-2-130153-0 (web pdf).

ILO, (2017b). *Ending Child Labour by 2025: A Review of Policies and Programmes.* Geneva: International Labor Organization. ISBN: 978-92-2-131400-4 (web pdf), https://reliefweb.int/sites/reliefweb.int/files/resources/2017_World_Report_on_Child_Labour_EN_Web.pdf accessed July 26, 2018.

ILO, (2018). *IPEC: Training Resource Pack on Elimination of Hazardous Child Labour in Agriculture.* Geneva: International Labor Organization, www.ilo.org/ipec/lang--en/index.htm accessed July 26, 2018.

Kiran, S., Unal, A., Ayoglu, F., et al., Effect of working hours on behavioral problems in adolescents: A Turkish sample. *Neurology Psychiatry and Brain Research* (2007) Vol. 14, pp. 103–110.

Lamar, S., O'Leary, P., Chui, C., Benfer, K., Zug, S., Jordan, L., Hazardous child labor in Nepal: The case of brick kilns. *Child Abuse & Neglect* (2017) Vol. 72, pp. 312–325.

Lee, J., Child survival: A global health challenge commentary. *The Lancet* (July 26, 2003) Vol 362. www.thelancet.com. Page 262 Jong-wook Lee WHO, CH 1211, Geneva 27, Switzerland (e-mail: Leej@who.int).

Lieten, G., deGroot, A., Ensing, A., Baum, N., de Graaf, M. (2010). *The Worst Forms of Child Labour in Asia—Main Findings from Bangladesh and Nepal*. Amsterdam, the Netherlands: International Research on Working Children (IREWOC). ISBN: 978-90-79078-24-0, https://resourcecentre.savethechildren.net/node/3990/pdf/3990.pdf accessed May 12, 2018.

Mardis, A., Pratt, S., Nonfatal injuries to young workers in the retail trades and service industries in 1998, *Journal of Occupational and Environmental Medicine* (2003) Vol. 45, No. 3, pp. 316–323.

McCurdy, S., Kwan, J., Ethnic and gender differences in farm tasks and safety practices among rural California farm youth, *Journal of Occupational and Environmental Hygiene* (2012) Vol. 9, No. 6, pp. 362–70.

NIOSH (July 2003). NIOSH ALERT preventing deaths, injuries, and illnesses of young workers. DHHS (NIOSH) Publication No. 2003-128.

NIOSH, Agricultural Safety (2017) www.cdc.gov/niosh/topics/aginjury/default.html accessed September 11, 2017.

Park, H., Reynolds, S., Kelly, K., Stromquist, A., Burmeister, L., Zwerling, C., Merchant, J., Characterization of agricultural tasks performed by youth in the Keokuk County Rural Health Study, *Applied Occupational and Environmental Hygiene* (2003) Vol. 18, No. 6, pp. 418–429.

Paruzzolo, S., The impact of programs relating to child labor prevention and children's protection: A review of impact evaluations up to 2007, *Working Paper, Understanding Children's Work*, ILO, UNICEF, (March, 2009) http://documents.worldbank.org/curated/en/872791468152387187/pdf/480220WP0UCW0p10Box338877B01PUBLIC1.pdf accessed May 12, 2018.

Pransky, G., Snyder, T., Dembe, A., Himmelstein, J. Under-reporting of work-related disorders in the workplace: A case study and review of the literature, *Ergonomics* (1999), Vol. 42, pp. 171–182.

Probst, T., Brubaker, T., Barsotti, A., Organizational injury rate underreporting: The moderating effect of organizational safety climate, *Journal of Applied Psychology* (2008), Vol. 93, pp. 1147–1154.

Rosenman, K., Kalush, A., Reilly, M., Gardiner, J., Reeves, M., Luo, Z., How much work-related injury and illness is missed by the current national surveillance system? *Journal of Occupational and Environmental Medicine* (2006) Vol. 48, pp. 357–365.

Rahman, A. et al. (2005). *Bangladesh Health and Injury Survey: Report on Children*, Dhaka: Ministry of Health and Family Welfare.

Salminen, S., Have young workers more injuries than older ones? An international literature review, *Journal of Safety Research* (2004) Vol. 35, pp. 513–521.

Santana, V.S., Villaveces, A., Bangdiwala, S.I., Runyan, C.W., Albuquerque-Oliveira, P.R., Workdays lost due to occupational injuries among young workers in Brazil, *American Journal of Industrial Medicine* (2012) Vol. 55, pp. 917–925.

Save the Children's Resource Center, About the Resource Center (2018) https://resourcecentre.savethechildren.net/about/resource-centre accessed May 12, 2018.

Sturrock, S., Hodes, M., Child labor in low- and middle-income countries and its consequences for mental health: A systematic literature review of epidemiologic studies, *European Child & Adolescent Psychiatry* (2016) Vol. 25, pp. 1273–1286.

Tucker, S., Diekrager, D., Turner, N., Kelloway, E., Work-related injury underreporting among young workers: Prevalence, gender differences, and explanations for underreporting, *Journal of Safety Research* (2014) Vol. 50, pp. 67–73.

UI, Child Labor Education Project. University of Iowa (2018) https://docuri.com/download/welcome-to-the-child-labor-public-education-project_59c1e50af581710b286b7d6d_pdf accessed June 25, 2018.

US Census Bureau U.S. and World Population Clock (2017) www.census.gov/popclock/ accessed August 24, 2017.

UN, General Assembly Forty Fourth Session, 44/25. Convention on the Rights of the Child (2018) www.un.org/ga/search/view_doc.asp?symbol=A/RES/44/25 accessed December 27, 2018.

USHS, Homeland Security, Human Trafficking (2018) https://search.usa.gov/search?utf8=%E2%9C%93&affiliate=dhs&dc=&query=human+trafficking accessed July 26, 2018.

WHO. (2008a). *World Report on Child Injury Prevention*. Geneva: World Health Organization. ISBN 978-92-4-156357-4.

Windau, J., Sygnatur, E., Toscano, G., Profile of work injuries incurred by young workers, *Monthly Labor Review* (June 1999) Vol. 122, pp. 3–10..

16 Modern Slavery and Occupational Health

Thomas P. Fuller
Illinois State University

CONTENTS

16.1 INTRODUCTION

Despite extensive evidence to the contrary, many people in the world today believe that slavery and forced labor are occurrences fixed in past civilizations and societies.

They seem to adhere to several myths about the extent of modern slavery, and the forms that it takes. One myth is that slavery no longer exists in the U.S. or modern European or Western civilizations. Another myth is that if it does exist in a developed nation, it is only in the sex industry. Yet another myth is that slavery only exists as one form and does not include the trafficking of people into slavery. Unfortunately, none of these common assumptions made by many people in our modern and civilized societies are true.

Modern slavery, in all its forms, represents a gross violation of human rights and international law, and it is considered a crime against humanity. It is also a major threat to global economic stability, democracy, political development, and world peace (Ngwe, 2012). The lack of rights and protections, normally afforded to workers, such as the ability to voice concerns or refuse dangerous jobs, makes slaves exceedingly more vulnerable to workplace risks.

16.2 DEFINITIONS OF MODERN SLAVERY

The most basic definition of slavery is found in Article 1(1) of the Slavery Convention of 1926 as "the status or condition of a person over whom any or all of the powers attaching to the right of ownership are exercised."

In recent years, the concept of "modern slavery" has arisen and has broadened the definition to include forced labor, forced marriage, forced sexual exploitation, human trafficking, and state-imposed labor. The International Labor Organization (ILO) in the Forced Labour Convention, 1930 (No. 29), Article 2.1, defines forced labor as "all work or service which is exacted from any person under the menace of any penalty and for which the said person has not offered himself voluntarily" (ILO, 1930).

Forced labor includes six basic conditions:

1. Threat or actual physical harm to the worker
2. Restriction of movement and confinement
3. Debt bondage
4. Withholding wages and wage reductions
5. Retention of passports and identity documents
6. Threat of denunciation of illegal status to authorities (Craig, 2007a; ILO, 2005).

Slavery was prohibited by Article 4 of the Universal Declaration of Human Rights (UN, 1948). On paper, slavery is illegal in every nation. Modern forced labor differs from historic slavery in a few basic ways. In contemporary slavery it costs less to acquire slaves. Some slaves can be purchased for as little as 10 U.S. dollars (Bales, 2008). Unlike in the past, slaves are not owned for life, but for limited durations of a few years, or even months. Slaves cost less and are easy to get rid of, and replace. This is a particular concern because there is less interest in keeping them healthy and safe. Lastly, slavery today is globalized, meaning that it is occurring in numerous countries and in broad swaths of geographic and economic regions on a regular and persistent basis. Some of the basic differences between old slavery and modern slavery are provided in Table 16.1.

TABLE 16.1
Basic Differences between Old Slavery and Modern Slavery

Old Slavery	Modern Slavery
Slaves were owned and traded as a commodity	Legal ownership is not always part of the process
High purchase costs	Low purchase costs
Low return on investment	High return on investment
Long-term ownership	Short-term usage and exploitation
Slaves were of limited commodity and supply	Unlimited supply of slaves, easy to transport
Slaves maintained	Slaves used and disposed of

Source: Adapted from Bales (2000).

In an article entitled "Badges of modern slavery," Amir Pas-Fuchs helps to identify and define modern-day slavery by a series of basic working and living conditions called "badges" (Pas-Fuchs, 2016). The first of these is the lack of dignity or state of humiliation in the social order. Slaves have no rights as citizens or members of society for a variety of logistical and legal reasons. But this lack of status can be used to identify and indicate when modern slavery is occurring or is likely to result in exploitative conditions. Another badge is ownership of a person. A person may willingly pay to be transported to a new employer in a foreign land, and in the end be coerced or tricked into a life of bondage or indebtedness. The term "ownership" is not always clearly applicable but needs to be applied more broadly to identify conditions where the level the employer maintains is tantamount to possession. Control, and thus possession, can be maintained through violence, sexual abuse, threats of legal action, excessive debt, withholding pay, psychological abuse, and physical restraint.

Another badge of modern slavery that is common to the practice is that it feeds on the otherness, race, or sex of the persons. Differences, weaknesses, and disabilities are exploited in order to seek some benefit. These slaves are often children or even family members who are unlikely to protest or even be aware initially of the degree to which they are being abused. In many parts of the world, race is still the key feature that determines who is enslaved by whom. Differences are exploited and accepted as part of the social landscape and order.

The last badge of slavery is the lack of consent or free choice. The choice between starvation and work rationally puts the employer in a position of exploitation of the worker. They can wield complete control and offer the most meager of wages. Working excessively long hours in hazardous conditions, in isolation, for meager wages could be considered exploitative and thus modern slavery. It is common for employers to hold workers' hostage on boats or isolated locations, withholding their passports, and restricting their departure. These activities should be considered modern slavery and called out as such.

The Palermo Protocol defines trafficking in persons as the recruitment, transportation, harboring, or receipt of persons, by means of coercion, abduction, deception, or abuse of power or of vulnerability, for the purpose of exploitation. It goes on to specify that exploitation shall, at a minimum, include sexual exploitation, forced labor, slavery, and slavery-like practices. The full definition of trafficking in persons in Article 3 of the protocol is as follows: "the recruitment, transportation, transfer, harbouring or receipt of persons, by means of the threat or use of force or other forms of coercion, of abduction, of fraud, of deception, of the abuse of power or of a position of vulnerability or of the giving or receiving of payments or benefits to achieve the consent of a person having control over another person, for the purpose of exploitation. Exploitation shall include, at a minimum, the exploitation of the prostitution of others or other forms of sexual exploitation, forced labour or services, slavery or practices similar to slavery, servitude or the removal of organs." (Art. 3 (a)). It also specifies that "The recruitment, transportation, transfer, harbouring or receipt of a child for the purpose of exploitation shall be considered 'trafficking in persons' even if this does not involve any of the means set forth in subparagraph (a) of this article" (Art. 3 (c)).

- *Smuggling* is defined as explicit consent to be taken illegally across national borders. The relationship between smuggler and migrant typically ends when the destination is reached.
- *Trafficking* is a form of ongoing exploitation: even if the person has at some stage consented, this is meaningless because of the deception and coercion involved. Trafficking occurs within as well as across national borders (Craig, 2007a)(2007b).

Smuggling is the act of moving people or objects secretly and illegally. Trafficking is an act of transporting people or migrants to employers. Trafficking takes a more insidious connotation in that the process includes recruitment, transportation, transfer, and harboring persons by means of the use of force or coercion, abduction, fraud, the use of power or position, giving or receiving benefits to achieve consent of the person for the purpose of exploitation (UN Protocol 3A, 2000a).

16.3 TYPES OF SLAVERY

16.3.1 Human Trafficking

Human trafficking includes the recruitment, transportation, transfer, harboring, or receipt of people by means of threat or use of force or other forms of coercion, abduction, fraud, deception, abuse of power or position, or receiving payment or benefit to achieve consent of a person with control over another person for the purpose of exploitation (UN, 2000b). Depending on the terms of the trafficking activities, international courts have made different interpretations of law for trafficking cases. In general, the trafficking process and traffickers are conducting slavery despite the consent of the victim, in some cases, to be exploited. Trafficking is basically the transport of people into slavery. In a review by Siller (2016) however, in the international judicial use, the terms "slavery" and "trafficking" have become mostly indistinguishable. In many cases, "the law of trafficking is encompassed under the umbrella of enslavement as a crime against humanity" (Siller, 2016).

Since persons are often brought into trafficking by means of deception, the traffickers are often initially trusted by the victim. They may be the same sex, race, religion, or come from the same country of origin (Simmons, 2013). These commonalities are used to form a bond with the victim. Many of these bonds remain in the victims' mind even as they are suffering the exploitation and bondage (Aronowitz, 2010).

Contrary to common thought, traffickers are not always affiliated with organized criminal groups, but often acted through close knit cultural or family groups. Many of these groups did not see their activities as criminal but as a means to make profit on the lives of their victims (Simmons, 2013). Trafficking has often rightly been associated with immigration fraud and the corruption of federal agents (David, 2010). Trafficking has been shown to be associated with the illegal transfer of money including money laundering (Simmons, 2013). In Vietnam, Thailand, and Cambodia, two-thirds of the trafficked labor were men working in fishing, agriculture, and factories (Pocock, 2016).

16.3.2 Debt-Based Slavery

Debt-based slavery is one that forms when the victim enters an arrangement to work to pay off prior debts. However, wages are not enough to pay the debt, related interest, and living expenses during the work, and as a result, the victim enters an increasing spiral of uncontrollable debt. In some cases, these workers willingly take employment and sign extended contracts to work that they would not normally sign due to their vulnerability and desperation as a form of **contractual bondage**. In colonial times, this form of debt bondage was also known as **indentured servitude**. In many cases, workers enter into these agreements even knowing, or sensing, the possible conditions they will be living or working under, yet feel they cannot refuse the opportunity. In other cases, workers given the opportunity to escape the extreme conditions (including violence and degradation) refuse to leave until the end of their next pay period (Phillips, 2012).

In Southeast Asia, debt bondage is practiced as a form of slavery where poor parents sell their children or themselves as collateral for loans. Between 15 and 20 million people have entered slavery in this way (iAbolish, 2018). Unfortunately, few are able to earn enough to ever repay their debts and never regain their freedom. Enslavement of children in this form is common in Haiti and other Latin American countries where children live with a family and in return for food and shelter they work excessively, are mentally and physically abused, and never receive an education (Issa, 2017).

16.3.3 Contract Slavery

A significant portion of people entering into slavery do so on their own volition in search of work due to economic desperation (ILO, 2009). Workers migrate to distant regions or nations in order to avoid poverty and hunger. They are enticed by offers of money and security from landowners and businesses. However, it often turns out that the costs for room and board for workers exceed what they earn, and they enter into an endless spiral of debt, without contact with their home countries, and without any way to pay to return to home.

16.3.4 State-Imposed Forced Labor

State-imposed forced labor is a special type of modern slavery whereby the government abuses its citizens in the form of abuse of conscription or compulsory labor under the guise of economic development. The abuse of internationally accepted standards for prison labor also falls under this category.

16.3.5 Forced Marriage

Women make up more than 70% of forced labor. This figure includes women and girls who comprise 84% of people into forced marriage globally. More than 90% of all forced marriage takes place in Asia and Africa. More than 36% of those forced into marriage were children at the time of the marriage (ILO, 2017).

16.3.6 Descent-Based Slavery

Descent-based slavery describes the conditions and cases where people are born into slavery because their parents were slaves (Quirk, 2009).

16.3.7 Forced Recruitment for Armed Services

In a number of regions, children as young as six are forcibly recruited to fight in armed conflict and combat. Children are often used as informers and trained to handle crude weapons and improvised devices. According to a recent report by the United Nations (2016), child recruitment is common in Afghanistan, India, Central African Republic, Democratic Republic of Congo, Iraq, Palestine, Lebanon, Libya, Bali, Myanmar, Somalia, South Sudan, Sudan, Yemen, Columbia, Nigeria, the Philippines, and Thailand (UN, 2016).

16.4 NUMBERS OF SLAVES

The ILO estimates that between 19.5 and 22.3 million people were victims of forced labor or human trafficking at any given time in the period between 2002 and 2011q (ILO, 2012). By 2016, there were 40.3 million people living in modern slavery, including forced marriage, of which 24.9 million were in forced labor. There are 5.4 slaves for every 1,000 people living in the world (ILO, 2017). A breakdown of global estimates of forced labor in 2011 by activity is shown in Figure 16.1.

By region, the ILO has identified the numbers of forced laborers in million as (ILO, 2012):

Asia and the Pacific—11.7
Africa—3.7

Latin America and the Caribbean—1.8
Central and southeastern Europe—1.6
Developed economies—1.5
The Middle East—0.6

Data show that most victims of human trafficking come from Asia, Central and Eastern Europe, and Africa. The most desireable destinations for traffickers are the more developed countries of Western Europe and North America. Fifty-eight percent of people living in slavery are in the five countries of India, China, Pakistan, Bangladesh, and Uzbekistan (Walk Free Foundation, 2016). Figure 16.2 shows a geographic distribution of slavery in the world.

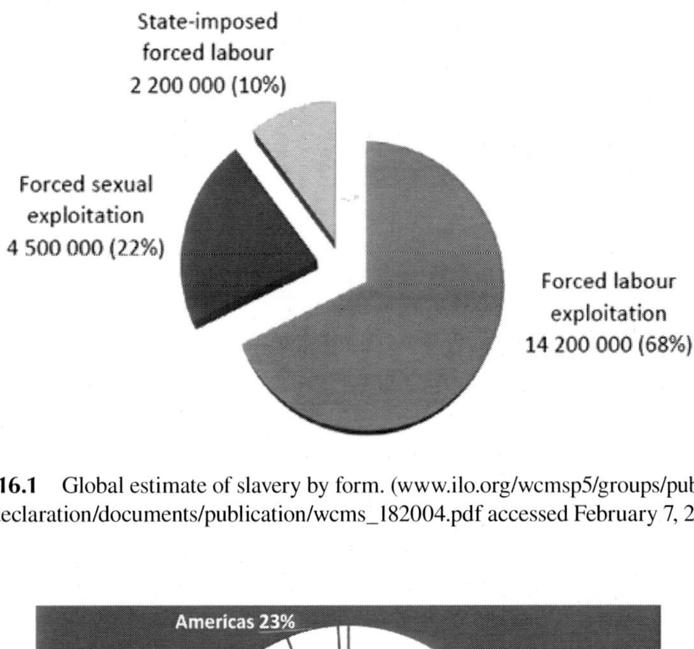

FIGURE 16.1 Global estimate of slavery by form. (www.ilo.org/wcmsp5/groups/public/---ed_norm/---declaration/documents/publication/wcms_182004.pdf accessed February 7, 2018.)

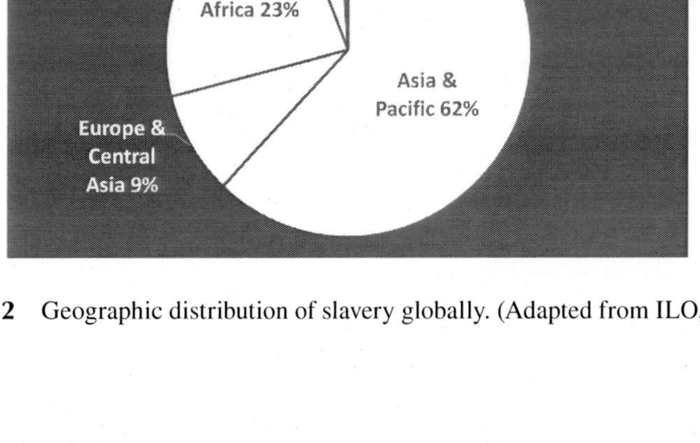

FIGURE 16.2 Geographic distribution of slavery globally. (Adapted from ILO, 2017.)

Although there are numerous resources for data on modern slavery, there is an immense amount that remains unknown. Human trafficking and slavery remain hidden crimes in an underground economy. Neither perpetrators nor victims of slavery are likely to report slavery, and thus, reliable data on the exact numbers of victims are impossible (Ngwe, 2012; Bales, 2015). About 42% of all those trafficked work in the sex industry (Craig, 2007a). In a study by Brown (2011), 70% of trafficking victims were women and 50% were children. Other common slave industries include brickmaking, mining, fish processing, gem production, fireworks, and carpet production (Craig, 2009). The remaining 58% of trafficked workers go to other occupational sectors in a broad range of categories. In 2016, the victims of exploitation were working in the following sectors (ILO, 2017):

Domestic work—24%
Construction—16%
Manufacturing—15%
Agriculture and fishing —11%
Wholesale trade—9%
Personal services—7%
Mining and quarrying—4%
Begging—1%.

Modern slaves tend to be placed into industries with large staffing shortages, where there is a high demand for cheap labor. These are often the most arduous or dangerous jobs in hazardous industries. These sectors are often conductive to trafficked workers because they have little government oversight and workers can be easily hidden from authorities, such as in homes, restaurants, hotels, farms, logging, and boats (Lindley, 2011).

16.5 CONSEQUENCES OF SLAVERY

16.5.1 Economic and Social Consequences of Slavery

Much of the negative impact of slavery is in the form of economics. Employers and even multinational corporations and governments could be considered to be encouraging and facilitating modern slavery when they force the cost of labor into unrealistically low levels. Human trafficking and modern slavery are valued at $32 and $44 billion per year. It is the third largest illicit trade globally, falling just behind arms and drug sales (Bales, 2007; Belser, 2005). Trafficking is expected to surpass illegitimate arms and drug sales within the next few years (Wheaton, 2010). Home countries suffer the loss of the youngest, healthiest, and most capable workers. In host countries, slaves often create economic and social burdens and lead to increased legal, medical, and social costs that were unforeseen, and for which the host country is ill prepared.

Slavery creates unfair competition between businesses and nations. A neighboring country that uses slaves for a significant portion of an industry gains an unfair economic business advantage against their rival nations. Similarly, a global business that includes slaves within portions of their global supply chain creates an unfair advantage over their competitors who need to pay more for labor of free workers. It is in the interest of businesses to know whether their competitors use slave labor, in addition to consumers wishing to follow basic legal, moral, and ethical principles.

The money going into the hands of slave owners is a significant loss of tax revenue for governments. Taxes lost from slaves performing labor, at minimal or no wages, rather than legitimate workers in legitimate business represent significant financial losses to governments. These also represent losses to other competing businesses who create products and services without the use of forced labor at much lower labor costs. There are great financial reasons for legitimate businesses to care that their competitors are not using slavery in significant portions of their supply chains.

The societal costs of supporting trafficked victims are also significant. Although victim support is lacking in many countries, whether rescued or not, these slaves at some point end up in hospitals, social services, mental health facilities, or prisons. All of these represent significant potential costs. A person who sells a kidney to send money home to Somalia where their family is starving is inevitably going to end up in the hospital receiving tens of thousands of dollars of medical services that will likely never be repaid. Currently trafficked victims are supported by charities and nongovernmental organization (NGO) efforts. But with greater recognition of the problems, governments will be called to play a greater role in victim support, even if only in the form of judicial prosecution and deportation of illegal immigrants (Craig, 2017a).

16.5.2 Social and Health Consequences to Slaves

Slavery victims suffer a variety of psychological and physical consequences. Besides direct physical and psychological violence, victims often work in unsafe environments, live in conditions of poor sanitation, receive poor nutrition, and receive little or no medical care when it is needed (Rezaeian, 2013). Slaves tend to have and receive very low levels of schooling. Despite continuous work for many years, slaves tend to remain in poverty, which continues to be the principal cause of their vulnerability (ILO, 2009). Slaves seldom achieve long-term economic or social security, they maintain low levels of literacy and education, they tend to be excluded from political activity or discourse, and they seldom interact or receive benefits from social infrastructure or services (Phillips, 2012).

In a study of 387 survivors of trafficking of the ages 10–17 in Thailand, Vietnam, Cambodia, Myanmar, and Laos, 56% showed symptoms of depression, 33% had symptoms of anxiety, and 26% had symptoms of post-traumatic stress disorder (PTSD) (Kiss, 2015). In another study of sex-trafficked workers in Nepal, 85.5% reported symptoms of depression, 87% had clinical symptoms of anxiety, and 27% reported symptoms of PTSD. In one study from Southeast Asia, 53.8% of trafficked males working in the fishing industry had been subject to severe violence (Pocock, 2016). This same study indicated extended working hours (mean 132.3 h per week) without breaks, and excessive lifting and bending and other exposure hazards.

Slaves often suffer occupational illnesses and injuries from hazardous machinery, exposure to pesticides, poor sanitation, temperature extremes, inadequate shelter, excessive work durations, and physical abuse. Injuries and illnesses due to the treacherous occupational exposures and living conditions lead to incapacities that reinforce the precarious position of the slave (Harriss-White, 2006; Kabat, 2017).

Perhaps one of the most impressive demonstrations of the lack of empathy for slaves is in the Morecambe Bay tragedy in 2004 when 21 Chinese illegal immigrant laborers were drowned by an incoming tide off the coast of England. Men and women between the ages of 18 and 45, who spoke little English and knew little about the area, were picking cockles in the sands off the coast when the tide came in suddenly. They were part of a larger total group of 36 workers. The Chinese gangmaster of the group was eventually sentenced to 12 years for manslaughter, but British natives purchasing the cockles from the illegal group were not prosecuted (BBC, 2004). A prime example of the racially biased nature of modern slavery where a jury of British natives found the British men purchasing the cockles were not guilty and 36 Chinese workers without proper cockling permits were working and living under the noses of immigration officials and local cockling licensing authorities.

If modern slaves are not kept in shackles by force, there are many other means of coercion to keep them in place. The leading means of coercion include withholding wages and physical violence. Other methods include threats against other family members, the need to repay debts, physical isolation and barriers, withholding passports, and threats of legal action or imprisonment (ILO, 2017). Debt bondage occurs when workers arrive at a job and find that they cannot make enough to pay for their lodging and food and fall into an endless chain of debts they can never repay.

Many victims that become part of the sex industry come from homes with extreme physical, sexual, and psychological abuse (NISMART, 2002). These children, often living on the street, are extremely vulnerable and common targets for pimps, sex offenders, and pornographers.

16.6 FACTORS THAT FOSTER MODERN SLAVERY

The various factors that lead and contribute to the growth of modern slavery are often overlapped and intertwined. Modern slavery exists due a large part, to apathy and ignorance of government officials, the general public, and corporations. Several of the major factors that affect the likelihood of slavery occurring in a particular nation are the basis of country ratings in the Global Slavery Index (WFF, 2016). Some key features of the likelihood of slavery are included in the following sections:

16.6.1 Overpopulation

The rapid increase in population after World War II, particularly in developing nations, led to further breakdown of already fragile economic and social conditions. When countries cannot economically secure the well-being of their people, it can lead to poverty and often higher infant mortality rates. As a means to compensate, populations tend to rise overall. Subsequent overpopulation and the lack of food or work in poorer countries rendered many of the most able-bodied vulnerable to trafficking as a means to escape their status (Bales, 2008; Population Matters, 2010). Overpopulation and lack of local opportunity is still a main reason today why people willingly give themselves up to migrate or be "taken" to a new and potentially better life.

16.6.2 Poverty

Poverty is often the reason why people seek to migrate or move to a different geographic region. The causes of poverty can be many. They include famine, war, overpopulation, economic collapse, government corruption, and injustice in legal and social systems. When faced with a choice of continuing on in poverty or migrating to a new region, even at great risk, many people will accept those risks. They may pay for transport directly to a trafficker and without any control or recourse, become the victim of the trafficker.

Migrants from Haiti, Mexico, and Central America seeking to escape poverty and violence come to America to work in agriculture and domestic service; many of these people fall into various forms of modern slavery and exploitation. Originally trafficked with the promise of "good" work and pay, they barely make enough for mere subsistence and continue to work and live in harsh and dangerous conditions without the freedom to get away (Bales, 2009).

16.6.3 Government Instability, Weakness, Corruption, and Lack of Interest

The growth of modern slavery must also be linked with the governments in the countries where it occurs. Government institutions and the rule of law are difficult to develop and maintain when a nation is in economic and social turmoil. Corruption undermines existing government policies and destroys the morale of the population (Ngwe, 2012). When democracy, justice, and the rule of law breakdown, it leads to general discontent and violence. Criminal traffickers use the cover of chaos, conflict, and disaster to inflict violence and trickery on desperate people to enslave them (Bales, 2008). The lack of civil, political, and social protections for all or a portion of a population increases the likelihood that various forms of modern slavery will occur.

Modern slavery occurs in developed countries when governments do not promulgate or enforce laws that prohibit it. International treaties and laws clearly set the standards for the prohibition of slavery, yet it goes on in many advanced countries throughout the world. This occurs for a variety of reasons from lack of funding and resources, lack of interest, or weak and corrupt government agencies. Slavery is allowed to occur where there is complicity between slave owners and lax enforcement by government officials (Craig, 2017a).

Since it is often an international problem with traffickers working across borders and continents, there is a need for multiple nations to work together and coordinate activities, yet there is relatively little or no formal cooperation between governments and the various international antihuman trafficking agencies (Laczko, 2005).

Slavery is also a result of global capitalistic economic and monetary policies that require maximum financial efficiency and profit for corporations. Slavery is clearly linked to global supply chains through the global economy and not exclusive to any one industry or geographic region (Issa, 2017). Ever-increasing global competition requires that labor be very cheap. The reduction in wages, and the unavailability of other choices, forces workers into a type of "wage slavery" doing appalling work in hazardous conditions just to survive. Economic and business models and policies that force nations and people to lower standards for environmental care, occupational safety and health (OSH), and the basic rights of workers to refuse hazardous work are a form of slavery. Businesses in developed countries that do not maintain supply chains free of slaves and other hazardous working conditions are supporting systems where nations and people accept work that would be illegal in developed countries.

In regions of Latin America, a high demand for labor in remote areas where there is little supply leads to trafficking and slavery where government oversight is difficult (Issa, 2017). Urban workers may agree to a contract and be taken to a remote area where they are cheated out of their wages and physically confined. This form of slavery that occurs in remote areas without government oversight in industries such as mining, agriculture, and logging often coincides with various other illegitimate operations, hazardous working conditions, and environmental destruction.

16.6.4 War and Social Disruption

Factors that disrupt social stability have been associated with victims entering into human trafficking and slavery. This has been notably demonstrated in the precarious positions of Syrian refuges who in order to escape civil war allow themselves to be trafficked out of the region, often ending in modern slavery-like conditions in Western Europe.

16.6.5 Cultural Values

Many cases of modern slavery come about due to cultural and religious beliefs and practices. In northern African countries such as Mauritania, the Republic of Sudan, and Mali, racially based **chattel slavery** is thriving. This is perhaps the worst form of slavery where people are captured and forced into bondage, and becomes the "property" of their captors. Children can be born into this slavery and become the property of the slave owners. People including children can be traded or bartered for property or payment of debts (Fight Slavery Now, 2018).

In some nations, the historical context of slavery itself is an impetus to continue the practice. In Brazil for example, wealthy landowner families may have owned slaves, in some form, for several generations. It is an accepted and expected practice and almost considered a "right." There are even cases as late as 1998 where slaves that have escaped from landowner estates are captured by local law enforcement and returned to the plantation by the police (ILO, 2009). Many of these wealthy landowners and business people exercise power and influence over state, federal, and municipal authorities and practice slavery with impunity.

16.6.6 Environmental Destruction

Wherever conditions lead to the loss of the natural environment, the impacts on local society can be profound in terms of food and water supply, availability of agricultural jobs, loss of fuel for cooking or heating, and other factors. Environmental destruction is often closely linked to the human population that it supports through overpopulation, overharvesting or fishing, and financial gain through poaching, illegal mining, and deforestation. This loss of the environment then leads to poverty and the loss of paying work. People move towards various forms of slavery and forced labor to survive.

Interestingly, many of the illegal activities destroying the environment that lead to increased levels of slavery due to financial hardship and desperation also use slaves to perform the work. In Brazil, where slavery has been entrenched for 300 years, slavery still exists in every region (Antero, 2013). And today, the deforestation of the tropical rain forests in Brazil is primarily conducted by slave labor in order to create more land for ranching and livestock production, production of charcoal, and other agricultural development (sugarcane, coffee, cotton) (ILO, 2009; Phillips, 2012).

16.7 HOW IS SLAVERY REPORTED OR IDENTIFIED?

Considering the magnitude of the modern slavery problem, there is still a lot unknown about how it occurs, and what the impacts are. The phenomenon is understudied on many fronts from the economic impacts on nations and corporations, the costs to societies in loss of people and provision of services, and the long-term impacts on victims. Health impacts on slaves that result from working conditions are mostly unknown and undocumented. This is an area where OSH academics could make a large impact and provide a benefit.

16.8 WHAT CAN ORGANIZATIONS DO TO COMBAT SLAVERY?

The discovery and public identification of modern slavery in a company's supply chain can have serious economic and public image consequences (Noble, 2014). Corporate social responsibility (CSR) tools and programs have been developed to help large companies identify a broad range of ethical and environmental conditions and challenges (New, 2004). Tools that can be used to identify and remove modern slavery in supply chains are less well developed to this point. Companies that allow slavery within their supply chains are a major reason nations and workers accept this type of work.

In recent years, retailers such as Wal-Mart and Carrefour established trade restrictions on meatpacking groups connected to producers in Brazil known to use slaves in their livestock production. Nike also placed restrictions on the use of leather coming from the Amazon region of Brazil due to the lack of reliable governance of illegal operations such as deforestation, but also including slavery (Phillips, 2012).

Supply chain management is often seen to fall under a larger umbrella of corporate sustainability initiatives (Gladwin, 1995). Supply chain ethics, environmental evaluation, and even occupational health and safety concerns are often covered by the same professionals (Rasche, 2013; Epstein, 2014; New, 2015; Bartley, 2007). Several firms have prohibited slavery and forced labor in their supply chains through the use of standard wording in their corporate codes of conduct and CSR statements. In general, these firms place the ethical responsibility to ensure that slavery is not part of their supply chain within environmental and worker health and safety management systems and consider the problem resolved. It turns out, however, that finding and eliminating modern slavery from supply chains is more difficult than thought, and often the corporate business models themselves are part of the reason slavery begins and flourishes. In an extensive review by Crane (2013), he shows how corporate actions that require suppliers to cut costs to minimum levels in order to survive, leave suppliers to cut costs in the only part of the chain they control, wages of workers. Workers who may have entered into a working arrangement suddenly have their wages cut and can no longer cover their living expenses supplied by the employer, and fall into a form of debt bondage. Geographically isolated and destitute, the workers must continue on in the position.

Often, it is difficult for a CSR program to see or evaluate all the parts of the supply chain. The electricity used to produce a product in China may use a form of state-imposed forced labor. It would first be difficult to identify this form of labor as an independent corporation, and secondly, it would be problematic to try to influence change in a state-run operation without other larger influences.

Other types of workers who are often not covered by CSR programs would be contracted, temporary, and agency-supplied workers that might be used within the supply chain. These workers may be indistinguishable from other factory workers, but they may be subject to some of several aspects of the definitions of modern slavery (Barrientos, 2008). Ultimately, these workers have lower visibility within CSR programs. Even victims of slave labor have reason to hide their plight from CSR evaluators since their discovery would probably mean the end of their position. These people are also often bound to their lives by complex social and psychological ties (Bales, 2002).

The illegality of slavery brings another challenging dimension to CSR programs. Due to the illegality, suppliers go to great lengths to hide slavery, or " slavery-like" practices. If they are caught, they not only may lose an important customer but also could be fined or even go to prison. So there is a known criminality to modern slavery which can put CSR auditors or inspectors in harm's way if they are doing their job effectively. Unlike environmental conditions that might or might not meet the goals of the CSR program, or that may be overlooked for a period where new systems or controls are put in place, the identification of slavery is often more clear and explicitly illegal. A company would be expected to cut ties with the supplier, the CSR evaluator would be expected to report the slavery to authorities, and the supplier would be expected to take immediate actions to rectify the situation, which may not be possible. Evaluators or auditors could be subject to bribery or grave physical harm.

Some of the best advances in CSR supply chain management in regard to identifying and policing slavery have come from the nongovernment sector. NGOs have increased the visibility of trafficking activities through websites and other media attention. One example is the Walk Free Foundation, a nonprofit that was formed in

2001. The group performs independent studies and published a Global Slavery Index and is working with numerous global nonprofit organizations, religious groups, and governments to communicate conditions and measures to be used to combat slavery through advocacy and policy change (Walk Free Foundation, 2018).

In one study, it was shown that organizations that include human rights due diligence in their CSR programs are four times more likely to uncover actual or potential human rights violations (McGregor, 2017). Companies and organizations that use due diligence processes such as those to evaluate workplace violence or harassment could effectively apply similar methods to identify and prevent modern slavery in their supply chains (Olsen, 2017).

NGOs can play a leading role in the development of antislavery policy in governments, in addition to corporations. It has been shown that NGOs played a significant role lobbying for and shaping the UK Modern Slavery Act (Craig, 2017a). Much of the work and output of NGOs in the United Kingdom was in the form of education and public awareness of the existence of slavery amidst their communities, despite ongoing inattention and lack of action by the government. One NGO called "unchosen" has created a series of short films on various aspects and conditions of modern slavery based on actual case studies (Unchosen, 2017; Craig, 2014). The films are valuable tools for building awareness in the community and can be shown as educational tools in schools to create empathy.

The Walk Free Foundation includes ratings of national government regulations to curb slavery in their annual Global Slavery Index. The listing reviews existing legislation in each country and rates them on a scale of 12 different categories (WFF, 2016). The index also rates nations on their level of response to and support of victims of slavery once they are identified and rescued. Measures in this index include such categories as survival support and criminal prosecution of perpetrators.

16.9 LEGISLATION TO IMPROVE GOVERNMENT OVERSIGHT AND CONTROL

As modern slavery is primarily a criminal activity, it would follow that creation and strict enforcement of laws and regulations would be a good starting point for combatting it. But despite the fact that international laws outlawing slavery have been in place for nearly 100 years, it still abounds. In 2015, the United Kingdom passed a Modern Slavery Act. This is a comprehensive series of laws that clarifies the various actions that are considered slavery, gives the government the means and funding to identify wrongdoing and enforce the laws, and structures judicial prosecution and sentencing for perpetrators (UK Government, 2015). Unfortunately, this Act does not go far enough to ensure that slavery does not occur, or to enforce and apply penalties if it does.

Sex trafficking and prostitution of minors has been an extremely difficult phenomenon for governments to effectively address. Where it is illegal for a minor to consent to sex, trafficked children have been arrested and charged for prostitution by local law enforcement agencies. Rather than being treated as victims, these children earn criminal records and receive no assistance to change their predicament (Jordan, 2013). In the United Kingdom, as soon as a prostituted minor who has a record of prostitution turns 18, they are deported for being an illegal immigrant.

The United Nations has begun to address illegal trafficking in two relevant protocols: The first protocol addresses the smuggling of migrants, and the second is related specifically to trafficking (UN, 2000a, 2000b). These protocols provide guidance and standards for governments to follow to assist in the eradication of transnational human trafficking. In order for countries to be signatories of the protocols, they must first be signatories of the United Nations Convention against Transnational Organized Crime and the Protocols Thereto (UN, 2004).

The United Nations Global Compact is a voluntary system designed to align strategies and operations with universal principles on human rights, labor, and the environment. The system calls for the "elimination of all forms of forced and compulsory labor" (UN, 2018). The global compact has more than 9,500 participating companies in more than 160 countries that have committed to the sustainable development goals and a reporting framework for progress on human rights and sustainability. The elimination of forced labor and development of decent work in global supply chains is a main focus of the compact goals. It also promotes communication and education on the risks of slavery in supply chains.

In 2017, France passed a law that requires French-based companies employing 5,000 workers in the country or 10,000 employees worldwide, to develop a "plan of vigilance" to identify and report possible environmental, health, and human rights violations within their supply chains. This includes the use of child labor and slavery within corporate supply chains. The program requires companies to develop programs and procedures to evaluate their partners and subsidiaries and to take actions to alleviate potential violations. They must also arrange a system for reporting and responding to potential violations. Companies that are found to be in violation of the law can be fined up to 10 million euros (France, 2017).

In 2012, California began enforcing the California Transparency in Supply Chains Act of 2010. This law requires certain large companies to disclose their efforts to ensure that slave labor was not any part of their product supply chain to the public. The law provided detailed requirements for the content and format for the information to be made available. Disclosures must address by whom and how the supply chain was evaluated, how audits of material suppliers were conducted, forms of supply chain certification, accountability standards and procedures for addressing employees or contractors that do not meet company standards, and training for employees and supply chain managers (CA, 2010; Harris, 2015). The California Occupational Safety and Health Administration (Cal/OSHA) enforces State OHS regulations without any consideration of the immigration status of the worker. This California law provides a basic structure to guide companies to improve transparency and public disclosure of their supply chain activities, but this law has been criticized for not going far enough and leaving enforcement to other federal and state laws. Statements on a website regarding corporate policies and programs denouncing slavery may look good to the public, but may not actually be addressing the actual issues (New, 2015).

Much of human trafficking is preventable through immigration policy reform. Immigration policies that punish victims through racial profiling, and threats of imprisonment or deportation, force the endeavor further underground and exacerbate the problems (Brennon, 2017). Policies that encourage fair and decent wages and provide social supports for migrants who may be uneducated, unaware of available services, and afraid to report abuse and exploitation to authorities would automatically reduce the trafficking trade.

One of the closest links between OSH and slavery is through the regulation of health and safety in the workplace. Trafficked workers and migrants have been shown to routinely work in environments dangerous to their health (Benach, 2011). If governments had regulations on health and safety, and they conducted regular worksite inspections and enforcement, then in theory working conditions for migrants, or otherwise exploited laborers, would become apparent. If working conditions are unsafe, then the workers are clearly being exploited and the safety conditions may just be the tip of the iceberg indicating other deeper exploitation issues. Dangerous and unhealthy working conditions are one of the key exploitations inflicted by slave owners. Additional research regarding migrant and slave work exposures will work to expand understanding of the problems and issues, and work towards job safety improvements and safer working conditions.

Several legislative actions have been taken in Brazil over the past two decades to combat slavery. In 1995, a federal labor inspectorate system was established to create legislation. In 2003, a National Plan for the Eradication of Slave Labor was established that identified employers that kept workers in conditions that were "slave-like." As a result of negative publicity to these businesses, and other factors, between the years of 2003 and 2010, more than 21,000 workers were released from these "slave labor" conditions (Phillips, 2012). In 2005, a National Pact for the Eradication of Slave Labor in Brazil was created that asks firms to voluntarily commit to the antislavery effort. In 2010, the pact had more than 130 full signatories representing 20% of the gross domestic product of Brazil (Phillips, 2012).

16.10 OTHER RESPONSES TO SLAVERY—EDUCATE, COMMUNICATE, AND COLLABORATE

The fact that there are more than 40 million slaves in the world today does not mean that it is a well-known or understood phenomenon. Employers and enslavers, and even slaves themselves, have serious interests in keeping their activities hidden from society. As we have discussed, even people perpetrating the crimes of trafficking and slavery do not always understand the criminality of their behaviors, as they are only exploiting willing workers. The importance of researching modern slavery and trafficking for the purpose of disseminating information and communicating the issues and outcomes to a broader society cannot be understated. In a study of the narration of slave stories, Laura Murphy has documented how telling stories about slavery in various literary genres can be used to elevate the social awareness and discussion of the great injustices ongoing in our societies right below our noses. She shows how powerful personal stories of suffrage can connect an otherwise impartial reader with those who have suffered under bondage and exploitation. The strategic use of real-life experiences can instill a sense of world citizenship and responsibility that "exceeds the bounds of the personal identification with the solitary narrator who suffers" (Murphey, 2015). Individuals and organizations that would not normally feel that slavery is their problem or take an interest in stopping it, can be influenced to take action by creating a sense of duty and promote a sense of social justice activism.

Similar to the use of narrative to communicate stories of exploitation and injustice, researchers and other OSH professionals can report and disseminate information about exploitation of workers regarding health and safety issues. In many cases, exploitation may be evident by unsafe working conditions. Everyone globally deserves a safe workplace. It is the OSH professionals' responsibility to make the unsafe conditions known to officials and the public. This may include reporting conditions to their own managers and directors regarding unsafe workplaces of suppliers or others in the supply chain.

Case

As my well-meaning colleague stated, she has a small farm in Pennsylvania and she hires the nicest group of migrants to help her harvest pumpkins every autumn. When I asked what form of healthcare do they have? And what is available for medical response to the workers in the field? Or what types of safety training do they receive for the operation of the heavy equipment, she said "none". What differentiates this small operation from one that might be termed "modern slavery"?

In his article about supply chain management and CSR, Stephan New concludes that modern slavery is in many cases created by external social and economic forces that are ultimately created by economic and business models and practices. Policies that push to improve corporate profits do so at the expense of suppliers who are often pushed to reduce costs of labor to the point where they can only hire workers at the most meager of wages, taking advantage of other economic and social conditions and factors (New, 2015). The place of modern CSR programs is to butt against the profit incentives to increase transparency and root out slavery despite the economic incentives to utilize it, from within the corporations. Unfortunately, CSR programs are doing little to actually provide safe working conditions for workers, prevent the use of child labor, or eliminate the use of slavery. OSH professionals involved in CSR programs need to ensure that they are not part of a company or CSR program that condones slavery for ethical and legal reasons.

Modern slavery is a fact of our global society, and every person, company, and organization plays a role in its existence. The complexity of the relationships between business, government, academia, and every other aspect of our daily social lives makes addressing the underpinning issues and causes of modern slavery difficult. All civilized institutions need to communicate and collaborate to work together to combat slavery at the root causes, wherever possible. Collaboration between different agencies and social and scientific disciplines will be essential to solving the problem of slavery (Jordan, 2013). Conditions that allow slavery to exist are a long-term threat to global political and economic stability.

In his assessment of the UK Modern Slavery Act, Craig (2017) noted that training on the topic was lacking at many levels. Legislators, judges, and law enforcement needed better training on recognizing and defining illegal human labor conditions and activities. This is also true for other members of the community including educators and service providers.

Health-care and social workers who may eventually come in contact with victims of modern slavery, whether in an emergency room or courtroom, need to be trained on the signs and symptoms of modern slavery. Once a victim is identified, these workers need to know how to report the crime without bringing attention to the victim or the possible perpetrator who may be accompanying, and even assisting the victim. Reporting processes need to be created and used so that victims can be recognized and managed to address their immediate needs and work with them as they escape their captors and return to normal society. Medical, nursing, and emergency responder education should include the means to recognize and support victims of modern slavery, and how to effectively work with local law enforcement to report illegal conditions (Domoney, 2015; Stoklosa, 2015).

16.11 CONCLUSIONS/RECOMMENDATIONS

Governments need to adhere to international conventions against human trafficking and create associated laws and border controls. Border controls and immigration policies need to be improved and equipped to detect and prevent human trafficking. There should be increased cooperation and information exchange between nations regarding perpetrators and victims of trafficking. States should improve training on the issues and identifiers of trafficking for law enforcement, immigration officials, and others associated with the transfer and movement of people including airlines and shipping companies.

The failure of CSR programs to identify and eliminate slavery from global supply chains demonstrates an inherent fault with the approach. Businesses do not have an interest in social justice if it goes against corporate profits. Although the public face of CSR programs is to demonstrate interest in social and environmental concerns, the majority of companies do not include social justice or slavery parameters within their CSR programs or reporting. And corporations will practice CSR to the extent that the public takes an interest in slavery or is aware of and care about the negative ramifications, which to a large extent they are not and do not. To improve compliance with international regulations on the use of slavery in supply chains, governments need to become more involved in CSR reporting.

OSH professionals and their professional organizations need to advocate for the rights of workers in all situations. This includes slaves in all of the various forms, industries, and geographic regions described in this chapter. OSH professionals' ethical responsibilities do not stop at the doors of their employer or border of their nation. OSH professionals have an ethical responsibility to improve working conditions globally and in all workplaces.

As consumers, we need to demand that the supply chains of our retailers are free of slave labor. CSR information is becoming more readily available in most industries. Consumers can read the reports and demand more action and transparency regarding the use of slaves within the supply chains of the products they purchase. As citizens, we need to demand that our governments do more to protect our economies and the fair trade of our businesses to exclude the use of slaves in supply chains. As the government itself is a huge consumer, this must include transparency in governmental supply chains to ensure slaves are not used in government activities, in addition.

OSH professionals need to take an active role in educating businesses, the public, the government, and colleagues on the breadth and depth of the global slave problem that exists today. The provision of healthy and safe conditions for workers must include vulnerable and defenseless slaves, who have no resources or capacities to defend themselves. Modern-day slavery needs to be understood and explained to a broader population, and OSH professionals must take an active role in this process.

REFERENCES

Antero, S., Understanding the eradication of slave labour in contemporary Brazil—An implementation perspective, *Policy Studies* (2013) Vol. 34, No. 1, pp. 89–111.

Aronowitz, A., Theuermann, G., Tyurykanova, E. (2010). *Analysing the Business Model of Trafficking in Human Beings to Better Prevent the Crime.* Vienna: OSCE Office of the Special Representative and Co-Ordinator for Combating Trafficking in Human Beings. ISBN: 978-92-9234-430-6, www.osce.org/cthb/69028?download=true accessed February 13, 2018.

Bales, K. (2000). *Disposable People: New Slavery in the Global Economy.* Berkley: University of California Press.

Bales, K., The social psychology of modern slavery, *Scientific American* (2002), Vol. 286, No. 4, pp. 80–88.

Bales, K. (2007). *Ending Slavery.* Berkeley: University of California Press.

Bales, K., Cornell, B. (2008). *Slavery Today.* Berkeley: Ground-Work Books.

Bales, K., Soodalter, B. (2009). *The Slave Next Door: Human Trafficking and Slavery in America Today.* Berkley: University of California Press. ISBN 978-0-520-25515-9.

Bales, K., Hesketh, O., Silverman, B., Modern slavery in the UK, *Significance* (June, 2015), pp. 16–21.

Barrientos, S., Contract labour: The 'Achilles heel' of corporate codes in commercial value chains, *Development and Change* (2008) Vol. 39, No. 6, pp. 977–990.

Bartley, T., Institutional emergence in an era of globalization: The rise of transnational private regulation of labor and environmental conditions, *American Journal of Sociology* (2007) Vol. 113, No. 2, pp. 297–351.

BBC, Man guilty of 21 cockling deaths. BBC News (2004) http://news.bbc.co.uk/2/hi/uk_news/england/lancashire/4832454.stm accessed February 16, 2017.

Belser, P. (2005). *Forced Labor and Human Trafficking: Estimating the Profits.* Geneva: International Labour Office.

Benach, J., Muntaner, C., Deicios, C., Menendez, M., Ronquillo, C., Migration and "low skilled" workers in destination countries, *PLos Medicine* (June, 2011) Vol. 8, No. 6, p. e1001043

Brennon, D., Fighting human trafficking today: Moral panics, zombie data, and the seduction of rescue, *Wake Forest Law Review* (May 23, 2017) Vol. 52, pp. 477–496.

Brown, G., Women and children last: The prosecution of sex traffickers as sex offenders and the need for a sex trafficker registry. *Boston College Third World Law* (2011) Vol. 1, pp. 1–40.

CA (2010). California Transparency in Supply Chains Act of 2010 (Senate Bill No. 657 Chapter 556 2010, enacted 2011).

Craig, G., Gaus, A., Wilkinson, M., Skrivankova, K., McQuade, A. (2007a). *Contemporary Slavery in the UK: Overview and Key Issues*. New York: Joseph Rowntree Foundation, www.jrf.org.uk/sites/default/files/jrf/migrated/files/2016-contemporary-slavery-uk.pdf accessed December 12, 2018.

Craig, G., Gaus, A., Wilkinson, M., Skrivankova, K., McQuade, A., Modern Slavery in the United Kingdom. Joseph Rowntree Foundation (February 26, 2007b) www.jrf.org.uk/report/modern-slavery-united-kingdom accessed February 9, 2018.

Craig, G., (ed.) (2009). *Child Slavery Now*. Bristol: Policy Press.

Craig, G., Modern slavery in the UK: The contribution of research, *Journal of Poverty Social Justice* (2014) Vol. 22, No. 4, pp. 159–164.

Craig, G., The UKs modern slavery legislation: An early assessment of progress, *Social Inclusion* (2017) Vol. 5, No. 2, pp. 16–27.

Crane, A., Modern slavery as a management practice: Exploring the conditions and capabilities for human exploitation, *Academy of Management Review* (2013), Vol. 38, No. 1, pp. 45–69.

David, F. (2010). *Labour Trafficking*. Research and public policy series no. 108. Canberra: Australian Institute of Criminology. ISBN 978 1 921532 62 7 (online), http://aic.gov.au/publications/current%20series/rpp/100-120/rpp108.html accessed February 13, 2018.

Domoney, J., Howard, L.M., Abas, M., Broadbent, M., Oram, S., Mental health service responses to human trafficking: A qualitative study of professionals' experiences of providing care, *BMC Psychiatry* (November 17, 2015) Vol. 15, p. 289.

Epstein, M.J., Rejc Buhovac, A. (2014). *Making Sustainability Work: Best Practices in Managing and Measuring Corporate Social, Environmental, and Economic Impacts*. New York: Berrett-Koehler Publishers.

Fight Slavery Now, Chattle slavery (2018) https://fightslaverynow.org/why-fight-there-are-27-million-reasons/otherformsoftrafficking/chattel-slavery/ accessed February 10, 2018.

France, National Assembly, Proposition de loi relative au devoir de vigilance des sociétiés mères et des entreprises donneuses d'orde, Text 924, (February 21, 2017), www.assemblee-nationale.fr/14/pdf/ta/ta0924.pdf accessed July 7, 2018.

Gladwin, T.N., Kennelly, J.J., Krause, T.S., Shifting paradigms for sustainable development: Implications for management theory and research, *Academy of Management Review* (1995) Vol. 20, No. 4, pp. 874–907.

Harris, K. (2015). *The California Transparency in Supply Chains Act of 2010—A Resource Guide*. Sacremento: California Department of Justice.

Harriss-White, B., Poverty and Capitalism. Economic and Political Weekly. pp. 1241–1246. iabolish, Modern Slavery 101-Types of slavery, (April 1, 2006) www.iabolish.org/index.php?option=com_content&view=article&id=25:modern-slavery-101&catid=4:-modern-slavery-101&Itemid=7)

iAbolish, Modern Slavery – 101 (2018) www.iabolish.org/index.php?option=com_content&view=article&id=25:modern-slavery-101&catid=4:modern-slavery-101&Itemid=7 accessed December 12, 2018.

ILO. (1930). *Forced Labour Convention (No. 29)*. Geneva: International Labor Organization, www.ilo.org/dyn/normlex/en/f?p=NORMLEXPUB:12100:0::NO::P12100_ILO_CODE:C029 accessed December 12, 2108.

ILO (2005). Human trafficking and forced labour exploitation: Guidelines for legislators and law enforcement. In B. Anderson, B. Rogaly (Ed.), *Forced Labour and Migration to the UK* (p. 16). London: TUC/COMPAS.

ILO, Fighting Forced Labor: The Example of BRAZIL. ISBN: 978-92-2-122293-4 (web PDF) (2009) www.ilo.org/wcmsp5/groups/public/---ed_norm/---declaration/documents/publication/wcms_111297.pdf accessed July 5, 2018.

ILO, ILO Global Estimate of Forced Labor. ISBN: 9789221264125; 9789221264132 (web pdf) (2012) www.ilo.org/wcmsp5/groups/public/---ed_norm/---declaration/documents/publication/wcms_182004.pdf accessed February 7, 2018.

ILO (2017). *Global Estimates of Modern Slavery: Forced Labour and Forced Marriage.* Geneva: International Labour Office. ISBN: 978-92-2-130132-5 (web pdf).

Issa, D., Modern slavery and human trafficking in Latin America, *Latin American Perspectives* (November, 2017) Vol. 44, No. 6, pp. 4–15.

Jordan, J., Patel, B., Rapp, L., Domestic minor sex trafficking: A social work perspective on misidentification, victims, buyers, traffickers, treatment, and reform of current practices, *Journal of Human Behavior in the Social Environment* (2013) Vol. 23, pp. 356–369.

Kabat, M., Desalvo, A., Egan, J., Media coverage of "Slave Labor" in Argentina, *Latin American Perspectives* (2017) Vol. 44, No. 217, pp. 50–62.

Kiss, L., Yun, K., Pocock, N., Zimmerman, C., Exploitation, violence, and suicide risk among child and adolescent survivors of human trafficking in the Greater Mekong Subregion. *JAMA Pediatrics* (September 2015) Vol. 169, No. 9, p. e152278.

Laczko, F., Data and research on human trafficking, *International Migration* (2005) Vol. 43, No. 1–2, Blackwell Publishing Limited, http://onlinelibrary.wiley.com/doi/10.1111/imig.2005.43.issue-1-2/issuetoc

Lindley, J., Beacroft, L., Vulnerabilities to trafficking in persons in the Pacific Islands. Trends & Issues in Crime and Criminal Justice no. 428. Canberra: Australian Institute of Criminology (2011) www.aic.gov.au/publications/current%20 series/tandi/421–440/tandi428.aspx

McGregor, A., Smit, J., Human rights due diligence in corporate global supply chains, *Governance Directions* (February 2017) Vol. 69, No. 2, pp. 16–21.

Murphey, L., The new slave narrative and the illegibility of modern slavery, *Slavery and Abolition* (2015) Vol. 36, No. 2, pp. 382–405.

New, S.J. (2004). The ethical supply chain. In New, S.J., Westbrook, R. (Eds.), *Understanding Supply Chains: Concepts, Critiques and Futures* (pp. 253–280). Oxford: Oxford University Press.

New, S., Modern slavery and the supply chain: The limits of corporate social responsibility? *Supply Chain Management: An International Journal* (2015) Vol. 20, No. 6, pp. 697–707.

Ngwe, J., Elechi, O., Human trafficking: The modern day slavery of the 21st century, *African Journal of Criminology and Justice Studies* (November 2012) Vol. 6, No. 1 & 2, pp. 103–119.

NISMART, National Incidence Studies of Missing, Abducted, Runaway & Throwaway Children (2002). Runaway/thrownaway children: National estimates and characteristics, www.ncjrs.gov/html/ojjdp/nismart/04/index.html accessed February 16, 2018.

Noble, D., 11% of UK Businesses Say Slavery in Their Supply Chains is 'likely', The Guardian, 18th August (2014) www.theguardian.com/sustainable-business/2014/aug/18/11-of-uk-businesses-say-slavery-in-theirsupply- chains-is-likely accessed February 10, 2015.

Olsen, L., Sullivan, C., Moving beyond modern slavery statements: Performing human rights due diligence, *Risk Management* (November 2017) pp. 597–601.

Pas-Fuchs, A., Badges of modern slavery, *Modern Law Review* (September 2016) Vol. 79, No. 5, pp. 757–785.

Phillips, N., Sakamoto, L., Global production networks, chronic poverty and 'Slave Labour' in Brazil, *Studies in Comparative International Development* (2012) Vol. 47, pp. 287–315.

Pocock, N., Kiss, L., Oram, S., Zimmerman, C., Labour trafficking among men and boys in the greater Mekong subregion; Exploitation, violence, occupational health risks and injuries, *Plos One* (December 2016) Vol. 11, No. 12, pp. 1–16.

Population Matters, Poverty (2010) https://populationmatters.org/wp-content/uploads/D15Poverty.pdf accessed July 25, 2018.

Quirk, J. (2009). *Unfinished Business*. Paris: UNESCO.

Rasche, A., De Bakker, F.G., Moon, J., Complete and partial organizing for corporate social responsibility, *Journal of Business Ethics* (2013) Vol. 115, No. 4, pp. 651–663.

Rezaeian, M., The association between natural disasters and violence: A systematic review of the literature and a call for more epidemiological studies, *Journal of Research in Medical Sciences* (2013)Vol. 18, pp. 1103–1107.

Siller, N., Modern slavery, *Journal of International Criminal Justice* (2016) Vol. 14 pp. 405–427.

Simmons, F., O'Brien, B., David, F., Beacroft, L., Human Trafficking and Slavery Offenders in Australia, Trends and Issues in Crime and Criminal Justice, Australian Institute of Criminology (November 2013) No. 464.

Stoklosa, H., Grace, A.M., Littenberg, N., Medical education on human trafficking, *AMA Journal of Ethics* (October 1, 2015) Vol. 17, No. 10, pp. 914–921.

UK Government, Modern Slavery Act 2015 (2015) www.legislation.gov.uk/ukpga/2015/30/section/3/enacted accessed February 13, 2018.

UN (1948). *Universal Declaration of Human Rights*. New York: United Nations, www.un.org/en/documents/udhr/index.shtml#a4

UN, Protocol Against the Smuggling of Migrants by Land, Sea and Air, Supplementing the United Nations Convention Against Transnational Organized Crime (2000a) www.unodc.org/documents/middleeastandnorthafrica/smuggling-migrants/SoM_Protocol_English.pdf accessed July 6, 2018.

UN, Protocol to Prevent, Suppress and Punish Trafficking in Persons, Especially Women and Children (November 2000b) https://treaties.un.org/doc/Treaties/2000/11/20001115%20 11-38%20AM/Ch_XVIII_12_ap.pdf accessed July 6, 2018.

UN, United Nations Convention Against Transnational Organized Crime and the Protocols Thereto, U.N. Doc. A/55/383/Add.1, at 62, 87, (2004) www.unodc.org/documents/middleeastandnorthafrica/organised-crime/UNITED_NATIONS_CONVENTION_AGAINST_TRANSNATIONAL_ORGANIZED_CRIME_AND_THE_PROTOCOLS_THERETO.pdf accessed July 6, 2018.

UN, United Nations, General Assembly Security Council, Agenda Item 68 Promotion and Protection of the Rights of Children—Children and Armed Conflict (April 20, 2016), A/70/836-S/2016/360.

UN, The Ten Principles of the Global Compact (2018) www.unglobalcompact.org/what-is-gc/mission/principles accessed June 21, 2018.

Unchosen, Unchosen (2017) https://encounters-festival.org.uk/unchosen-gallery/ accessed December 12, 2018.

Walk Free Foundation, The Global Slavery Index (2016) www.globalslaveryindex.org/ accessed December 12, 2018.

Walk Free Foundation, Overview of Global Slavery (2018) www.globalslaveryindex.org/2018/findings/regional-analysis/overview/ accessed December 12, 2018.

Wheaton, E., Schauer, E., Galli, T., Economics of human trafficking, *International Migration* (2010) Vol. 48, No. 4, pp. 114–141. doi:10.1111/j.1468–2435.2009.00592.x.

17 Occupational Health and Informal Work

Marianne Levitsky
Workplace Health Without Borders

CONTENTS

17.1 INTRODUCTION

According to the International Labor Organization (ILO), more than 60% of the world's employed population earn their livelihood in the informal economy. Informal workers generally lack the protections that may be afforded those in the formal economy: access to social security programs, workers' compensation, government regulation, and sick/holiday time and pay. This has profound implications for occupational health and safety (OHS). In a recent report on the informal economy (ILO, 2018), the ILO states that "informality has a harmful effect on workers' rights, including fundamental principles and rights at work, social protection, decent working conditions and the rule of law."

17.2 DEFINITION OF INFORMAL WORK

The "informal economy" has a different meaning than "informal work" or "informal employment." "Informal economy" refers to an economic sector at the enterprise level, whereas "informal work" and "informal employment" refer to individual jobs. Everyone who works in the informal economy does informal work, but the converse is not true: some informal workers work in the formal economy.

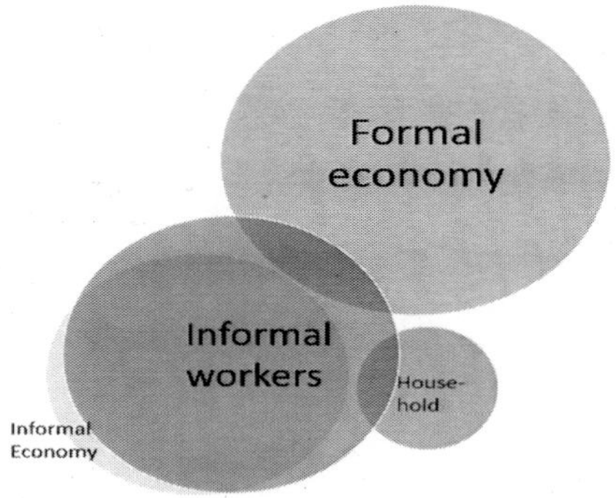

FIGURE 17.1 Overlap of informal work with formal/informal economies.

The ILO definition of the informal economy (ILO, 2018) encompasses enterprises that

- Produce goods only for household use.
- Are not registered with government institutions or programs.
- Do not have formal bookkeeping systems.
- Have employees, if any, who do not pay into social pension or insurance programs.
- Have fewer than six workers, or do not have a fixed location.

Workers in informal employment include those who own and/or work in an informal enterprise, as described above. Other workers considered to work in informal employment (even if they work for a formal enterprise) are those who do not pay into social insurance programs or do not have paid vacation and sick leave. Types of informal workers include domestic workers, waste pickers, casual day laborers, street vendors, and workers in cottage (home-based) industries.

Figure 17.1, while not to scale, illustrates the concept that informal workers are mainly those who work in the informal economy but also include some who work for a household and some who work in the formal economy. A visual depiction of formal and informal economy relationships is shown in Figure 17.1.

17.3 THE GLOBAL DIMENSIONS OF INFORMAL WORK

While more than 60% of the world's workforce is in informal work (50% if agriculture is excluded), the informal economy is not equally distributed. Figure 17.2 [Panel A reproduced from the ILO report (ILO, 2018)] shows the global distribution of work in the informal economy (including agriculture), ranging from less than 20% in developed countries like the United States to more than 90% in some parts of Africa. Figure 17.3 (Panel C from the ILO report) shows the different components of informal work by region.

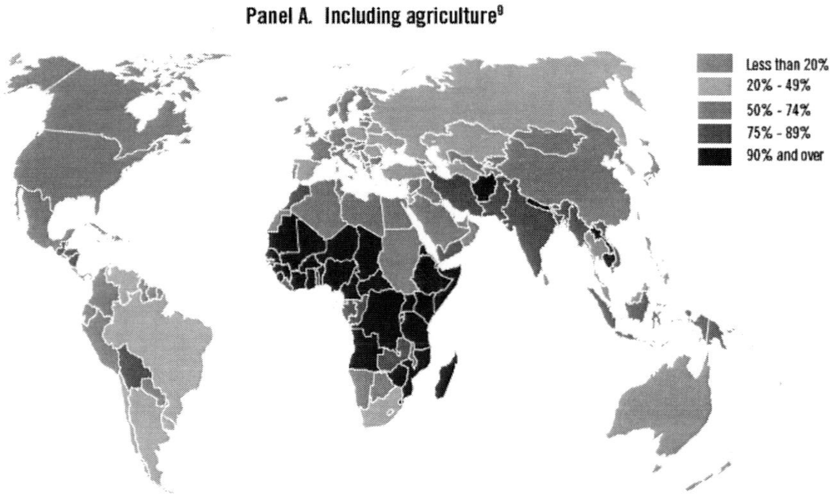

FIGURE 17.2 The global distribution of work in the informal economy. [International Labour Organization (ILO). Women and men in the informal economy. 2018. 156 pp. ISBN 978-92-2-131581-0 (web pdf), www.ilo.org/wcmsp5/groups/public/---dgreports/---dcomm/documents/publication/wcms_626831.pdf.]

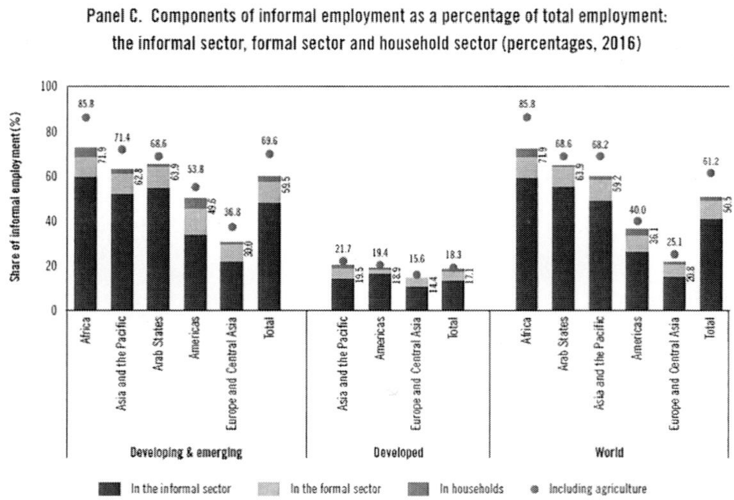

FIGURE 17.3 Components of informal work by region. [International Labour Organization (ILO). Women and men in the informal economy. 2018. 156 pp. ISBN 978-92-2-131581-0 (web pdf).]

Not only does informal work account for most of the world's workforce, but in developing countries, it may account for the majority of economic activities. It has been estimated that informal work in developing countries may contribute up to 60% of the gross domestic product (DCPP, 2007).

17.4 HEALTH AND SAFETY AND ADVERSE IMPACTS OF INFORMAL WORK

As stated by the ILO (2018),

> Although not everyone in the informal economy is poor and there is also poverty in the formal economy, ample empirical research has shown that workers in the informal economy face a higher risk of poverty than those in the formal economy, while informal economic units face lower productivity and income. Indeed, most people enter the informal economy not by choice but as a consequence of a lack of opportunities in the formal economy and in the absence of any other means of earning a living.

According to the ILO, "the 2 billion women and men who make their living in the informal economy are deprived of decent working conditions."

The nature of informal work makes it very difficult to obtain reliable overall data on the health and safety of informal workers. As observed in a report on a South African seminar on informal work, "it is still very difficult to collect reliable epidemiological data, which means that the health risks that informal workers face often remain invisible to policy makers" (London, 2018).

However, various studies conducted in the past two decades, reported by Lund and Marriott (Lund, 2011), cite the following estimates of the dimensions of hazardous conditions faced by informal workers:

- Exposures to dangerous chemicals with no protection, ranging from 6% of casual laborers and wage workers in Gujarat, India, to 74% of workers in Indonesia.
- High percentages of workers reporting that their general working conditions are unsafe, including nearly 50% of all wage workers in Bangladesh, 30% of workers in Russia, 24% of workers in Argentina, and 17% of workers in Chile and Brazil.
- In Tanzania, 40% of all workers but 80% of casual agricultural workers reported unsafe conditions.
- In Bangladesh, 16% of rural workers had suffered a work-related injury that required them to miss a week of work.
- A survey of 1585 informal workers in rural and urban Zimbabwe found similar occupational injury and mortality rates to those found in the formal economy, but higher rates of occupational illness.

While hazards faced by informal workers are similar to those encountered in the formal economy, their effects on informal workers are exacerbated by factors related to the nature of informal work. These factors include the following:

- Absence of workers' compensation, health insurance, or other forms of social protection to support workers incapacitated by injury or illness
- Lack of access to health care to treat injury or illness
- Lack of health and safety resources, including knowledge and expertise to assess and control hazards, and lack of control equipment such as ventilation and personal protective equipment

- No access to government regulatory and enforcement regimes. As noted by the ILO, "OSH [Occupational Safety and Health] legislation often does not apply to [informal] workplaces or, if it does, it is not effectively implemented and enforced. Individual workers do not have access to sound OSH advice and training, and are often exposed to risks to their safety and health without adequate protection" (ILO, 2011)

Surveys by the ILO conducted early this century found that the burden of the costs of occupational injury and illness fell on the informal workers themselves and their households (ILO, 2004). These surveys found that

- The majority of workers in Bangladesh, Brazil, Chile, China, India, Indonesia, Moldova, Pakistan, the Philippines, Tanzania, and Ukraine themselves bear the costs of work-related injuries or illness.
- Workers in Africa are among the least likely to have insurance against accidents or injury at work.
- In Gujarat, India, 93% of workers had no insurance against wage-work risks. Where employers do pay the medical costs of work accidents, payment only covers 22% of work injury costs for male workers and only 7% of such costs for women.
- Sixty percent of workers in Hungary are entitled to employer-provided medical services for work injuries and illness but more than 20% never actually receive such services.

Predictably, the ILO also found that in the absence of workers' compensation or health insurance, workers continue to work if at all possible, despite injury and illness. As noted by Lund and Marriott (Lund, 2011), "if workers do not take time off to recover or to seek necessary health care, their illness or condition is likely to deteriorate, possibly causing more long-term productivity declines or more long-term absenteeism in the future. Workers who continue to work while unwell and infectious also increase the risk of occupational illness for other workers with whom they come into contact."

The negative correlation of informal work with well-being is demonstrated by the ILO in Figure 17.4 [Figure 17 from the ILO report (ILO, 2018)]. As described by the ILO, "a measure of social development is the Human Development Index (HDI), which combines the indicators of long and healthy lives, knowledge, and a decent standard of living. Comparing national data on informal employment as a share of total employment with HDI values shows that countries with higher informality also have a lower HDI value."

Low occupational health standards in the informal economy can have spillover effects in the formal sector, as observed by a participant in the South African seminar on informal work. As cited in the seminar report, this participant "argued that the political and economic forces that have resulted in the trend towards sub-contracting, outsourcing and the casualisation of labour … have also led to the undermining of OHS regulation, so that it is not only the informal economy that now suffers from a lack of appropriate regulation and standards" (London, 2018).

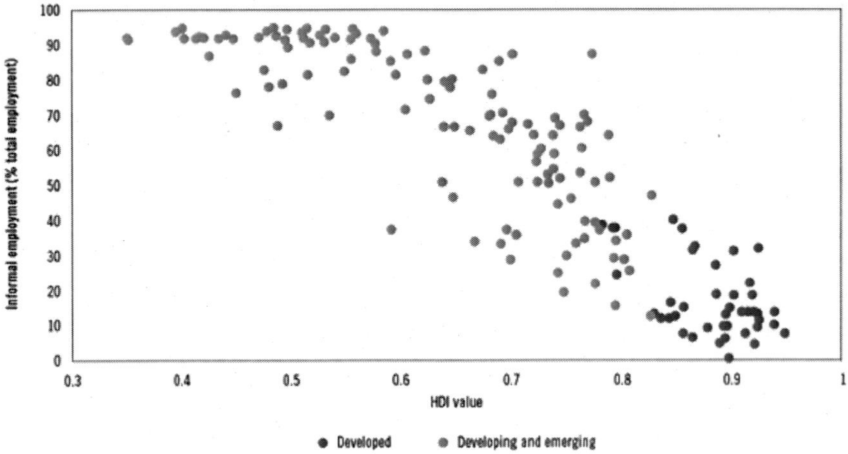

FIGURE 17.4 Relationship between informal employment and HDI values. [International Labour Organization (ILO). Women and men in the informal economy. 2018. 156 pp.]

17.5 EXAMPLES OF HAZARDS IN INFORMAL WORK

The hazards of informal work are similar to those often encountered in the formal sector, though as noted, these are often exacerbated by lack of knowledge and resources to ensure adequate assessment and control. There is considerable research literature on specific health and safety hazards in informal work. Examples include the following:

- A report on Zanzibar seaweed farmers documented numerous health and safety problems, including rashes and eye irritation from seawater contact, stings by poisonous fish, and ergonomic issues resulting from heavy physical demands (WIEGO, 2011).
- A study of informal construction workers in Pakistan revealed that both employers and workers lack knowledge of OHS laws/standards, resulting in the absence of safe work practices and nonreporting of injuries (Ahmed, 2018).
- A number of studies reported on the health of agate processors in India (NIOSH, 2015; Chaudhury, 2012), who usually work from their homes and have high rates of silicosis and tuberculosis due to silica exposure from processing gemstones.

17.6 THE GROWING GLOBAL INFORMATION GRID ECONOMY

While international development agencies have concentrated on the problems of informal work in developing economies, the growth of the Global Information Grid (GIG) has given rise to a "gig economy," which results in similar problems in more developed countries. While Figure 17.3 indicates that the percent of informal work in developed countries is less than 20%, the growth of companies like Uber suggests that this percent is growing.

In her 2018 book *Gigged*, Sarah Kessler (2018a) says that "20% to 30% of the working-age population in the United States and European Union had engaged in freelance work. Add part-time work to the mix, and some estimates put the percentage of the US workforce that did not have a full-time job as high as 40%." Articles by Kessler and others illustrate how gig workers lack the social and workplace protections that are often enjoyed by workers in the formal economy. Citing the example of Pablo Avendano, a bicycle courier killed on the job, Thomas Fox Parry (2018) points out that "in this gig economy, liability for work injuries, including death, falls on the worker and their family."

An article by Kessler (2018b) describes the "crazy hacks" that a Canadian woman used to earn a living when her husband lost his job. Kessler documents the experiences of this woman who managed to support her family by sourcing work on Mechanical Turk, Amazon's online crowdsourcing marketplace. As described by Kessler,

> No matter where Milland was in her house, if she heard the alarm go off, she would run to her computer. There were thousands of other Mechanical Turk workers competing with each other to grab the high-paying work, which was assigned to whoever could claim it first. Milland would sleep in her office so that she could listen for the alarm to go off at night without waking her husband. When she spotted good tasks, often through her alarm system, she used an automated tool to keep her queue full with the maximum 25 tasks that could be assigned to her at one time, and then worked furiously to finish them and grab more before they were snatched by other people.

Unsurprisingly, this woman developed ergonomic injuries due to this work. Kessler notes that

> There is no paid sick leave in the GIG economy. And among US workers who rely on sites like Mechanical Turk for their entire income, almost 40 percent don't have health insurance. Milland lived in Canada, with universal health care, but she couldn't afford the break. She wore a wrist brace and an elbow brace and kept on clicking.

17.7 ADDRESSING THE PROBLEMS OF INFORMAL WORK

The ILO has stated that in view of the dimensions and consequences of informal work, "transition from the informal to the formal economy is of strategic significance for hundreds of millions of workers and economic units around the world that are working and producing in precarious and vulnerable conditions." Therefore, it concludes, "there is an urgent need to tackle informality" (ILO, 2018).

Addressing problems of the informal economy has been identified as crucial to Sustainable Development Goal #8, (UN, 2016) to promote sustained, inclusive, and sustainable economic growth, full and productive employment, and decent work for all. (Sustainable Development Goals, or SDGs, are global priority-setting goals established by the United Nations in 2015.)

A target set under this goal (Target #8.3) is to "Promote development-oriented policies that support productive activities, decent job creation, entrepreneurship, creativity and innovation, and encourage the formalization and growth of micro-, small-, and medium-sized enterprises, including through access to financial services." Complementing this SDG, in 2015 the International Labour Conference adopted Recommendation No. 204, Recommendation Concerning the Transition from the Informal to the Formal Economy (International Labour Conference, 2015). This recommendation sets out a wide-ranging agenda for countries to promote the transition from informal to formal enterprises, and ensure decent work for those currently employed in the informal economy. Among the recommendations is that nations should "promote and extend occupational safety and health protection to employers and workers in the informal economy."

While the ILO emphasizes transition out of informal work as its highest priority in addressing these problems, efforts are also being made to improve the health of informal workers by extending occupational health services. In a 2014 workshop on Health Coverage and Occupational Health and Safety for the Informal Workforce in Developing Countries, participants from Asia, South America, and Africa reported on a variety of initiatives to offer health services to informal workers (Taylor, 2016).

Other efforts have been strongly promoted by Women in Informal Employment: Globalizing and Organizing (WIEGO), a network focusing on conditions of women in informal work. A section of their website (WIEGO, 2018) focuses on OHS, documenting experiences of informal workers and efforts to improve their health. For example, WIEGO has worked with informal street vendors in Accra, Ghana, to advocate for a variety of protections against health and safety hazards, including the following:

- Better waste disposal systems to avoid clogged drains and gutters, which lead to often intolerable smells and disease vectors that can cause food poisoning and diarrhea
- Improved cooking equipment and access to fire extinguishers
- Better lighting and security measures to prevent traders from being preyed upon by criminals (Cities Alliance, 2014)

Other examples of successful advocacy cited by WIEGO include the following:

- The provision of safety equipment to rag pickers in Pune, India
- The Self-Employed Women's Association (SEWA) in India, which has worked with local designers to develop equipment with better ergonomic features for waste workers, garment workers, and bakers in the informal sector (WIEGO, 2013)
- The provision of seaweed farmers with protective equipment, such as boots, gloves, and hats, by the companies to which they sell (WIEGO, 2011)

In citing this last example, WIEGO's report notes that "one of the most practical ways to start extending OHS to these seaweed farmers may be to put pressure on the end sellers and consumers to enforce health and safety standards through ethical trade initiatives."

In a 2014 paper, Valentina Forastieri of the ILO called for a variety of measures aimed at integrating OHS of informal workers with general OHS services and data gathering (Forastieri, 2014). "In designing or strengthening national OSH policies and programmes," she states, "promoting safe and healthy working conditions should aim not only at the formal but also at the informal economy." She goes on to call for "a self-sustainable health insurance scheme and a referral system for the extension of occupational health services using the existing public health structure and a community health approach to prevent and control injuries and communicable, endemic and occupational diseases."

17.8 CONCLUSIONS

Protecting the health of informal workers will require the involvement of institutions that have not traditionally carried lead responsibilities for OHS. As noted by Forastieri, these include the existing public health infrastructure and community groups, which could be enlisted to provide occupational health services that in a formal economy may be provided by employers, unions, regulatory agencies, and workers' compensation systems. As pointed out by WIEGO, traders (the buyers and sellers of the products of informal workers) could be motivated to provide protective measures for their suppliers, fulfilling responsibilities that employers in the formal sector would be expected to meet. Through ethical trade initiatives, consumers are in a position to pressure such traders to protect the health and safety of their suppliers.

Addressing the occupational safety and health problems of informal work is critical to overall social and public health. This is not only because informal workers account for the majority of the global workforce but also because poor working conditions in the informal sector can have spillover effects by driving down standards in the entire economy. The growth of outsourcing and the gig economy threatens to counteract international efforts to transition economies from the informal to formal sectors. Improving global health and well-being for everyone will therefore depend to a great extent on affording to informal workers the health and safety protections and social supports enjoyed in the world's formal workers in the most advanced industries and economies.

REFERENCES

Ahmed, I., Shaukat, M.Z., Usman, A., Nawaz, M.M., Nazir, M.S., Occupational health and safety issues in the informal economic segment of Pakistan: A survey of construction sites, *International Journal of Occupational Safety and Ergonomics* (2018) Vol. 24, No. 2, pp. 240–250.

Chaudhury, N., Phatak, A., Paliwal, R., Co-morbidities among silicotics at Shakarpur: A follow up study, *Lung India [Internet]* (2012) Vol. 29, No. 1, p. 6, www.lungindia.com/text.asp?2012/29/1/6/92348

Cities Alliance, Accra's Street Vendors Collaborate to Create A Safer, Healthier Workplace [Internet], pp. 1–3 (2014) www.citiesalliance.org/WIEGO-AccraTraders

Disease Control Priorities Project (DCPP), Developing Countries Can Reduce Occupational Hazards, pp. 2–5 (October 2007) http://pria-academy.org/pdf/OHS/DCPP-OccupationalHealth.pdf accessed December 12, 2018.

Forastieri, V. (2014). Economia Informal Y Salud Ocupacional, Occupational Safety and Health (OSH) in the informal economy, pp. 1–16.

International Labour Conference, Recommendation 204. Recommendation concerning the transition from the informal to the formal economy (2015), *Adopted by the Conference at Its One Hundred and Fourth Session*, Geneva, June 2015.

ILO, Economic Security for a Better World. Socio-Economic Security Program [Internet], p. 478, (2004) www.ilo.org/ses

ILO, XIX World Congress on Safety and Health at Work, Global Trends and Challenges on ILO Introductory Report: Occupational safety and health, pp. 11–15 (2011) www.ilo.org/wcmsp5/groups/public/@ed_protect/@protrav/@safework/documents/publication/wcms_162662.pdf accessed December 12, 2018.

ILO, Women and Men in the Informal Economy—A Statistical Picture, 156 pp, ISBN 978-92-2-131581-0 (web pdf) (2018) www.ilo.org/wcmsp5/groups/public/---dgreports/---dcomm/documents/publication/wcms_626831.pdf accessed September 7, 2018.

Kessler, S. (2018a). *Gigged*. New York: St. Martin's Press, 304 p.

Kessler, S., The crazy hacks one woman used to make money on Mechanical Turk. *Wired [Internet]* (2018b), pp. 1–14, www.wired.com/story/the-crazy-hacks-one-woman-used-to-make-money-on-mechanical-turk/ accessed December 12, 2018.

London, L., Lund, F., Policy, E.H., Seminar on new approaches to occupational health and safety in the informal economy and extractives industry. *Occupational Health Southern Africa* (2018) Vol. 24, No. 2, pp. 56–57.

Lund, F., Marriott, A., Occupational Health and Safety and the Poorest, pp. 1–84 (2011) www.inclusivecities.org/wp-content/uploads/2012/07/Lund_WIEGO_WP20.pdf accessed December 12, 2018.

National Institute of Occupational Health, Prevention Control and Treatment of Silicosis & Silico Tuberculosis in Agate Industry [Internet] (2015) www.lungindia.com/text.asp?2012/29/1/6/92348

Parry, T.F., The death of a gig worker. *The Atlantic* (June 1, 2018) pp. 1–11 www.theatlantic.com/technology/archive/2018/06/gig-economy-death/561302/ accessed December 12, 2018.

Taylor, R. (2016). *Approaches to Universal Health Coverage and Occupational Health and Safety for the Informal Workforce in Developing Countries, Workshop Summary [Internet]*. Washington, DC: National Academies Press. ISBN: 978-0-309-37406-4, www.nap.edu/catalog/21747

United Nations, Sustainable Development Knowledge Platform, Sustainable Development Goal 8 [Internet] (2016) https://sustainabledevelopment.un.org/sdg8

WIEGO, Occupational Health and Safety for Informal Workers (2011), *Newsletter* (2: February), pp. 1–18.

WIEGO, Occupational Health and Safety of Informal Workers, (7: June), pp. 1–18 (2013) www.wiego.org/sites/wiego.org/files/resources/files/OHS_Newsletter_june_2013_English.pdf

WIEGO, Occupational Health and Safety in the Informal Economy [Internet] [cited 2018 Jul 31] (2018) www.wiego.org/ohs

18 Travel Safety and Security for the Global Worker

Thomas P. Fuller
Illinois State University

CONTENTS

18.1 INTRODUCTION

The increased globalization of business, industry, and agriculture requires an ever-increasing number of world workers and employees to travel internationally. The number of workers traveling abroad is expected to increase by 50% by 2020 (PWC, 2010). These workers on extended duty beyond the borders of their homelands are subject to many risks unlike and greater than most that they see in their daily lives in their home countries. The average total cost of a 1-year international assignment is US$311,000. The average cost of a failed international assignment in terms of productivity, reputation, medical evacuation, and other costs is up to US$950,000 (SOS/IOSH, 2016). These workers represent a valuable asset to the organization sending them abroad, but it is argued that employers also have a duty to care for these workers' security, health, and safety beyond the actual workplaces in these foreign countries while they are on assignments. This chapter describes many of the risks that traveling workers may be subjected to, and the responsibilities that employers have to protect them while working abroad. It also provides sources of accurate information regarding travel risks. It identifies programs and procedures to minimize and control risks, in addition to responding to emergencies.

Whether it is for a short visit to a foreign manufacturing site, or a 2-year overseas assignment, employers have an ethical responsibility to ensure the safety of travelers and ex-patriots while they are abroad. Not only should the home country occupational exposure limits and job safety working conditions be met, but traveling abroad brings a whole new set of security and safety concerns that may not be part of the normal occupational safety and health rubric at home. Workers abroad may be subjected to potentially violent social settings and workplaces. They may encounter a variety of infectious agents, or wild and venomous animals or poisonous insects. Differences in languages, food safety customs, availability of safe drinking water, and standards of medical care each warrant a thorough review for every employee traveling to any country internationally.

18.2 INTERNATIONAL WORK HAZARDS

18.2.1 Violence/Kidnapping

Being the victim of crime or violence when traveling abroad for work is not unusual in many regions of the world. International travelers may appear racially different, they may dress differently, they may not know the language, and they behave differently from local populations. They are not aware of local conditions and appropriate security measures; it is easy to become a target of petty or serious crimes.

There are more than 100,000 kidnapping cases annually, but only 10% are reported to local authorities for fear of revenge and repercussions of talking to the police (Smith, 2010). Kidnapping hot spots include Mexico, Brazil, India, Venezuela, Nigeria, Pakistan, Afghanistan, Columbia, Iraq, Ecuador, Somalia, and the Philippines.

But they also happen in economically developed and western/European countries. International workers and jobsites are common kidnapping targets due to their known and seen wealth. Workers or family members may be taken off the street, in hotels, and at jobsites.

18.2.2 Transportation Hazards

If transportation is a significant workplace hazard in the developed countries, then transportation is especially hazardous in economically underdeveloped ones. In addition, the employer should consider travel to and from work, as part of the job assignment for the international worker. According to the Aid Worker Security Database, 48.8% of violent attacks on workers occur in transit on roads (AWSD, 2016).

Approximately 1.25 million people die each year in road traffic accidents. And 90% of road fatalities happen in low- and middle-income countries even though these countries have only half the vehicles in the world. Nearly 50 million people suffer nonfatal injuries and disability. In less developed countries, road accidents equal up to 5% of their gross national product (WHO, 2016).

Since a significant portion of road fatalities occur to workers, employers should be mindful of the impact these accidents have on their organizations. And although companies can control such factors as worker driving and vehicle maintenance to reduce their direct risks, many risk factors that impact their worker's safety depend on social factors of the host country. Factors including road conditions, and laws that affect other drivers' behaviors such as speed limits, drinking and driving, and distracted driving, can play a large role in determining the level of risk of road travel for workers on assignments internationally. Organizations with workers abroad have an interest to become politically active on these issues as a means to improve their own worker's safety.

In addition, in low-income countries, half of the road deaths are of pedestrians, cyclists, and motorcyclists. Although being injured or killed while commuting to and from work is not typically considered a workplace occurrence, it would still behoove employers to consider the risks to their workforce and take action to reduce the likelihood of losing this valuable commodity of human resource. Employees might be encouraged to take public transportation, or special training might be provided to help workers understand the risks and possible precautions that could be taken on a personal level.

18.2.3 Political Instability

Many countries are under some forms of political instability or strife, or even civil war. Gunfire and artillery are common occurrences in many countries. Several sources of information regarding the political conditions and safety for international travel are available:

U.S. Department of State—Travel (www.state.gov/travel/)
European Commission—Consular Protection (https://ec.europa.eu/consular-protection/content/home_en)

UK Government—Foreign Travel Advice (www.gov.uk/foreign-travel-advice)
Australian Government—Smart Traveller (http://smartraveller.gov.au/Pages/default.aspx)

18.2.4 Infectious Diseases

A wide variety of infectious diseases that have been eradicated in the western world are still endemic in large regions of the world. Malaria, typhoid, tuberculosis, and yellow fever still are endemic in Africa, India, and large parts of China. Much of the world still has polio and small pox. Huge parts of Africa are still being ravaged by human immunodeficiency virus and acquired immunodeficiency syndrome.

Workers traveling to and through various countries need to consider all the pathways of exposure that they may encounter during their travels. Travel notices are provided at a variety of sources including the Center for Disease Control and Prevention and the World Health Organization (WHO). It cannot be assumed that the family physician will be aware of the appropriate travel vaccines and endemic diseases in foreign countries. It is best to verify the recommendations with a variety of sources and reference points well before travel begins.

Case

An engineering firm sent a worker to tend to a project for one week on an oil-drilling platform well off the coast of Gabon in western Africa. After checking with the CDC travel information, the company did not recommend or provide malaria medicine for the worker since he would be working far out at sea and there would not be mosquitos there. A few days after returning home from his trip, the worker began getting headaches and diarrhea. After a couple days, he saw his doctor and they ran some tests. The worker had contracted malaria and after a few days had an extreme case of liver failure and died. Upon investigation of the case, it was found that the day the worker arrived in Gabon, the helicopter service to the oil platform was not working and he spent the night in a hotel on land. It was concluded that he was bitten by a mosquito at this time, and contracted the disease. In a lawsuit brought by the employees' family, the company was found to be responsible for the workers' death by not providing appropriate protection for the worker.

Foodborne infectious diseases can also be problematic for many workers traveling abroad in many parts of the world. Combined with long airline travel, lack of sleep, poor diet, and dehydration, infectious organisms in the food can be catastrophic. Elderly workers, and employees with other medical conditions, can be especially vulnerable to foodborne illnesses.

18.2.5 Natural Disasters

Natural phenomenon such as earthquakes, volcanoes, typhoons, hurricanes, and others may occur any time and disrupt travel and the well-being of international workers. Some regions of the world may be more prone to various natural disasters. Some natural disasters may be more likely to occur in certain seasons or regions. Others may provide no warnings whatsoever.

Some countries may not only be prone to natural disasters, such as earthquakes, but the infrastructure may not be designed to support the occurrences. Not only are parts of Mexico more prone to large-scale earthquakes, many of the buildings are not built to withstand them and often tumble to the ground, where more modern buildings built to higher safety specifications would not be affected.

Some regions of countries may be more susceptible to natural disasters. Low-lying regions may be more susceptible to flash flooding or storm surges. Coastal regions are more likely to suffer greater wind damage from hurricanes and large storms.

18.3 LEGAL PERSPECTIVES

The International Labor Organization (ILO) Promotional Framework for Occupational Safety and Health (2006) is a collective agreement between countries to develop acceptable programs and regulations to prevent occupational accidents, illnesses, and diseases. This framework requires signatory countries to develop national policies, systems, programs, and cultures that ensure safe and healthy work environments for all workers. Legal debate and discussion about the scope of the framework has led to interpretations that include employees working in foreign destinations (Mathiason, 2013). That is, workers traveling to another country on assignment also have the right to have safe and healthy working conditions.

The ILO 1981 Occupational Safety and Health Convention's Preventative Approach to Occupational Safety and Health and the Duty of Care also described the responsibilities for governments to ensure a broad interpretation of the legal concept "duty of care" to require employers to provide occupational health and safety to all workers. The convention applies the scope to all branches of economic activity in which workers are employed, including public service. And Article 3 of the convention defines "workplaces" as "all places where workers need to be or to go by reason of their work and which are under the direct or indirect control of the employer" (ILO, 1981).

Article 2 of the 2006 ILO framework requires continuous improvement of programs for worker health and safety to reflect relevant ILO guidelines and industry standards as they are created. This can be interpreted to include the assessment of new risks as they emerge, and development of ways to control those risks. This now includes protections for risks to business travelers, international assignees, and all employees of the company working abroad (Mathiason, 2013). This has expanded to include such foreseeable location-specific risks to international workers such as kidnapping because the hazard is incurred during the line of duty (Claus, 2009).

Risks to these international employees are considered unique but not exceptional and therefore need to be identified by the employer and any actions available to control risks need to be set in motion by the employer (Berkowitz, 2011). If a risk is foreseeable, then the employer must take action to control and minimize it. These controls could include ensuring workers have appropriate vaccines before travel, security measures are in place, and workers have received appropriate safety and security training (ILO, 2006). Overall, business traveler risk assessment and mitigation programs are now considered the standard of care for corporate health and safety program managers (Claus, 2009, 2010).

Case law in the European Union (EU), the United Kingdom, the United States, and Australia has already begun to reflect the industry standards in international travel responsibilities for employers (Berkowitz, 2011). Employers have already been held responsible for employees who have been injured in accidents, victims of crime, kidnapping, and gotten sick with infectious diseases while working abroad. The UK Corporate Manslaughter and Corporate Homicide Act of 2007 specifically imposes criminal liability on employers for gross breaches of the duty of care for the death of employees working abroad (UK, 2007).

The UK duty of care requires employers to provide for the safety of workers when there is a foreseeable risk and a causal link between the work required and an injury. When working abroad, this duty of care extends beyond the worksite to include all aspects of life while on assignment including travel and safety at their accommodations. Additionally, the European Directive 89/391 also outlines responsibilities for employers to provide and promote occupational safety and health for employees traveling abroad throughout the EU (EU, 1989).

In the United Kingdom, the Health and Safety at Work Act of 1974 requires employers to provide a duty of care to all workers. In addition, the UK Corporate Manslaughter and Corporate Homicide Act of 2007 can impose criminal liability on organizations where there is a gross breach of duty that results in the death of an employee. Both of these laws can be applied to workers extraterritorially even when the negligent decisions causing the death were made by managers remaining in the United Kingdom. The EU and Australia each have similar laws that place responsibility on employers to provide for worker safety while on international assignments (Berkowitz, 2011).

In an in-depth analysis of laws associated with dangerous international employee assignments, Berkowitz (2011) showed that the determination of the extent of legal responsibility for employers to provide for all aspects of worker safety while they are abroad is a complex issue. The laws involved in each international case depend on the laws in the host country and specific details about each case, including whether the worker was "at work," whether the injured worker is covered by workers' compensation, and whether they can sue their employers.

18.4 BUSINESS PERSPECTIVES

Separate from the legal concerns and implications for provision of safety for international workers, there are fundamental business reasons for having global worker security programs. By promoting worker travel safety, organizations will

- Reduce the disruptions to business due to unnecessary incidents.
- Improve worker moral and peace of mind while on international assignments.
- Increase return on investments.
- Improve public relations and reduce the likelihood of media attention.
- Reduce the number of premature worker returns from assignments.
- Reduce the costs associated with repatriation and medical responses.
- Lessen legal liabilities associated with international travel (SOS Gold, 2013).

18.5 IDENTIFYING AND ASSESSING RISKS

Before sending employees to international assignments, the employer has a responsibility to thoroughly assess all the various risks the worker will be exposed to during their travels. It may not be possible to eliminate all the risks that an employee may encounter abroad, but legal liabilities can be reduced by providing adequate warning of the risks that the workers may encounter. A comprehensive risk assessment should follow predetermined protocols and be documented for future reference and revision, when necessary.

Basic risk assessment for international travel should include at a minimum the following:

- Individual and detailed aspects of the travel to and work within the host country
- Identification of specific health, safety, and security risks for each activity
- Determining which individuals are at risk
- Evaluating what risk control measures are available and most practicable
- Measuring residual risks that remain after controls are in place
- Reviewing and updating the risk assessment periodically to reflect environmental, social, political, and economic changes

As risk is greatly dependent on the country or region the worker will be traveling to, this is a good place to begin the risk assessment. Political institutions and local laws should identify any potential problems for individuals working abroad. The online database of the State Department of the United States is an excellent resource to identify the political systems and stability of hundreds of countries and is updated frequently (USSD, 2018). The State Department lists travel advisories for numerous countries that identify particular hazards, crime reports, and even areas to avoid. Requirements for passports or visas for each country are also provided. In addition, general guidance for traveling abroad is provided on the United States State Department website.

The ILO also provides information for workers going abroad on assignments. Much of the information is country specific regarding working conditions, national regulations, and ratification of ILO treaties and conventions (ILO, 2018). Travel advisories and notices for several countries are provided at the ILO website.

In addition to political stability, the economic stability of the country should be assessed. Methods for employee payment and financing of their work and living expenses abroad need to be identified prior to assignment. Trustworthy financial institutions must be identified before travel, and the legal means of payment and movement of money must be understood in order to avoid any possible illegal activities. Tax liabilities need to be evaluated to ensure an employee working abroad complies with all tax and financial obligations during their stay abroad.

Workers need to be trained on how and where to attain cash, and how to secure the cash that they have on hand. Cash can be obtained at hotel automatic teller machines (ATMs) and at currency exchanges on the secure side of airport clearance systems. ATMs on the street should never be used in many cities and regions.

Social systems and cultural norms must be analyzed and assessed. Religious and cultural differences can present a form of "risk" if the traveling employee is unfamiliar with social norms and practices. Consumption of alcohol may be illegal. Certain prescription drugs such as pain relievers, medical marijuana, other medicines, and even poppy seeds may be illegal. These need to be identified before travel to preclude complications from arising at immigration checkpoints. Certain clothing or attire may be inappropriate or even illegal. Other clothing may be required, such as head of face coverings for women in public.

Security systems within the region of assignment should be analyzed. Private armed services may be needed in addition to local law enforcement. In fact, the validity and conduct of local law enforcement may also warrant explicit evaluation by professional consultants prior to assignment. Weaknesses and inefficiencies, and even corruption of law enforcement in the area, may preclude placing workers in the area, or at least warrant increasing private security systems to offset the concerns.

Medical incidents that occur during worker travel can be very disruptive to business activities and the lives of workers. Thousands of international trips are canceled, delayed, and negatively affected by injury and illness during the trips (Druckman, 2014). Many of the illnesses in workers are caused by communicable and infectious diseases. Workers who become dehydrated and tired from travel become susceptible to these diseases. Illnesses are often exacerbated and prolonged due to limited and inadequate health-care facilities and lack of appropriate medicines (Allegranzi, 2011).

Medical risks may include the lack of access to acceptable emergency or routine health-care services. The WHO provides a comprehensive list of endemic diseases throughout the world. The WHO lists various countries and now has a crisis-rating system that grades the various outbreaks of infectious diseases in different countries. In this system, Grade 3 countries indicated events with substantial public health consequences that require substantial WHO response. Grade 2 conditions or events represent outbreaks that represent moderate public health consequences and require moderated WHO response. Grade 1 conditions are events that represent minimal public health consequences and require minimal WHO response (WHO, 2018).

Routine and emergency medical care sources and hospitals should be identified before travel. Services should include the availability of psychiatric treatment and support if the need arises. Emergency medical travel capabilities should be identified. The means to pay for both local emergency medical services and emergency medical travel should be arranged prior to appointment and arrival.

The medical conditions of employees and their families traveling with them should be evaluated prior to assignment. Workers with known preexisting medical conditions (e.g., diabetes, hypertension, pacemakers/defibrillators) should be considered especially. Special medical treatments and routine examinations or treatments should be considered. Sources for examinations and treatment should be identified and arranged prior to travel, if possible. Employees with special medical conditions may not be appropriate for assignments that would put them or their families at excessive risk due to inadequate medical facilities or capabilities in the assigned country.

In a study of 48 large U.S. multinational companies with employees traveling abroad, it was observed that contrary to common perceptions, most serious medical problems that require aggregate services occur in what are considered to be "low-risk" countries that are economically developed rather than " high-risk" economically developing countries (Druckman, 2014). This study also showed that business travelers in what would be considered "safe" working environments such as finance or business services actually used a greater proportion of medical services than more hazardous industries such as mining or construction. This is partially explained by larger numbers of workers from those sectors and the injuries being related to the employees themselves. In addition, many of the injuries and illnesses in "low-risk" countries occur when employees are on personal time, or traveling within the host country to local tourist attractions, and partaking in non-work-related activities. This same study showed, however, that the "high-risk" countries did show a clear "gradient of risk when viewed from the perspective of the individual trip" (Druckman, 2014). On an individual trip basis, a trip to a "high-risk" country is much more likely to require hospitalization than travel to a "low-risk" host country.

The need for vaccinations and ongoing prophylactic treatments, such as malaria medications, should be evaluated for employees prior to travel. Malaria medications, for example, must begin prior to travel, and continue for a period after travel, in order to be fully effective. Some countries require proof of vaccination prior to admission. This may be particularly true when returning from countries with ongoing outbreaks of highly contagious diseases, such as yellow fever.

Additional information on vaccines and vaccination programs can be found at the following websites:

www.who.int/immunization/en/—World Health Organization Immunization, Vaccines, and Biologicals

www.cdc.gov/vaccines/index.html—United States Centers for Disease Control and Prevention

https://ecdc.europa.eu/en/home—European Center for Disease Prevention and Control

www.dh.gov.uk—United Kingdom Department of Health and Social Care

Transportation risks include a broad variety of modes of travel. Airline travel today is fairly standardized and includes control of terrorism activities on the planes themselves. Airline travel security, including aircraft maintenance, is fairly well regulated and controlled by international standards of conformance.

Once inside a nation, however, transportation safety and security can vary widely. Generally, train and bus travel will be safer and more secure that traveling in individual vehicles. Personal vehicles are significantly more susceptible to damage and injury from collision and road accidents. Vehicular accident rates due to different traffic and road conditions vary greatly between countries, and traveling to work by car can present significant hazards to workers and should be considered and minimized when possible.

In addition to the hazards of vehicular accidents, workers traveling in a vehicle by themselves may become a victim of kidnapping, crime, or terrorism. Armored vehicles and traveling with security staff may reduce the overall risks but may also draw additional attention to the worker.

Special worker risks are characteristics that might put workers at particular risk in a given nation or region. Women may be especially vulnerable in certain cultures, and special precautions may be warranted. They may be likely subjects of violence or segregation from certain areas or facilities. In many parts of the world, women need to adhere to strict social norms, and if they do not, they are subject to criminal charges.

Lesbian, bisexual, gay, and transgender employees may be at particular risk in a number of countries. In several countries, sexual relations with members of the same sex are subject to fines, imprisonment, or death. It is important that workers, and particularly family members, understand the laws and ramifications of violations.

Religious proselytization is also considered a crime in many countries. Symbols of certain religions are illegal in many countries, and laws should be followed to the greatest extent possible. In France, it is illegal to wear face coverings or hijabs and other conspicuously religious symbols in public. In many Muslim-oriented countries, wearing a crucifix or cross could be construed as proselytizing and subject to criminal charges. Wearing religious symbols of a counter religion is not a sensitive or wise activity in any culture and easily leads to resentment and hostility, and should be avoided whenever possible for both legal and security reasons.

18.6 BEFORE WORKERS TRAVEL

Organizations with workers traveling abroad should have clearly developed policies and procedures to guide travel activities and responses to unusual conditions or situations. Safety practices, policies, and limits should follow a combination of regulations and standards of both the home and host countries, with a slant towards the requirements that are most stringent and conservative. Organizational policies should be sure to include the individual responsibilities of employees too. Sample organizational travel policies and programs are provided in the Appendix.

Traveling abroad should begin with an analysis of the host country including political and social stability, visa requirements, embassy locations, health conditions, and cultural norms. Special support organizations are typically available both within host countries and from sources abroad before travel. An employee selection process should be used to identify existing employees who may be willing to travel abroad. This process would also identify language capabilities, experience traveling abroad, understanding of and sensitivity to other cultures, and possible limitations to traveling internationally.

Factors to consider before sending workers to a foreign country include currency stability and possible expenses, religious and cultural differences, employment laws, and the availability of reliable communications systems. Driving laws and access to medical care should also be determined before travel begins. The political and social conditions of the country should be considered too, including terrorist activity, civil unrest, and lack of infrastructure, such as grocery stores (Berkowitz, 2011).

Useful sources of information regarding traveling abroad include the following:

- ILO Country Profiles on Occupational Safety and Health (www.ilo.org/safework/countries/lang--en/index.htm)
- ILO Global Database on Occupational Safety and Health Legislation (LEGOSH) (www.ilo.org/dyn/legosh/en/f?p=14100:1:0::NO:::)

- CIA, Central Intelligence Agency World Fact Book—The Fact book provides information on the history, people, government, economy, energy, geography, communications, transportation, military, and transnational issues for 267 world entities. (www.cia.gov/library/publications/resources/the-world-factbook/)
- WHO (www.who.int/en/ and www.who.int/ihr/IVC200_06_26.pdf)
- UK Department of Health (www.dh.gov.uk)
- U.S. Centers for Disease Control (www.cdc.gov/travel)
- European Centre for Disease Control (www.ecdc.europa.eu)
- U.S. Department of State Bureau of Consular Affairs—Country Information (https://travel.state.gov/content/passports/en/country.html)

Prior to travel abroad, the means by which employees will communicate should be determined and verified. Special cellular telephone adaptions might be necessary through the service provider. If they are not available, it may be necessary to make arrangements to purchase service once the new country is entered. Means to charge existing or new phones should be available or adaptable to the country's power sources. E-mail should be available at the new country, and power adapters for portable computers may be necessary. Charging systems and adapters should also be available while en route to and from the host country. Backup communications systems should be considered and identified prior to travel.

Employees traveling abroad should be in good physical condition and have routine medical checkups throughout the travel period. Physical and electronic copies of medical and dental records, blood groups, vaccines, and prescriptions should be available. Medical and dental insurance providers should be contacted prior to travel to determine existing coverage and whether additional insurance or premiums are required. Workers should be sure to take an adequate amount of prescription and nonprescription medicines to last for the duration of travel. The legality of prescription, and even over-the-counter medications, should be verified prior to travel.

Medical services available to the worker in the host country should be evaluated and verified prior to travel. Some countries need to have proof of medical insurance coverage for certain minimum monetary values prior to issuing a visa.

Special potential medical problems at host sites such as extreme temperatures or altitudes should be considered for travelers. There may be some period of acclimatization associated with the assignment that should be accommodated in the planning of the itinerary. Older workers or those with special medical conditions may need more time, or special support in adjusting to the new environment. Medical kits may be warranted depending on the locations to be visited. These could include regular first-aid materials, water purification tablets and filters, sunscreen, diarrhea treatment, insect repellent, mosquito nets, and antihistamines. In extremely impoverished locations with minimal access to health care, sterile medical supplies such as syringes, dressings, and sutures may be useful (IOSH, 2015).

In areas where hospital standards are extremely low, additional medical equipment and supplies such as blood plasma and single-use thermometers may be warranted (IOSH, 2015). Workers or family members with any special nutritional needs should consider where they will acquire the food. Special allergies to foods or other materials should be evaluated prior to travel.

Financial services and measures should be in place before traveling abroad. Adequate amounts of cash should be on hand, and the means to send payments to employees should be known and verified before travel, if possible. Credit card companies and banks should be contacted to ensure that the credit and debit cards will work properly internationally. Copies of all card numbers, access codes, and passwords should be kept in a separate document when traveling. Access to credit card and bank account information through the Internet cloud should be considered for continued access from remote locations. The data stored as a PDF and sent via e-mail is a good backup measure to ensure access to the bank and card information from anywhere.

If possible, it is a good idea to open a local bank account in the host country. An account in a local bank is a good way to ensure access to money when it might be needed. Local laws should be investigated to ensure there are no violations, but local bank managers should also be familiar with requirements. A local credit/debit card from a national bank could also be useful for long-term assignments.

Employees should learn as much as they can about their host country before traveling by reading travel journals, looking at websites, asking people who have been there, and in some cases by professional consultants who can teach them specifics about host country culture and customs.

Workers should be familiar with the seasons of the host country and bring appropriate clothing. If necessary, they may need to obtain special clothing or equipment before they travel. Whenever possible, dress should be casual and not draw attention. Expensive jewelry and watches should be minimized and kept out of sight.

The appropriate luggage to be used may also vary by host country and the route of travel. Be sure to use suitcase locks so it is easy to see whether someone has tampered with your luggage. Keep a tag with your name and address attached inside the luggage as a backup to the outer tag. And keep a list of all the items inside the suitcase. Avoid using soft-sided luggage since it can be easily compromised. Be sure to keep a reasonable supply of necessary daily prescription medications in your hand luggage in case your luggage is lost or delayed.

18.7 SECURITY MEASURES

Each country presents a different type of personal security threat to business travelers. Employers need to be made aware of potential threats and how to deal with or avoid them. Employees with little or no experience traveling abroad are at particular risk. Even the most-simple aspects of packing and preparation can affect the likelihood of being victim. Certain materials and important medications should not be left in luggage and should be in carry-on bags along with applicable copies of prescriptions.

Strict guidelines for itineraries and timelines should be adhered to, and specific check-in times should be predetermined. Accommodations should be selected carefully and in safe areas with easy access to public transportation and hotel shuttles and taxis. Fares to destinations should be agreed upon before entering vehicles, whenever possible. Even the route to the destination can be determined before you get in the car.

Case

In Mexico City, there are taxis that are vetted by western hotels that are accepted as respectable and "safe" through a system of registration or recognition by such characteristics as age and company emblems. Other less reputable taxis frequent the streets and are taken by locals at a much lower fare, and slightly higher risk. Taxis have been known to target foreigners and then take them for ransom. This can be especially true if the fee for service to the destination is not agreed upon before entering the vehicle. Many of the "non-approved" taxi drivers do not speak English. You need to know what the taxi etiquette is in whatever country you are travelling in, and if you plan on taking the taxi from the airport to your hotel when you get there, you not only need the right money, but which taxis to take and what to say. What is taxi etiquette in Johannesburg, South Africa?

If workers are driving a car in the foreign country, then the rules of the road should be learned prior to travel. Carjackings are common in some countries. In some countries, drivers are known to cause intentional "accidents" in order to try to solicit bribes or other forms of payment.

It goes without saying that foreign visitors should keep their defenses up at all times. When approached on the street by people wanting money or directions, travelers should keep their distance and discourage interaction even though it appears the person is in need, lost, needs money for a bus ticket, or found what appears to be a valuable ring or wallet. Travelers should always keep at least arms-length away from people on the street, if possible. Opulent visible jewelry should not be worn, even if costume, as it attracts the eye of would-be thieves. Valuable jewelry is sometimes best left at home altogether for the duration of the travel overseas.

Whether a natural disaster, social or political unrest, disease outbreaks, or economic instability, there should always be a contingency plan for each event. The plan and the strategy may be different for each type of event and for each country. It may be useful and safer to have an adequate amount of cash on hand in the event that electronic banking goes down. Exit strategies and timeframes should be determined prior to travel in the event that management cannot communicate with their employees. Emergency meeting locations for multiple staff in a country should be determined ahead of time. Emergency responses and exit strategies should be practiced periodically.

Travelers should avoid carrying large sums of money, but enough cash for emergencies should be kept on hand. Money should be kept in more than one place to prevent losing it all at once. Travelers' checks should be kept in addition to various international currency. Whenever possible, store cash and other valuables in available safes.

Keep important phone numbers in more than one location. Numbers should include emergency contacts and the numbers of the nearest embassy or consulate.

18.8 TRAVEL INSURANCE

Actions need to be taken that ensure travelers have adequate medical, dental, and emergency response insurance before traveling. Many corporate or state government worker medical insurance policies do not cover extended travel outside the home country. Most policies only cover emergency services or emergency repatriation in some instances. Some policies require prior notification before international travel in order to be validated.

When medical insurance policies do not provide reasonable coverage for workers and their families traveling abroad, they need to be upgraded or supplemented with additional policies. In some instances for extended visits, certain countries will require proof of adequate medical insurance prior to administering a visa or travel permit in the country. Employees need to know how to access the medical insurance, and which facilities or health-care systems in the country should be used with the carrier policies. Dental and ophthalmic insurance coverage should also be verified and supplemented, if needed, for the employees and traveling family members.

Insurance for other emergencies is highly recommended, depending on the region or country visited. Insurance for security incidents can include responding to kidnapping and ransom situations. Security incidents may also include blackmail and extortion, and other legal responses to criminal situations. Separate from ransom payments themselves, professional security and legal consultant advice and response can cost several tens of thousands of dollars. The advice of security and legal professionals can go a long way when dealing with criminals and complex legal and political situations in a foreign country. Workers charged with crimes, actual or fabricated by local law enforcement, need competent professional legal representation that can be extremely costly.

18.9 CONTRACTED SERVICES

Organizations working with external consulting services and contracted agencies should have clear arrangements for delineation of responsibilities. Contractor and consultant competencies should be vetted thoroughly by independent sources. Lines of normal and emergency communications should be set up and tested. Insurance and funding sources must be clearly identified and put in place before travel begins.

18.10 PROGRAM EVALUATION AND CONTINUOUS IMPROVEMENT

The travel safety and security program should be reviewed and evaluated periodically. Deficiencies and weaknesses should be documented, reported, and corrected as soon as possible. A problem identified in one country or worker arrangement may be likely to exist in another worker's assignment. Problems identified should not be hidden, but reported to other workers, as a way to improve safety and security. Open communications in the area of program evaluation is the best means to ensure worker safety is continuously moving forward. All levels of management must be made aware of the status of travel security at all times. A list of all workers on assignment should be readily available to management and all departments with responsibilities for worker travel safety and security.

18.11 LIVING ABROAD

18.11.1 Food and Drink

Many infectious diseases are transmitted via food and drink, so precautions should be taken continuously during the time in the host country. Between 30% and 70% of travelers have traveler's diarrhea, and the majority are believed to be caused by bacterial pathogens in food or drink (Connor, 2018). To be safe, only cooked food that is served hot should be eaten. Raw fruits, salad, and vegetables should be avoided. Water should be boiled before drinking or brushing teeth. If boiling is infeasible, disinfectant tables or bottled water should be used. The use of ice should be avoided unless made from boiled, chlorinated, or treated water.

Travelers with food allergies should practice extreme caution. Many countries do not require food labeling, of if there is labeling, it may be in a foreign language, incomplete, or incorrect. Antidote medication for allergic reactions may be unavailable, so travelers should carry their own, whenever possible.

18.11.2 Communicable Diseases

If all medical preparations were completed properly prior to travel, then issues with communicable diseases while in the host country should be reduced. Ongoing vigilance is important, however, as conditions may change over time and in different regions. Outbreaks that occur in nearby countries or areas may overflow into the host country with little warning or notice. Additional vaccines, social distancing, or other precautions may be warranted.

18.11.3 Hotel Safety

Prior to selecting a hotel, learn as much as possible about it and the surrounding area. Find out local transportation availability and what other institutions are in the surrounding area. Inquire whether they have a written hotel safety program and onsite security. Is the hotel compound gated to vehicles and pedestrians and secured by armed guards? Do visitors go past a checkpoint to get to the elevators, or are they operated by key card?

If possible, avoid staying in hotel ground floor rooms to reduce public access and improve security. For fire safety, rooms on the second to fourth floors are preferable to higher floors.

When you arrive in your room, check for other access points (balconies, patios, adjoining rooms) and ensure they are secure. Test the room telephone to be sure it is functional. Never answer the room telephone unless you have made prior arrangements to accept a call at a certain time. If the phone rings and you are not expecting a call, wait until after the phone stops ringing and call the operator to see who called, and if someone in the hotel was looking for you. Using the phone to check on your whereabouts is a key way that criminals can track your activities. Recently, hotel telephones have been increasingly used for extortion scams, kidnapping, and other criminal activities.

If someone knocks on your hotel door, use the spyhole. Never open the door to anyone you do not know. If someone is at your door that you are not expecting and do not know, call the hotel reception desk to send someone to inquire who it is. If someone from the hotel arrives unexpected to make repairs in your room, check with the hotel reception before letting them in. While they are in your room, maintain control of all your valuables.

Always keep the deadbolt or chain on your hotel room door when you are there. If the door does not have a chain or deadbolt, ask to change to another room. If second door locks are not available, use a wedge of wood to secure the door from the inside. When you are not in the room, leave the television on low.

Do not linger in the hotel lobby and be cautious about who you talk to there. Do not have loud conversations about your trip, work, room, or personal life with others in the lobby, restaurant, or bar or the hotel.

Locate the nearest fire alarm and extinguishers. Learn the available fire escape routes, and follow them to the final exit points to ensure they are clear.

Avoid keeping large amounts of cash in your room, and use the room safe for small items and small amounts of cash. Keep laptop computers and other cash at the hotel safe.

18.11.4 Driving Safety

Be sure that your driver's license from your home country will not expire while you are traveling. If it will, get it renewed before you leave on your trip. Always carry your license and insurance information in the vehicle when driving.

Learn about official local driving rules before driving. Learn about driving culture and practices from travel books or local residents. Be aware of local police traffic practices, including frequency and types of roadblocks. Carry cash in order to pay traffic fines you may receive while driving.

Never depend solely on electronic navigation systems for travel, but carry hard copy maps in the vehicle at all times. Be aware of, and avoid, restricted travel areas. Be sure you have a spare tire and know how to change it prior to traveling to remote areas. Carry additional emergency supplies in the vehicle including first-aid kits, fire extinguishers, toolkits, reflective vests, emergency flares, and warning triangles.

Keep your car in good working condition and check tires, fluid levels, and gasoline levels before traveling. For long trips, check on the location of gas stations along the route. Keep vehicle doors locked at all times, even when exiting for short times, such as to refuel. Do not leave valuables in the car.

If you are not familiar with your surroundings, do not get out of the vehicle. If you are involved in an accident and are suspicious of the cause, do not exit the vehicle. If people are pointing out problems with your car or tires, be particularly wary. Drive on to a separate distant public and busy area to stop and inspect the vehicle. Be wary of people wishing to assist you in your auto repair before you have asked them.

If you are ever stopped by local law enforcement and are then asked to report to the local police station, contact your emergency contact numbers to let them know your situation.

Before leaving on a trip, check traffic and road conditions. Avoid peak travel times and avoid traffic congestion when possible. Consider weather, air pollution, and driving terrain, and make appropriate adjustments in your expected trip duration. In low-economic countries, avoid driving at night, if possible.

18.11.5 Communications

During the worker's time abroad, systems should allow for 24/7 two-way communications between the organization and the traveler. Mechanisms should be in place to ensure that workers can be informed of changing risk levels either in their host country or surrounding regions or in the home countries. Backup systems should also be identified and tested periodically.

Systems for workers to communicate with family members' traveling/non-traveling should be developed to ensure safety and peace of mind. Protocols should require normal contact times and response activities in the event contact cannot be made.

18.11.6 Using Taxis

In many countries, taxis represent a significant risk from both traffic accidents and criminal activity. Local conditions should be investigated to determine the safest practices before a taxi is engaged. Hotels will typically know which taxi companies and car services are safe and reputable. In many countries, hailing a taxi in the street is not a safe option. It is always preferable to hire a cab by telephone or from the hotel. Never get into a taxi without verifying that it is the one you hired.

Once in the taxi, confirm your route and the expected fare with the driver. Always sit in the back of the cab and always wear a seat belt. Do not be overly familiar with the driver or share personal information such as where you are from, your profession, employer, or title. If you are uncomfortable with the driver, their driving, or the route being taken, consider having the driver take you to a secure and public location to hire a new taxi for the remainder of the trip.

18.11.7 Environmental Conditions

Extreme weather conditions may not unduly strain workers in top physical shape, but they may stress other workers who may be less physically fit. Extreme weather can put additional stresses on workers of family members and exacerbate preexisting medical conditions sensitive to them. High altitude or high levels of air pollution can put stress on individuals with respiratory illnesses.

18.12 SPECIAL CONCERNS FOR FEMALE TRAVELERS

Depending on the host country, women may be at particular risk. Many countries have gender-specific cultural and legal restrictions. In many countries, it is illegal for a woman to drive, or even ride in a car with a man that is not her husband. In some countries, a woman alone in public is frowned upon and a likely target for crime. Women may be required to wear certain clothing in public. In some host countries, women's magazines depicting western attire and partial nudity may be considered pornographic and lead to arrest. In some countries, bare skin shown on women is frowned upon and goes against some cultural and religious norms. To minimize unwanted sexual advances and garner greater acceptance by the local residents, it is prudent to dress with discretion and show respect for the culture you are visiting.

18.13 PSYCHOSOCIAL ADAPTATION

Even the most seasoned traveler can eventually be worn down by extended travel and long duration assignments. Lack of family and social support, foreign cultures and languages, hostile social or political situations, and dangerous security conditions can make life abroad stressful and lead to a variety of psychological problems such as depression. Psychological issues can be even more pronounced and common in family members who relocate with the assigned worker. Spouses and children who accompany the worker may be exposed to different, and possibly worse, conditions of exclusion and social separation. Mental issues should be anticipated and accepted as normal psychological responses to stressful situations. Employers should be proactive in helping workers and their families in their work-life and work-family issues for the psychological well-being of the employee, and success of the travel assignment (Black, 2007). The need for professional support should be recognized as soon as possible, and arrangements should be made to get assistance as soon as possible.

18.14 EMERGENCY PREPAREDNESS

Emergency plans and procedures should be developed to cover every possible foreseeable emergency situation. Meeting points and evacuation routes should be planned and practiced ahead of time. Resources to respond to all types of emergencies should be readily available, and/or deployable. Access to security providers and other consultants should be available 24/7, and funds should be available as necessary. Workers and their families need to be trained on all aspects of emergency preparedness and response. Periodic drills and exercises should be used to test emergency preparedness, and corrective actions should be taken when needed.

Workers should carry medical and security travel assistance contact information in case they do not have the capacity to communicate. In countries where the worker is not fluent in the host language, arrangements should be in place for interpreters to respond in all emergencies. Care should be taken to assure that the response to an emergency affecting one worker does not put other workers at the jobsite at risk.

18.15 CONCLUSIONS/RECOMMENDATIONS

Globalization has led to ever-increasing numbers of workers traveling abroad. The responsibility of employers to provide for the safety of their employees extends beyond the borders of their nation of origin. A combination of education and awareness on the risks and hazards in each country, and the basic risks associated with international travel need to lead to programs and systems to ensure the ongoing well-being of workers abroad. The risks are broad and numerous, so individual workers on assignments cannot be expected to have all the tools and expertise needed to ensure their own security. A comprehensive and elaborate system should be in place before and during employee international travel.

APPENDIX: SAMPLE ORGANIZATIONAL POLICIES, PROGRAMS, AND PROCEDURES FOR INTERNATIONAL WORKERS

Policies

Policy statements generated at the top levels of management set the foundation for international travel safety. Leadership and commitment to the health and safety of workers traveling abroad provide guidance for the programs and procedures developed to ensure that goals and objectives are met.

Policy statements on intent provide guidance for how travel security should be developed to ensure meeting organizational objectives. Compliance with home and host government regulations should be part of the policy, in addition to a commitment to follow the accepted industry standards of practice. Well-designed policy statements regarding international travel will be sure to align with other corporate policies such as sustainable development and corporate social responsibility.

Programs

Once organizational policies regarding international travel have been developed, various programs should be created to address the identified goals. The organizational structure to address various travel security objectives should be developed to delineate roles and responsibilities of departments or individuals in the organization. Clear lines of communication and responsibility between different levels of management should be part of written travel safety programs.

One primary organizational department should be identified with the central role and responsibility for the overall travel security program. This person or department should have the experience and resources available to complete the required tasks that are assigned. Various programs that will need to be developed to support a comprehensive and thorough travel safety program will include at a minimum the following:

- Risk assessment
- Traveling worker/family assessment
- Pretravel protocols
- Medical evaluation/preparation
- Training for workers and traveling/non-traveling family members
- Communications
- Accommodations
- Transportation (to and from host country and within the host country)
- Security services
- Educational systems (for family members traveling with worker)
- Financial services/operations

- Insurance
- Legal services
- Emergency response
- Post-travel debriefing
- Auditing/reporting /continuous improvement

Procedures

Once the overarching programs are completed for the international worker travel safety and security goals and organization, detailed procedures can be created to support the individual program objectives. Clear and concise written procedures offer the opportunity for various departments to review other department's procedures and identify conflicts, inconsistencies, or redundancies before they occur in practice. Quality assurance and document control systems can be put in place to ensure consistency over time, and keep track of program changes and improvements as they are made. Procedures that are transparent and readily available to workers and managers help to clarify program objectives and set easy to follow tasks and steps for workers to follow at each step of the travel assignment.

Emergency response procedures should include initial notification, communication, risk assessment, response actions, ongoing event monitoring and analysis, continued support actions, and de-escalation/recovery. These procedures should be written and available onsite as hard copies and electronically as PDFs on computer systems, and also available through the Internet. Workers and managers should receive detailed training on emergency response procedures. Emergency drills and exercises should be held periodically.

Training programs lead to detailed procedures regarding how and when workers will receive training. The training procedures will identify the topics covered in each session and provide documentation of worker attendance.

Travel Security—Pretravel Checklist

Risk assessment
Political stability
Social stability
Economic stability
Religious requirements
Cultural requirements
Reliable telephone
Reliable Internet
Reliable electrical service
Criminal activity
Transportation systems
Weather conditions
Employee factor risks

(Continued)

Travel Security—Pretravel Checklist

Insurance
Medical
Repatriation
Dental
Ophthalmic
Ambulance
Auto
Medical preparation
Predeparture checkups and screenings
Copies of medical records and prescriptions
Necessary vaccinations
Prescription medications
Over-the-counter medications
Eyeglasses and contact lenses/solutions
Spare eyeglasses and contact lenses
Eyeglass and contact lens prescriptions
Finance
Employee payments
Local bank account
International credit/debit cards
Local credit cards

Security

Types of crime
Onsite security
Travel security
Emergency response team
24-h communications
Emergency meeting points
Emergency evacuation plans
Normal communications
High-risk areas identified
Travel itineraries provided
Employee and family member training
Accommodations clearance

Transportation

International driving permits required
Car service/security provided

REFERENCES

Allegranzi, B., Bagheri Nejad, S., Combescure, C., Graafmans, W., Attar, H., Donaldson, L., Pittet, D., Burden of endemic health-care-associated infection in developing countries: Systematic review and meta-analysis, *Lancet* (2011) Vol. 377, pp. 228–241.

Aid Worker Security Database (AWSD), Location of Attacks (1997–2016) (January 31, 2016) https://aidworkersecurity.org/incidents/report/location accessed June 26, 2018.

Berkowitz, P., Congiu, M., Managing the Global Workforce—A Legal and Practical Guide to Dangerous International Employee Assignments (February, 2011) 1 Littler Mendelson Publishers www.littler.com/files/press/pdf/WP_IntlAssignments_2-23-11.pdf accessed February 22, 2018.

Black, L., Jamieson, S., Up, up and fading away: The work and family life of executive international travelers, *Policy and Practice in Health and Safety* (2007) Vol. 5, No. 2, pp. 63–78.

Claus, L., International SOS, Duty of Care of Employers for Protecting International Assignees, Their Dependents, and International Business Travelers, Vol. 11, pp. 15–17 (2009) www.eisf.eu/wp-content/uploads/2014/09/0132-International-SOS-2009-Duty-of-care.pdf accessed December 12, 2018.

Claus, L., International assignees at risk: Employers have a duty of care for workers around the globe, *HR Magazine* (February 1,2010) Vol. 55, pp. 73–74. https://www.shrm.org/hr-today/news/hr-magazine/pages/0210legal.aspx accessed December 12, 2012.

Connor, B., Traveler's Diarrhea, CDC Yellow Book, Chapter 2 (2018) https://wwwnc.cdc.gov/travel/yellowbook/2018/the-pre-travel-consultation/travelers-diarrhea#4954 accessed March 3, 2018.

Druckman, M., Harber, P., Liu, Y., Quigley, R., Assessing the risk of work-related international travel, *JOEM* (November, 2014) Vol. 56, No. 11, pp. 1161–1166.

EU, Council Directive 89/391/EEC, Measures to encourage improvements in the safety and health of workers at work (June 12, 1989) http://eur-lex.europa.eu/legal-content/EN/TXT/PDF/?uri=CELEX:01989L0391-20081211&from=EN accessed February 21, 2018.

International Labor Organization (ILO), C155-Occupational Safety and Health Convention's Preventative Approach to Occupational Safety and Health and the Duty of Care (1981) *Convention concerning Occupational Safety and Health and the Working Environment (Entry into force: 11 Aug 1983) Adoption: Geneva, 67th ILC session (22 Jun 1981)—Status: Up-to-date instrument (Technical Convention).Convention may be denounced: 11 Aug 2023-11 Aug 2024* www.ilo.org/dyn/normlex/en/f?p=NORMLEX-PUB:12100:0::NO::P12100_ILO_CODE:C155 accessed February 21, 2018.

International Labor Organization (ILO), Promotional Framework for Occupational Safety and Health (2006) Report IV(1) ISBN 92-2-116608-2.

International Labor Organization (ILO), Homepage (2018) www.ilo.org/global/lang--en/index.htm accessed February 27, 2018.

Institute of Occupational Safety and Health (IOSH), Safety Without Borders: Keeping Your Staff Healthy and Safe Abroad, IOSH (May, 2015) http://www.iosh.co.uk/~/media/Documents/Books%20and%20resources/Guidance%20and%20tools/Safety%20without%20borders.pdf accessed June 21, 2016.

Mathiason, T., Are you part of the global workforce?: An examination of the "Duty of Care" to business travelers and international assignees under the ILO occupational health and safety conventions and as emerging international customary law, *American University International Law Review* (2013) Vol. 28, No. 3, pp. 873–904.

PWC, Price Waterhouse Coopers, Talent Mobility 2020—The Next Generation of International Assignments, s.l.: Price Waterhouse Coopers (2010) www.pwc.com/gx/en/managing-tomorrows-people/future-of-work/pdf/talent-mobility-2020.pdf accessed December 12, 2018.

Smith, R., For Employers Conducting Business Abroad, Kidnappings Remain an Ongoing Concern, *Human Resources Report (2010) Bureau of National Affairs, Inc.* ISSN 1095-6239 (2010) www.psfinc.com/wp-content/uploads/psfinc/2016/07/psf-human-resources.pdf accessed March 4, 2018.

SOS/IOSH, Gold, D. (2016). Managing the Safety, Health and Security of Mobile Workers: An Occupational Safety and Health Practitioner's Guide. Amsterdam, the Netherlands: SOS/IOSH, www.iosh.co.uk accessed August 2017.

SOS, SOS Foundation, Gold, D., Global Framework—Safety, Health and Security for Work Related International Travel and Assignments (September 26, 2013) www.internationalsos.com/newsroom/news-releases/global-framework-safety-health-and-security-for-work-related-international-travel-and-assignments-sep-26-2013 accessed December 12, 2018.

UK, Corporate Manslaughter and Corporate Homicide Act 2007, C19 § 1 (2007) www.legislation.gov.uk/ukpga/2007/19/pdfs/ukpga_20070019_en.pdf accessed June 26, 2018.

US State Department (USSD), Homepage (2018) www.state.gov/ accessed February 27, 2018.

World Health Organization (WHO), Humanitarian Health Action—Crises Webpage (2016) www.who.int/hac/crises/en/ accessed February 27, 2018.

World Health Organization (WHO), Road Traffic Injuries (February, 2018) www.who.int/news-room/fact-sheets/detail/road-traffic-injuries accessed June 26, 2018.

Index

D

E

F

G

U

V

W